I0056311

Least ◆ Squares
Support Vector Machines

Least Squares
Support Vector Machines

Johan A. K. Suykens, Tony Van Gestel,
Jos De Brabante, Bart De Moor and Joos Vandewalle

K U Leuven, Belgium

World Scientific

NEW JERSEY • LONDON • SINGAPORE • BEIJING • SHANGHAI • HONG KONG • TAIPEI • CHENNAI

Published by

World Scientific Publishing Co. Pte. Ltd.

5 Toh Tuck Link, Singapore 596224

USA office: 27 Warren Street, Suite 401-402, Hackensack, NJ 07601

UK office: 57 Shelton Street, Covent Garden, London WC2H 9HE

Library of Congress Cataloging-in-Publication Data

Least squares support vector machines / Johan A.K. Suykens ... [et al.].
   p. cm.
   Includes bibliographical references and index.
   ISBN-13 9789812381514 (alk. paper)
   ISBN-10 9812381511 (alk. paper)
   1. Machine learning. 2. Algorithms. 3. Kernel functions. 4. Least squares. I. Suykens,
Johan A. K.

Q325.5 .L45 2002
006.3'1--dc21   2002033063

British Library Cataloguing-in-Publication Data

A catalogue record for this book is available from the British Library.

First published 2002
Reprinted 2005

Copyright © 2002 by World Scientific Publishing Co. Pte. Ltd.

All rights reserved. This book, or parts thereof, may not be reproduced in any form or by any means, electronic or mechanical, including photocopying, recording or any information storage and retrieval system now known or to be invented, without written permission from the Publisher.

For photocopying of material in this volume, please pay a copying fee through the Copyright Clearance Center, Inc., 222 Rosewood Drive, Danvers, MA 01923, USA. In this case permission to photocopy is not required from the publisher.

Printed in Singapore

Preface

In recent years there have been many new and exciting developments in kernel based learning, largely stimulated by work in statistical learning theory and support vector machines. Our research in this area started in 1998 around the time that we organized the *International Workshop on Advanced Black-Box Techniques for Nonlinear Modelling* in Leuven, where Vladimir Vapnik presented his important breakthroughs in this area. Solving nonlinear modelling and classification problems by convex optimization without suffering from many local minima sounded indeed very appealing and interesting for a deeper investigation.

Driven by the dream to make the approach as simple as possible (but not simpler) led us to the formulation of least squares support vector machine classifiers, as a first contribution in this area. Many tests and comparisons showed great performance of LS-SVMs on several benchmark data set problems and were very encouraging for further research in this promising direction. At the ESAT-SISTA research division of the Electrical Engineering department of the Katholieke Universiteit Leuven a lot of expertise in the area of mathematical engineering, including neural networks, was available which largely motivated the study of least squares support vector machines from this perspective. Conceptually, the additional explicit *primal-dual* interpretations from the viewpoint of optimization theory turned out to be essential for further developments of least squares support vector machines. In the neural networks area the emphasis has always been on *universal models* with applications within a very broad context such as function estimation, recurrent modelling, classification, control, unsupervised learning, on-line learning and many more. Links between LS-SVMs and regulariza-

```
┌─────────────────────────────────────────────────────────────┐
│                    ┌──────────────────┐                       │
│                    │  neural networks │                       │
│                    └──────────────────┘                       │
│      ┌──────────────┐                   ┌──────────────┐      │
│      │  datamining  │                   │ linear algebra│     │
│      └──────────────┘                   └──────────────┘      │
│  ┌──────────────────┐      ╭───────╮    ┌──────────────┐      │
│  │ pattern recognition│    │       │    │  mathematics │      │
│  └──────────────────┘    │ LS-SVM  │    └──────────────┘      │
│  ┌──────────────────┐     ╰───────╯     ┌──────────────┐      │
│  │ machine learning │                   │  statistics  │      │
│  └──────────────────┘                   └──────────────┘      │
│      ┌──────────────┐           ┌──────────────────┐          │
│      │ optimization │           │ signal processing│          │
│      └──────────────┘           └──────────────────┘          │
│              ┌──────────────────────┐                         │
│              │  systems and control │                         │
│              └──────────────────────┘                         │
└─────────────────────────────────────────────────────────────┘
```

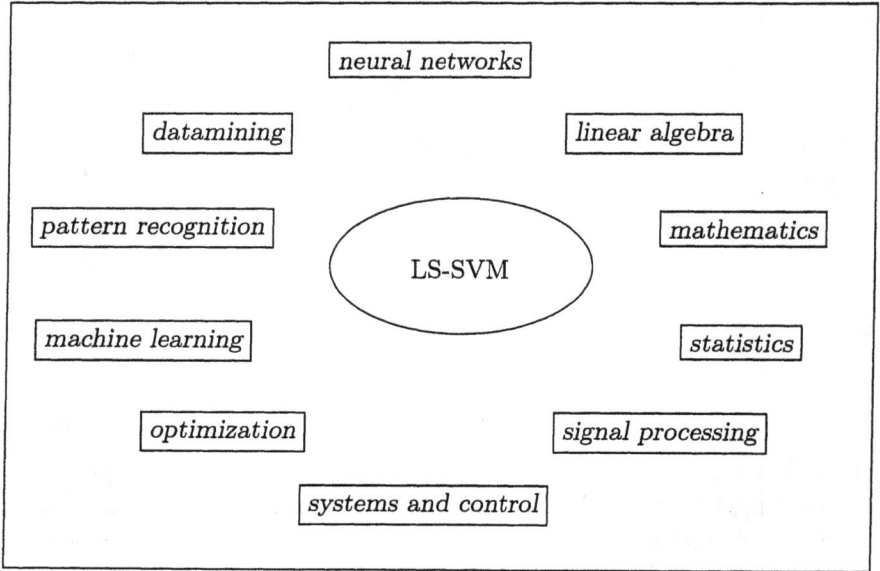

LS-SVM: an interdisciplinary topic.

tion networks, Gaussian processes and kernel Fisher discriminant analysis became clear. A Bayesian learning and a robust statistics framework were developed. Support vector machine formulations to principal component analysis and canonical correlation analysis and their kernel versions were made. Towards large scale problems the primal-dual formulations have been exploited in a fixed size LS-SVM algorithm. It turned out that several extensions of the SVM methodology are much easier to formulate in terms of least squares and equality constraints instead of other loss functions or inequality constraints. Therefore, one of the important motivations for this book is to present a *general framework* (in the sense of traditional neural nets) for a class of support vector machines towards supervised and unsupervised learning and feedforward as well as recurrent networks.

The topic of least squares support vector machines is also very interdisciplinary. The world in which we are living is characterized by fragmentation, but at the same time the emergence of several new technologies requires different fields to interact with each other. Given these two sometimes conflicting faces of reality, it is important to have common languages for transferring ideas between different fields and translating them into novel applications. Therefore another motivation for this book is to offer an *interdisciplinary forum* where different fields can meet, ranging from neural networks, machine learning, mathematics, statistics, optimization, pattern recognition, signal processing, circuits-, systems- and control theory to many applications areas. At this point we were stimulated by invitations and organizations of special sessions at international conferences and the opportunities for invited talks, tutorials and mini-courses such as for *IJCNN, ESANN, ECCTD, ISCAS, AMS, IMTC, ECC, FoCM, ICRM*. In parallel with the writing of this book, a *NATO Advanced Study Institute on Learning Theory and Practice* was organized in Leuven July 2002, which has been particularly motivating as well.

More specifically, for stimulating discussions, invitations and organization of joint meetings, we would like to thank Peter Bartlett, Sankar Basu, Jan Beirlant, Colin Campbell, Vladimir Cherkassky, Nello Cristianini, Felipe Cucker, Mark Embrechts, Lee Feldkamp, Martin Hasler, Simon Haykin, Gabor Horvath, Sathiya Keerthi, Anthony Kuh, Chih-Jen Lin, Lennart Ljung, Charlie Micchelli, Tommy Poggio, Massi Pontil, Danil Prokhorov, Johan Schoukens, Steve Smale, Stefan Vandewalle, Paul Van Dooren, Vladimir Vapnik, Mathukumalli Vidyasagar, Grace Wahba, Paul Werbos, Yu Yi, and many others.

All of us also highly appreciated the great efforts made by Bernhard Schölkopf and Alex Smola for setting up a website on kernel machines www.kernel-machines.org in recent years. For least squares support vector machines a Matlab/C toolbox called LS-SVMlab is available at

http : //www.esat.kuleuven.ac.be/sista/lssvmlab/.

We are especially grateful to Kristiaan Pelckmans for this development and to his colleagues Lukas, Bart Hamers and former master student Emmanuel Lambert.

Besides theoretical and algorithmical contributions on least squares support vector machines, several applications studies have been made with joint projects at K.U. Leuven. In the area of *bio-informatics* with applications to microarray data and textmining, we are grateful to Peter Antal, Tijl Debie, Frank De Smet, Patrick Glenisson, Bart Hamers, Kathleen Marchal, Janick Mathys, Yves Moreau and Gert Thijs. For *biomedical applications* we enjoyed our cooperation with Sabine Van Huffel in the projects on classification of brain tumours from magnetic resonance spectroscopy signals (with Andy Devos, Lukas, Rene Intzand, Leentje Vanhamme, and Rosemary Tate (University of Sussex)), detection of ovarian cancer (with Chuan Lu, and Dirk Timmerman, Ignace Vergote (K.U. Leuven Hospitals)) and prediction of mental development of preterm newborns (with Lieveke Ameye, and Hans Daniels, Gunnar Naulaers, Hugo Devlieger (K.U. Leuven Hospitals)). In the area of *nonlinear system identification and control* we are grateful to Luc Hoegaerts, Jeroen Buijs, several master students and Jakobus Barnard, Chris Aldrich (University of Stellenbosh) for our cooperation on the problem of prediction of air pollution. For *benchmarking studies and marketing applications* we want to thank Bart Baesens, Stijn Viaene, Jan Vanthienen, Guido Dedene (K.U. Leuven Applied Economic Sciences). For studies in *financial engineering* we are grateful to Dirk Baestaens (Fortis Bank Brussels) and several master students, in particular also Gert Lanckriet for contributions to the Bayesian LS-SVM framework.

Furthermore we are grateful to our many colleagues at other universities within the interuniversity poles of attraction in Belgium, the interdisciplinary center of neural networks of the K.U. Leuven, all the members of our (continuously growing) research group ESAT-SCD-SISTA and the interaction with its spin-off companies. Thanks to all of you for the great atmosphere!

Neural networks have often been presented as a universal solution to many real-life problems (sometimes even as a "miracle solution") and one has come in a stage now where it is important to understand the limits of intelligence, both artificial and human as stated by Steve Smale in his *mathematical problems for the next century* [218] (Problem 18). We are very grateful to him and others as Tommy Poggio and Felipe Cucker for invitations to special meetings on learning theory and hope that this book, together with several breakthroughs in this area, may further contribute towards understanding these exciting problems.

Johan Suykens
Tony Van Gestel
Jos De Brabanter
Bart De Moor
Joos Vandewalle
Leuven, June 2002

Acknowledgements

Our research is supported by grants from several funding agencies and sources: Research Council K.U. Leuven: Concerted Research Action GOA-Mefisto 666 (Mathematical Engineering), IDO (IOTA Oncology, Genetic networks), several PhD/postdoc & fellow grants; Flemish Government: Fund for Scientific Research FWO Flanders (several PhD/postdoc grants, projects G.0407.02 (support vector machines), G.0080.01 (collective intelligence), G.0256.97 (subspace), G.0115.01 (bio-i and microarrays), G.0240.99 (multilinear algebra), G.0197.02 (power islands), research communities IC-CoS, ANMMM), AWI (Bil. Int. Collaboration South Africa, Hungary and Poland), IWT (Soft4s (softsensors), STWW-Genprom (gene promotor prediction), GBOU McKnow (Knowledge management algorithms), Eureka-Impact (MPC-control), Eureka-FLiTE (flutter modeling), several PhD-grants); Belgian Federal Government: DWTC (IUAP IV-02 (1996-2001) and IUAP V-10-29 (2002-2006): Dynamical Systems and Control: Computation, Identification & Modelling), Program Sustainable Development PODO-II (CP-TR-18: Sustainibility effects of Traffic Management Systems); Direct contract research: Verhaert, Electrabel, Elia, Data4s, IPCOS. JS is a professor at K.U. Leuven Belgium and a postdoctoral researcher with the FWO Flanders. TVG has been a research assistant with the FWO Flanders and is presently a postdoctoral researcher with the FWO Flanders. BDM and JVDW are full professors at K.U. Leuven Belgium.

Contents

Chapter 1

Introduction

In this introductory Chapter we review some parts from the theory of neural networks which are of conceptual importance for the sequel of this book. A more extensive treatment of these topics can be found in [23; 41; 71; 110; 193; 230].

1.1 Multilayer perceptron neural networks

In the last decade many successful results have been obtained in different areas by applying neural network techniques. A major historical break-through was taking place after the introduction of multilayer perceptron (MLP) architectures together with the backpropagation method for learning the interconnection weights either off-line or on-line from given input/output patterns. Previously, one had realized that the perceptron was only able to realize a linear decision boundary in the input space by constructing a hyperplane. The perceptron was unable e.g. to learn the XOR problem which can only be solved by constructing a nonlinear decision boundary. In the sixties this caused some scepticism within the neural networks community about the general applicability of neural networks. However, thanks to the introduction of additional hidden layers more powerful multilayer perceptron architectures have been created in the age of backpropagation.

According to the McCulloch-Pitts model, a neuron is modelled as a simple static nonlinear element which takes a weighted sum of incoming signals x_i multiplied with interconnection weights w_i. After adding a bias term b (also called threshold) the resulting activation $a = \sum_i w_i x_i + b$ is sent through a static nonlinearity $h(\cdot)$ (activation function) yielding the output y such that

$$y = h(\sum_{i=1}^{n} w_i x_i + b). \qquad (1.1)$$

The nonlinearity is typically of the saturation type, e.g. $\tanh(\cdot)$. Biologically this corresponds to the firing of a neuron depending on gathered information of incoming signals that exceeds a certain threshold value. This McCulloch-Pitts model is a simple model for a biological neuron that has been frequently used within artificial neural network models. On the other hand one should also note that much more sophisticated models exist for biological neurons and the goal of artificial neural networks is not to mimic biology but rather to create a powerful class of mathematical models.

By means of a single neuron one can form a perceptron architecture as shown in Fig. 1.1. A more powerful model of a multilayer perceptron (MLP) is obtained by adding one or more hidden layers (Fig. 1.1). It is a static nonlinear model $y = f(x)$ which is described as follows in matrix-vector notation:

$$y = W \tanh(Vx + \beta) \qquad (1.2)$$

with input $x \in \mathbb{R}^n$, output $y \in \mathbb{R}^{n_y}$ and interconnection matrices $W \in \mathbb{R}^{n_y \times n_h}$, $V \in \mathbb{R}^{n_h \times n}$ for the output layer and hidden layer, respectively. The bias vector is $\beta \in \mathbb{R}^{n_h}$ and consists of the threshold values of the n_h hidden neurons. In these descriptions a linear activation function is taken for the output layer. Depending on the application one might choose other functions as well. For problems of nonlinear function estimation and regression, one takes a linear activation function in the output layer. Sometimes a neural network with two hidden layers is chosen.

One of the reasons for the neural networks success story is the fact that these are *universal approximators*. It has been mathematically proven that MLPs can approximate any continuous nonlinear function arbitrarily well over a compact interval to any degree of accuracy provided they contain one or more hidden layers. The history of these universal approximation theorems dates back in fact from the beginning of the previous century around 1900, when Hilbert formulated a list of challenging mathematical problems for the century to come. In his famous 13th problem he formulated the conjecture about the existence of analytical functions of three variables which

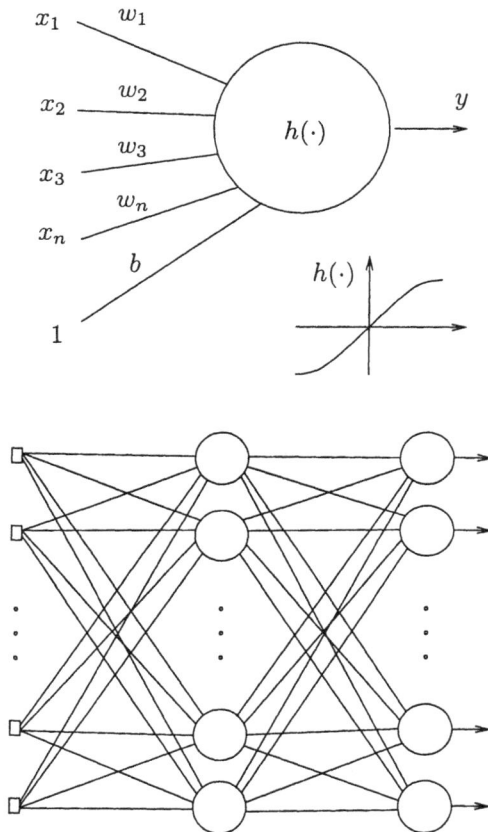

Fig. 1.1   *(Top) McCulloch-Pitts model of a neuron. A model consisting of a single neuron gives a perceptron. (Bottom) A more powerful model shown in this figure is the multilayer perceptron which is a universal approximator provided it contains one or more hidden layers.*

cannot be represented as a finite superposition of continuous functions of only two variables. This conjecture was refuted by Kolmogorov and Arnold in 1957. In 1987 Hecht-Nielsen made a connection between a refined version of the Kolmogorov theorem made by Sprecher in 1965 and showed that MLPs with two hidden layers can represent any continuous mapping. Later in 1989 it was shown by Hornik and others that one hidden layer is sufficient for universal approximation. The role of the activation function was further studied by Leshno in 1993, who showed that multilayer feedforward neural networks with locally bounded piecewise continuous activation function can approximate any continuous function to any degree of accuracy if and only if the network's activation function is not a polynomial.

Universal approximation is a nice property of neural networks. However, one can argue that also polynomial expansions possess this property. So is there any reason to prefer neural networks instead of polynomial expansions? This question brings us to a second important property of neural networks which is that they are better able to cope with the *curse of dimensionality*. In 1993 Barron showed that neural networks can avoid the curse of dimensionality in the sense that the approximation error becomes independent of the dimension of the input space (under certain conditions), which is not the case for polynomial expansions. The approximation error for MLPs with one hidden layer is of order of magnitude $\mathcal{O}(1/n_h)$, but $\mathcal{O}(1/n_p^{2/n})$ for polynomial expansions where n_h denotes the number of hidden units, n the dimension of the input space and n_p the number of terms in the expansion. Models that are based on MLPs will be able to better handle larger dimensional input spaces than polynomial expansions, which is an interesting property towards many real-life problems where one has to model dependencies between several variables.

1.2   Regression and classification

The first algorithm for the training of multilayer perceptrons and feedforward networks in general was the backpropagation algorithm. In the case of batch learning, this involves minimizing the residual squared error cost function in the unknown interconnection weights

$$\min_{\theta \in \mathbb{R}^p} J_{\text{train}}(\theta) = \frac{1}{N} \sum_{k=1}^{N} \|y_k - f(x_k; \theta)\|_2^2 \qquad (1.3)$$

where $\theta = [W(:); V(:); \beta] \in \mathbb{R}^p$ is a vector of p unknowns containing the weights and bias term. The training set of given input/output data is $\{x_k, y_k\}_{k=1}^N$ where N is the number of training data. The basic backpropagation (BP) algorithm is in essence a steepest descent local optimization algorithm for minimizing this objective function. Improvements with momentum term and adaptive learning rate have been developed as well. The BP algorithm is an elegant method for obtaining an analytic expression for the gradient of the cost function. The generalized delta rule gives recursive formulas for computing the gradient for an arbitrary number of hidden layers. The name backpropagation stems from the fact that the input patterns are propagated in a forward phase towards the output layer while the error made at the output layer is backpropagated in a backward phase towards the input layer. This is done by computing so-called delta variables based upon which the values of the interconnection weights are modified. BP can also be considered as an extension of the LMS algorithm which is well-known in the area of adaptive signal processing [109]. Several improved training methods have been developed both for off-line and on-line learning. For batch learning methods, insights from optimization theory have been taken e.g. in order to apply quasi-Newton and Levenberg-Marquardt algorithms. In these methods one usually makes approximations to the Hessian matrix. In comparison with BP, the convergence of these methods is much faster. Typically, for networks with more than about one thousand interconnection weights it might be better to apply conjugate gradient methods which don't require the storage of huge matrices. For efficient on-line learning extended Kalman filtering techniques have been applied.

A well-known problem which should be avoided with the training of neural networks is overfitting. This problem especially occurs when one optimizes a cost function of the form (1.3) in case one continues training until the local minimum is reached. Therefore, in daily neural networks practice one often divides the total number of data into a training set, validation set and test set. The validation set is used in order to decide about when to stop training (i.e. when the minimal error on this validation set is obtained). The test data are completely left untouched within the training and validation process (Fig. 1.2).

When using MLPs for function approximation one often works with the model (1.2) which contains one hidden layer of tanh activation functions and an output layer consisting of neurons with a linear characteristic. In the case of solving classification problems by MLPs one can also use tanh activation functions in the output layer as for the hidden layer. Nevertheless, often a linear characteristic is taken as in the function approximation case. The network is usually trained then in the form (1.2) and the classifier is then evaluated as $y(x) = \text{sign}[f(x)]$ or

$$y(x) = \text{sign}[W \tanh(Vx + \beta)]. \qquad (1.4)$$

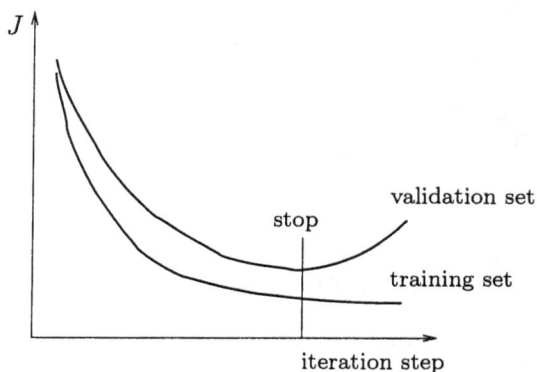

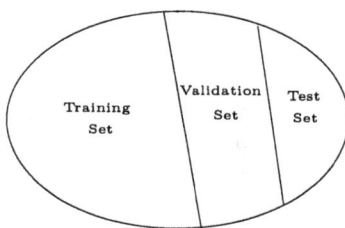

Fig. 1.2 *Training, validation and test set where the validation set is used for early stopping of the training process in order to obtain a good generalization. The test set is completely left untouched during the training and early stopping process and is used to check the performance of the trained model on fresh data.*

The network is trained then with desired output values $y_k \in \{-1, +1\}$ for binary class problems. In the multi-class case one takes additional outputs in order to represent the classes. One is confronted then with the question of how to choose the coding-decoding of the classes. One has experienced that it is good to employ as many outputs as the number of classes, but better schemes might exist as we will discuss later in this book. In Fig. 1.3 an illustration is given on the recognition of the letters of the alphabet.

7×5 pixels

input vector
dimension 35

output vector
dimension 26

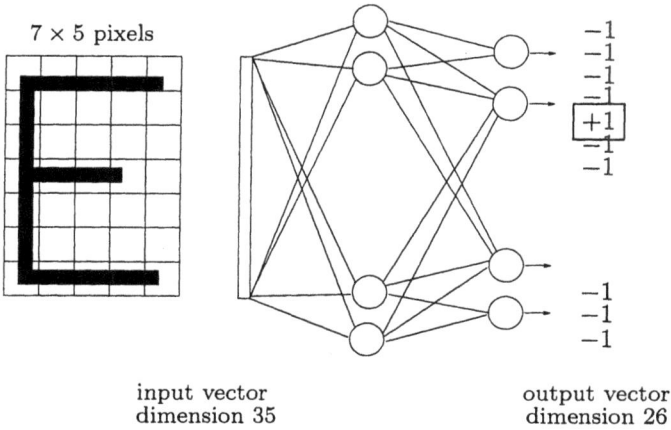

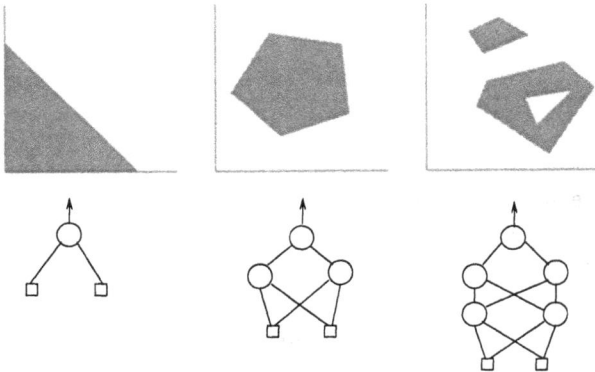

Fig. 1.3 *(Top) Illustration of a regression approach of MLPs to a multiclass classification problem on alphabet recognition; (Bottom) Complex decision boundaries can be realized by adding hidden layers. A perceptron which consists of a single neuron can only realize a linear separating hyperplane.*

1.3 Learning and generalization

1.3.1 *Weight decay and effective number of parameters*

After an initial boom of MLP applications with the BP algorithm in the early nineties, the problems have been investigated in an interdisciplinary manner with contributions from several fields such as statistics, engineering and physics. One of the important ideas was to add a weight decay term to the cost function and modify (1.3) into

$$\min_{\theta \in \mathbb{R}^p} J_{\text{reg}}(\theta) = \nu \frac{1}{2} \theta^T \theta + \frac{1}{N} \sum_{k=1}^{N} \|y_k - f(x_k; \theta)\|_2^2 \qquad (1.5)$$

which means that one should keep the values of the interconnection weights small in addition to obtaining a good fit of the data. Thanks to the regularization mechanism one can work with an effective number of parameters that is less than the number of interconnection weights depending on the choice of the regularization constant ν. This means that even when the number of neurons is large, it is still possible that one can implicitly work with a much smaller number of parameters and obtain a good generalization.

The *effective number of parameters* is given by

$$d_{\text{eff}} = \sum_{i=1}^{p} \frac{\lambda_i}{\lambda_i + \nu} \qquad (1.6)$$

where λ_i denotes the i-th eigenvalue of the (positive definite) Hessian matrix at a point in the interconnection weight space for the unregularized problem. When many of the values $\lambda_i \ll \nu$, the influence of many interconnection weights will be implicitly suppressed by the regularization mechanism [23; 149]. In this way the overfitting problem can be avoided without doing early stopping of the training process provided that a good choice is made of the regularization constant. Also one can show that an early stopping process implicitly corresponds to a certain form of regularization. The advantage of suppressing the influence of interconnection weights is clear from complexity criteria such as Moody's criterion which states that the generalized prediction error (GPE) equals

$$\text{GPE} = J_{\text{train}} + \frac{d_{\text{eff}}}{N} \sigma_e^2 \qquad (1.7)$$

where σ_e^2 denotes the variance of the noise. This can be considered as an extension of the well-known Akaike information criterion from linear to nonlinear models where the number of unknown parameters is replaced by the effective number of parameters.

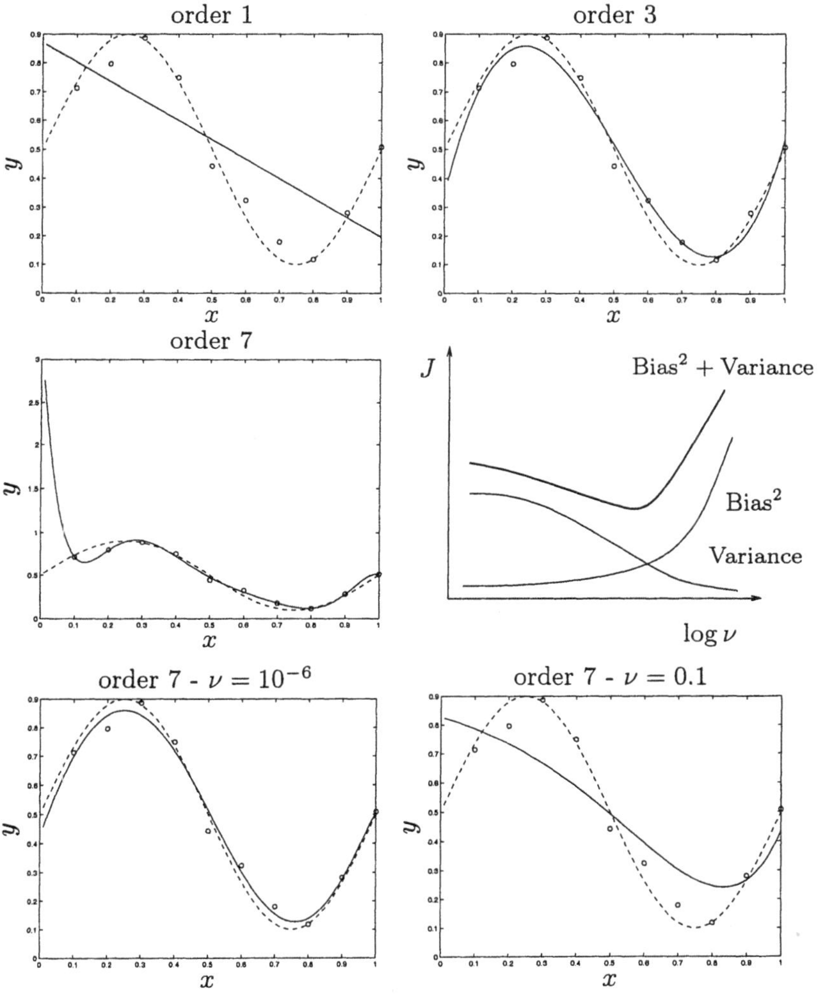

Fig. 1.4 *Estimation of a noisy sine function: polynomial models of order 1 (Top-left), order 3 (Top-right) and order 7 (Middle-left); (Middle-right) Bias-variance trade-off when applying regularization; (Bottom) ridge regression applied to the polynomial model of order 7 with $\nu = 10^{-6}$ (left) and for $\nu = 0.1$ (right).*

1.3.2 *Ridge regression*

The use of weight decay is a parametric form of regularization and is closely related to ridge regression. Let us illustrate this by means of a simple example. Assume that training data are generated from the true underlying function $f_0(x) = 0.5 + 0.4\sin(2\pi x)$ with training data x_k generated in the interval $[0.1, 1]$ with steps of 0.1 and zero mean Gaussian noise with standard deviation 0.05 is added to the output values, giving a training data set $\{x_k, y_k\}_{k=1}^{10}$. From the training data polynomials of different degrees are estimated. For example, for the polynomial model of degree 3

$$y = a_1 x + a_2 x^2 + a_3 x^3 + b \tag{1.8}$$

an overdetermined set of linear equations is constructed for the given data $\{x_k, y_k\}_{k=1}^{10}$

$$\begin{bmatrix} x_1 & x_1^2 & x_1^3 & 1 \\ x_2 & x_2^2 & x_2^3 & 1 \\ \vdots & \vdots & \vdots & \vdots \\ x_{10} & x_{10}^2 & x_{10}^3 & 1 \end{bmatrix} \begin{bmatrix} a_1 \\ a_2 \\ a_3 \\ b \end{bmatrix} = \begin{bmatrix} y_1 \\ y_2 \\ \vdots \\ y_{10} \end{bmatrix}. \tag{1.9}$$

This overdetermined system is of the form $\mathcal{A}\theta = \mathcal{B}$ with $\mathcal{A} \in \mathbb{R}^{q \times p}$ where $q > p$, $\theta = [a_1; a_2; a_3; b]$ and $e = \mathcal{A}\theta - \mathcal{B}$. By taking a cost function in a linear least squares sense, $\min_\theta J_{\mathrm{LS}}(\theta) = \frac{1}{2}e^T e = \frac{1}{2}(\mathcal{A}\theta - \mathcal{B})^T(\mathcal{A}\theta - \mathcal{B})$, the condition for optimality $\partial J_{\mathrm{LS}}(\theta)/\partial\theta = 0$ gives the solution [98]

$$\theta_{\mathrm{LS}} = (\mathcal{A}^T \mathcal{A})^{-1}\mathcal{A}^T \mathcal{B} = \mathcal{A}^\dagger \mathcal{B} \tag{1.10}$$

where $\mathcal{A}^\dagger$ denotes the pseudo inverse matrix. One can observe that for the higher order polynomial (order 7) the solution starts oscillating and overfitting is obtained because the polynomial interpolates the given training data but fails to generalize well in between the given training data points.

Let us now modify the least squares cost function by an additional term which aims at keeping the norm of the solution vector small. In ridge regression one solves $\min_\theta J_{\mathrm{ridge}}(\theta) = J_{\mathrm{LS}}(\theta) + \frac{1}{2}\nu\|\theta\|_2^2$, $\nu > 0$ which gives the following solution after taking the condition for optimality $\partial J_{\mathrm{ridge}}(\theta)/\partial\theta = 0$ [98]

$$\theta_{\mathrm{ridge}} = (\mathcal{A}^T \mathcal{A} + \nu I)^{-1}\mathcal{A}^T \mathcal{B}. \tag{1.11}$$

This technique is useful when $\mathcal{A}^T \mathcal{A}$ is *ill conditioned*. The results of ridge regression for the order 7 polynomial model is shown on Fig. 1.4 for different values of the regularization constant ν. Figure 1.4 conceptually shows the bias-variance trade-off in terms of the regularization constant ν. A large value ν decreases the variance but leads to a larger bias. This value of

ν is chosen as a trade-off solution by minimizing the sum of the variance and the bias square contributions. According to e.g. [107](pp.196-200) the bias-variance trade-off can be understood as follows. Assume a model of the form $Y = f(x) + e$ where $\mathcal{E}[e] = 0$ and $\mathrm{Var}(e) = \sigma_e^2$ with random output variable Y. The expected prediction error of a regression fit $\hat{f}(x)$ evaluated at input $x = x_0$ is then given by the following expression when using a squared error loss function:

$$
\begin{aligned}
\mathrm{PE}(x_0) &= \mathcal{E}[(Y - \hat{f}(x))^2 | x = x_0] \\
&= \sigma_e^2 + (\mathcal{E}[\hat{f}(x_0)] - f(x_0))^2 + \mathcal{E}[\hat{f}(x_0) - \mathcal{E}[\hat{f}(x_0)]]^2 \quad (1.12) \\
&= \sigma_e^2 + \mathrm{Bias}^2[\hat{f}(x_0)] + \mathrm{Var}[\hat{f}(x_0)].
\end{aligned}
$$

The first term is the variance of the target around its mean $f(x_0)$ and is an irreducible error no matter how well one estimates $f(x_0)$. The second term is the squared bias which characterizes the amount by which the average of our estimates differ from the true mean. The last term is the variance which is the expected squared deviation around its mean. The interpretation for the present example on ridge regression is that the more complex we make the model (higher degree of the polynomial) the lower the bias but the larger the variance.

In order to make a suitable choice for the regularization constant ν one has several options then. The following methods are popular in the neural networks area: (a) taking a value for the regularization constant that minimizes the error on a separate validation set; (b) instead of working with a single validation set one can work with a n_{CV}-fold cross-validation procedure (Fig. 1.5) and take ν in such a way that sum of the errors on the validation sets that are left out in the several runs is minimal; (c) applying methods of Bayesian inference which enable to do automatic selection of the hyperparameters.

1.3.3 *Bayesian learning*

In Bayesian learning [23; 149; 150; 151] a distribution is considered on the unknown parameter vector θ, rather than considering θ as a point in parameter space. For data $\mathcal{D}$ and parameter vector θ the application of Bayes' rule gives

$$
p(\theta | \mathcal{D}) = \frac{p(\mathcal{D} | \theta)}{p(\mathcal{D})} p(\theta) \quad (1.13)
$$

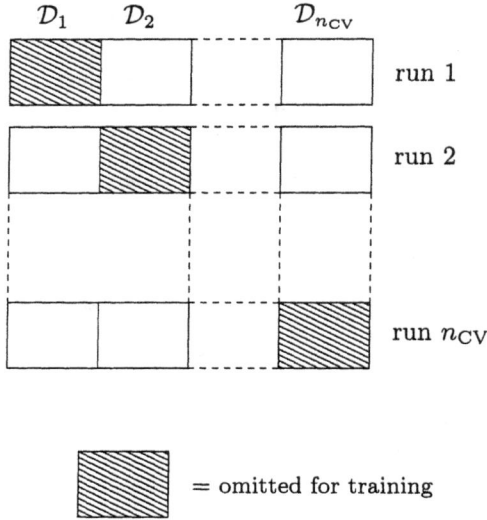

= omitted for training

Fig. 1.5 *In 10-fold cross-validation a number of $n_{CV} = 10$ runs are done where in each run a different part of the data set is omitted and the validation error is finally checked as the sum of error costs on the sets that were omitted.*

meaning that

$$\text{Posterior} = \frac{\text{Likelihood}}{\text{Evidence}} \times \text{Prior}. \qquad (1.14)$$

For i.i.d. data x_k ($k = 1, ..., N$) one has the likelihood $p(\mathcal{D}|\theta) = \prod_{k=1}^{N} p(x_k|\theta)$. The normalization factor $p(\mathcal{D}) = \int p(\theta') \prod_{k=1}^{N} p(x_k|\theta')d\theta'$ ensures $\int p(\theta|\mathcal{D})\, d\theta = 1$. Figure 1.6 illustrates the process of obtaining an estimate for the posterior $p(\theta|\mathcal{D})$ by combining the prior $p(\theta)$ with the data $\mathcal{D}$. In advanced methods for Bayesian inference of neural networks, this Bayes rule is applied at different levels. At a first level one considers a density in the unknown parameter values of the interconnection weights. The regularization weight decay term is related to the prior (keeping the weights small), while the likelihood corresponds to the training set error. Inference at the second level is done in the unknown hyperparameters which are the regularization constants that penalize the importance of the regularization term versus the training set error. Application of the Bayes rule at this level leads to implicit formulas for automatic selection of suitable regularization constants, which depend on the effective number of parameters. At the third level of inference model comparison can be done where the Bayes rule embodies the principle of Occam's razor which states that simple models are preferred.

More specifically, a Bayesian learning approach to regression starts from the following cost function with regularization

$$\min_{\theta \in \mathbb{R}^p} J_{\text{reg}}(\theta) = \mu \frac{1}{2} \theta^T \theta + \zeta \frac{1}{2} \sum_{k=1}^{N} \|y_k - f(x_k; \theta)\|_2^2 \qquad (1.15)$$

where $\theta \in \mathbb{R}^p$ denotes the unknown parameters of the neural network and μ, ζ are called *hyperparameters*. Bayesian inference is done at the following 3 levels:

- *Level 1* [inference of parameters]

$$p(\theta|\mathcal{D}, \mu, \zeta, \mathcal{H}_i) = \frac{p(\mathcal{D}|\theta, \mu, \zeta, \mathcal{H}_i)}{p(\mathcal{D}|\mu, \zeta, \mathcal{H}_i)} p(\theta|\mu, \zeta, \mathcal{H}_i) \qquad (1.16)$$

- *Level 2* [inference of hyperparameters]

$$p(\mu, \zeta|\mathcal{D}, \mathcal{H}_i) = \frac{p(\mathcal{D}|\mu, \zeta, \mathcal{H}_i)}{p(\mathcal{D}|\mathcal{H}_i)} p(\mu, \zeta|\mathcal{H}_i) \qquad (1.17)$$

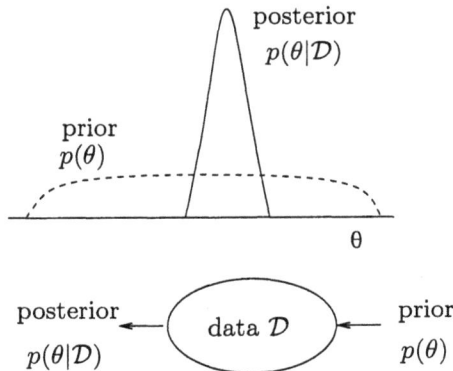

Fig. 1.6 *Bayesian inference: starting from a prior $p(\theta)$ with a large uncertainty on the parameter vector θ, the data $\mathcal{D}$ are used in order to generate a posterior $p(\theta|\mathcal{D})$ which becomes more accurate.*

- *Level 3* [model comparison]

$$p(\mathcal{H}_i|\mathcal{D}) \propto p(\mathcal{D}|\mathcal{H}_i)p(\mathcal{H}_i) \tag{1.18}$$

where $\mathcal{H}_i$ denotes the i-th model ($i = 1, ..., n_{\mathcal{H}}$) with $n_{\mathcal{H}}$ the number of models to be compared (this could for example be models with different numbers of hidden neurons). The connection between the posterior and prior on θ at Level 1 and the cost function (1.15) is made as follows:

$$\begin{aligned} p(\theta|\mathcal{D},\mu,\zeta,\mathcal{H}_i) &\propto \exp(-\zeta \tfrac{1}{2} \textstyle\sum_{k=1}^{N} \|y_k - f(x_k;\theta)\|_2^2) \\ p(\theta|\mu,\zeta,\mathcal{H}_i) &\propto \exp(-\mu \tfrac{1}{2} \theta^T \theta). \end{aligned} \tag{1.19}$$

The connection between Level 1 and Level 2 is made by observing that the Evidence at Level 1 for the inference of θ equals the Likelihood at Level 2 for the inference of the hyperparameters μ, ζ.

Other techniques that have been used in the neural networks area in order to improve generalization are *pruning methods* where one starts from a neural network with a large number of interconnection weights and gradually prunes the least relevant weights by computing so-called saliency values. Several versions have been developed such as optimal brain damage and optimal brain surgeon. One can also improve the generalization by combining several models and improve the results in view of the bias-variance trade-off. This is done by so-called committee networks (Fig. 1.7).

1.4 Principles of pattern recognition

1.4.1 *Bayes rule and optimal classifier under Gaussian assumptions*

The link between artificial neural networks and methods in pattern recognition and statistical decision theory has been quite well understood [23]. Suppose we consider a number of n_C classes and continuous variables x belonging to the input space. Application of the Bayes rule gives then

$$P(\mathcal{C}_i|x) = \frac{p(x|\mathcal{C}_i)}{p(x)} P(\mathcal{C}_i) \ , \ i=1,...,n_C \ , \ x \in \mathbb{R}^n \tag{1.20}$$

with normalization $\sum_{i=1}^{n_c} P(\mathcal{C}_i|x) = 1$ and unconditional density $p(x) = \sum_{i=1}^{n_c} p(x|\mathcal{C}_i)P(\mathcal{C}_i)$, where $P(\mathcal{C}_i)$ denotes the prior class probability of class

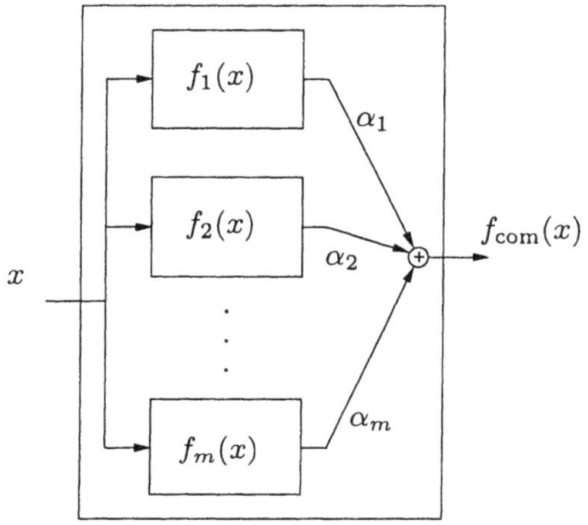

Fig. 1.7 *Combining models $f_i(x)$ for $i = 1, ..., m$ into a committee network $f_{com}(x) = \sum_{i=1}^{m} \alpha_i f_i(x)$ in order to improve the estimate in view of the bias-variance trade-off.*

$\mathcal{C}_i$, $P(\mathcal{C}_i|x)$ the posterior class probability of class $\mathcal{C}_i$ and $p(x|\mathcal{C}_i)$ the class conditional density. If one assigns a pattern x to a class i^* such that

$$i^* = \arg \max_{i=1,...,n_C} P(\mathcal{C}_i|x) \tag{1.21}$$

one can show that this classification rule which is based on the posterior class probability leads to a *minimal probability of misclassification*. In Fig. 1.8 it is illustrated for a binary class problem that this decision rule leads to a minimal shaded area determined by the curves $p(x|\mathcal{C}_1)P(\mathcal{C}_1)$ and $p(x|\mathcal{C}_2)P(\mathcal{C}_2)$. Choosing a threshold decision line at the intersection of the two curves yields a minimal misclassification.

In the context of discriminant functions the Bayes decision rule can be interpreted as assigning x to class $\mathcal{C}_i$ such that

$$d_i(x) > d_j(x) \ , \ \forall i \neq j \tag{1.22}$$

where $d_i(x), d_j(x)$ denote discriminant functions. The case

$$d_i(x) \ \propto \ p(x|\mathcal{C}_i)P(\mathcal{C}_i) \tag{1.23}$$

corresponds then to minimizing the probability of misclassification. Only relative magnitudes of discriminant functions are important. One may ap-

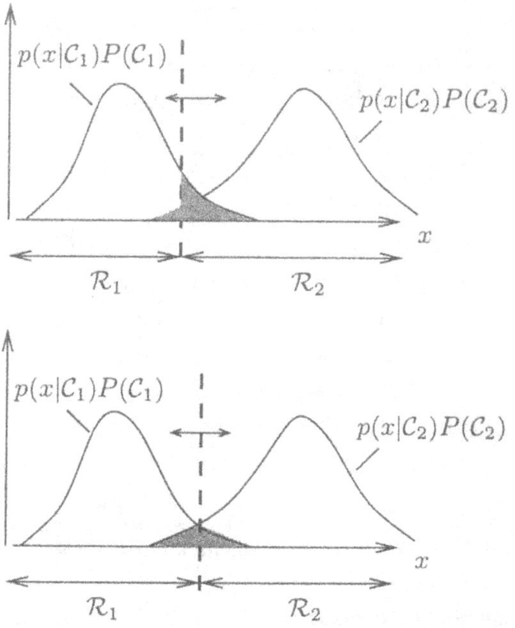

Fig. 1.8 *A minimal shaded area corresponds to minimizing the probability of misclassification (bottom figure), when considering a moving threshold line (dashed vertical line).*

ply a monotonic function $g(\cdot)$ as $g(d_i(x))$, e.g.: $d_i(x) = \log[p(x|C_i)P(C_i)] = \log p(x|C_i) + \log P(C_i)$. The decision boundaries $d_i(x) = d_j(x)$ are not influenced by the choice of this monotonic function. In the case of a binary classification problem one may take a reformulation by a single discriminant function $d(x) = d_1(x) - d_2(x)$ with class C_1 if $d(x) > 0$ and class C_2 if $d(x) < 0$. Instead of using two discriminant functions one can take a single one then.

It is often meaningful then to *assume* that the densities $p(x|C_i)$ are normally distributed for all classes $i = 1, ..., n_C$ and consider the discriminant functions [71; 255]

$$d_i(x) = \log p(x|C_i) + \log P(C_i). \qquad (1.24)$$

Recall that a multivariate normal density is given by $p(x) = ((2\pi)^n |\Sigma_{xx}|)^{-1/2}$ $\exp(-\frac{1}{2}(x - \mu_x)^T \Sigma_{xx}^{-1}(x - \mu_x))$ with $\int_{-\infty}^{\infty} p(x)dx = 1$, $\mu_x = \mathcal{E}[x] \in \mathbb{R}^n$

the mean, $\Sigma_{xx} = \mathcal{E}[(x - \mu_x)(x - \mu_x)^T] \in \mathbb{R}^{n \times n}$ the covariance matrix $(\Sigma_{xx} = \Sigma_{xx}^T > 0)$, and $|\Sigma_{xx}|$ the determinant of Σ_{xx}. Under this Gaussian assumption one obtains the following discriminant functions

$$d_i(x) = -\frac{1}{2}(x - \mu_{x_i})^T \Sigma_{xx_i}^{-1} (x - \mu_{x_i}) - \frac{1}{2} \log |\Sigma_{xx_i}| + \log P(\mathcal{C}_i). \quad (1.25)$$

The decision boundaries which are characterized by $d_i(x) = d_j(x)$ are quadratic forms in $\mathbb{R}^n$. For the binary classification case one obtains the following important special cases (Fig. 1.9):

- Case $\Sigma_{xx_1} = \Sigma_{xx_2} = \Sigma_{xx}$:
  For equal covariance matrices the decision boundary becomes linear, for distributions having a small overlap as well as a large overlap. The fact that a linear decision boundary is optimal in the case of overlapping distributions means in fact that one has to tolerate misclassifications. Allowing no misclassifications in such a case on training data, which would have been generated from such underlying distributions, would lead to an overfitting situation in this classification context. This result also gives a better insight between neural networks and overfitting for classification problems.

- Case $\Sigma_{xx_1} = \Sigma_{xx_2} = \Sigma_{xx} = \sigma_x^2 I$:
  Under this assumption one obtains

$$d_i(x) = -\frac{\|x - \mu_{x_i}\|_2^2}{2\sigma_x^2} + \log P(\mathcal{C}_i), \quad i \in \{1, 2\}. \quad (1.26)$$

If the prior class probabilities are equal, then the mean vectors μ_{x_1}, μ_{x_2} act as prototypes. Under these assumptions the classification can be done by a simple calculation of the Euclidean distance to μ_{x_i}. When applying this kind of classification rule in a more general context one should be aware of the fact that this classification rule is only optimal under these restrictive assumptions.

The link between these results and neural network architectures can be made now as follows. For normally distributed class-conditional densities, the posterior probabilities can be obtained by a logistic one-neuron network

$$y = h(w^T x + b) \quad (1.27)$$

Case $\Sigma_{\mathbf{xx}_1} = \Sigma_{\mathbf{xx}_2} = \Sigma_{\mathbf{xx}}$

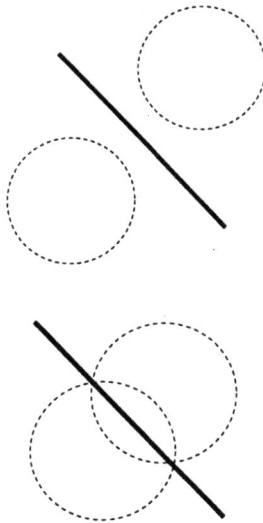

Case $\Sigma_{\mathbf{xx}_1} \neq \Sigma_{\mathbf{xx}_2}$

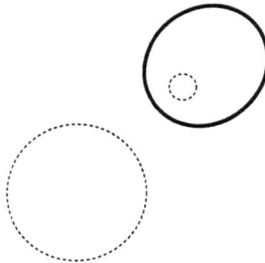

Fig. 1.9 *Optimal linear and quadratic decision boundaries under different assumptions for the covariance matrices of the classes in a binary classification problem.*

with logistic activation function $h(a) = 1/(1 + \exp(-a)) \in [0, 1]$ and activation $a = w^T x + b$. For class-conditional densities with $\Sigma_{\mathbf{xx}_1} = \Sigma_{\mathbf{xx}_2} = \Sigma_{\mathbf{xx}}$ we had $p(x|\mathcal{C}_i) = ((2\pi)^n |\Sigma_{\mathbf{xx}_i}|)^{-1/2} \exp(-\frac{1}{2}(x - \mu_{x_i})^T \Sigma_{\mathbf{xx}_i}^{-1}(x - \mu_{x_i}))$. This leads to the following posterior

$$P(\mathcal{C}_1|x) = h(w^T x + b) \tag{1.28}$$

with

$$w = \Sigma_{xx}^{-1}(\mu_{x_1} - \mu_{x_2})$$

$$b = -\tfrac{1}{2}\mu_{x_1}^T \Sigma_{xx}^{-1}\mu_{x_1} + \tfrac{1}{2}\mu_{x_2}^T \Sigma_{xx}^{-1}\mu_{x_2} + \log \frac{P(\mathcal{C}_1)}{P(\mathcal{C}_2)}.$$

(1.29)

Note that the bias term b depends on the prior class probabilities (which are often unknown in practice). Hence, different prior class probabilities will lead to a translational shift of the hyperplane or straight line in the case of a two dimensional feature space. In the other case when the above mentioned assumptions do not hold one may apply techniques for density estimation such as mixture models.

1.4.2 *Receiver operating characteristic*

For binary classification problems one can consider the following so-called confusion matrix shown in Fig. 1.10 with TP the number of correctly classified positive cases, TN the number of correctly classified negative cases, FP the number of wrongly classified positive cases, FN the number of wrongly classified negative cases. Negative/positive means class 1/class 2 (the meaning of negative or positive might depend on the specific application area, e.g. in biomedicine it means malignant/benign) and true/false means correctly/wrongly classified data, respectively.

It is convenient then to define

$$\text{Sensitivity} = \text{TP}/(\text{TP} + \text{FN})$$

$$\text{Specificity} = \text{TN}/(\text{FP} + \text{TN})$$

$$\text{False positive rate} = 1 - \text{Specificity} = \text{FP}/(\text{FP} + \text{TN}).$$

(1.30)

The receiver-operating characteristic (ROC) curve [249; 250] shows the sensitivity with respect to the false positive rate (Fig. 1.10). The larger the area under the ROC curve the better the classifier, which is achieved when the classifier has a high sensitivity for a small false positive rate. This ROC curve method has been applied since the second world war where it was used on radar signals. Later it became popular in biomedical application areas. These criteria should be evaluated not only on training data but also on test data. In Fig. 1.10 classifier C is better than B and the classifier with curve A has no discriminatory power at all. The points T_1, T_2, T_3 illustrated on a ROC curve B on Fig. 1.10 are examples of different operating

	Negative	Positive
True	TN	TP
False	FN	FP

Fig. 1.10 *(Top-Left) Confusion matrix; (Top-Right) ROC curve for three classifiers A,B,C. The classifier C has the best performance. It has the largest area under the ROC curve; (Bottom) An ROC curve is obtained by moving the threshold T of the classifier.*

points on the ROC curve. These correspond to varying decision threshold values of an output unit. Note that in the case of logistic discrimination this corresponds in fact to varying the prior class probabilities, because the bias term explicitly depends on these probabilities.

1.5 Dimensionality reduction methods

In view of complexity criteria it is often useful to reduce the dimensionality of the input space. In this way a smaller amount of interconnection weights will be needed for the neural network.

A well-known and frequently used technique for dimensionality reduction is linear PCA analysis. Suppose one wants to map vectors $x \in \mathbb{R}^n$

into lower dimensional vectors $z \in \mathbb{R}^m$ with $m < n$. One proceeds then by estimating the covariance matrix:

$$\hat{\Sigma}_{\mathbf{xx}} = \frac{1}{N-1} \sum_{k=1}^{N} (x_k - \overline{x})(x_k - \overline{x})^T \qquad (1.31)$$

where $\overline{x} = \frac{1}{N} \sum_{k=1}^{N} x_k$ and computes the eigenvalue decomposition

$$\hat{\Sigma}_{\mathbf{xx}} \, u_i = \lambda_i u_i. \qquad (1.32)$$

By selecting the m largest eigenvalues and the corresponding eigenvectors, one obtains the transformed variables (score variables)

$$z_i = u_i^T (x - \overline{x}), \quad i = 1, ..., m. \qquad (1.33)$$

One has to note, however, that these transformed variables are no longer real physical variables. The error $\sum_{i=m+1}^{n} \lambda_i$ resulting from the dimensionality reduction is determined by the values of the neglected components.

Another possibility for dimensionality reduction, but in a supervised learning context, is the use of Automatic Relevance Determination (ARD) [151]. In this method one assigns additional regularization constants to each set of interconnection weights associated with a certain input. These additional hyperparameters are inferred at the second level of inference within the Bayesian learning context. In this case one finds the relevance of the original input variables itself instead of transformed variables in the PCA analysis case.

The case of linear PCA analysis can also be extended to nonlinear PCA analysis making use of neural nets [23]. Consider patterns $x \in \mathbb{R}^n$ in the original space and transformed inputs $z \in \mathbb{R}^m$ in a lower dimensional space with $m \ll n$. Conceptually this problem can be understood as an encoding/decoding problem to minimize the reconstruction error. For the encoder mapping $z = G(x)$ and decoder mapping $\hat{x} = F(z)$ one has the following objective of minimizing the squared distortion error between the original variables $x \in \mathbb{R}^n$ and the reconstructed variables $\hat{x} \in \mathbb{R}^n$

$$
\begin{aligned}
\min J &= \frac{1}{N} \sum_{k=1}^{N} \|x_k - \hat{x}_k\|_2^2 \\
&= \frac{1}{N} \sum_{k=1}^{N} \|x_k - F(G(x_k))\|_2^2
\end{aligned}
\qquad (1.34)
$$

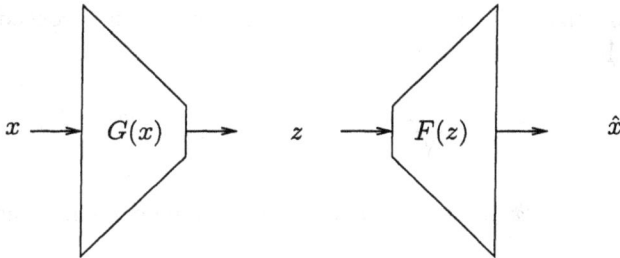

Fig. 1.11 *Information bottleneck in dimensionality reduction. Linear PCA is related to linear mappings $F(\cdot), G(\cdot)$. Nonlinear PCA analysis is obtained by considering nonlinear mappings that can be parameterized e.g. by MLPs.*

where the error measure is evaluated on a given training set of data points. An important special case is obtained when $F(\cdot), G(\cdot)$ are linear mappings. It can be proven that this corresponds to linear PCA analysis. However, one can take these mappings also nonlinear and parameterize it e.g. by means of MLPs. In that case it leads to nonlinear PCA analysis. This insight gives an interesting interpretation of linear and nonlinear PCA analysis in terms of regression.

1.6 Parametric versus non-parametric approaches and RBF networks

Besides multilayer perceptrons another popular class of neural networks are Radial Basis Function (RBF) networks. While parametric regularization is important for reliable training of MLPs, the optimality of RBF networks can be understood in terms of non-parametric regularization. In [183] RBF networks have been derived from a viewpoint of functional analysis and calculus of variations by Poggio & Girosi, in relation to ill-posed problems [252]. Given a training data set $\{x_k, y_k\}_{k=1}^{N}$ one considers the problem

$$\min_{f} J[f] = \sum_{k=1}^{N}(y_k - f(x_k))^2 + \nu\|Pf\|^2 \qquad (1.35)$$

where $\nu > 0$ denotes the regularization parameter, P is a differential operator and $\|\cdot\|$ usually corresponds to the l_2 norm defined on the function space. Loosely speaking, this regularization term means that one keeps higher order derivatives small for the function to be estimated. One has shown that for a specific choice of the differential operator one obtains RBF networks as the optimal solution (for other choices one obtains e.g. spline networks). The optimal network model is given by

$$f^*(x) = \sum_{k=1}^{N} \alpha_k \, G(\|x - x_k\|) \tag{1.36}$$

where $G(\cdot)$ denotes a Gaussian activation function. These are centered at each of the given points of the data set. The weighting coefficients follow from the solution to a linear system

$$(G + \nu I)\alpha = y \tag{1.37}$$

where $y = [y_1; y_2; ...; y_N]$, $\alpha = [\alpha_1; \alpha_2; ...; \alpha_N]$ and the kl-th entry of the matrix G equals $G_{kl} = G(x_k, x_l) = G(\|x_k - x_l\|)$. Note that the size of this system grows with the number of data. The form of the network is also related to methods in non-parametric statistics where one often employs kernel based models of the form

$$p(x) = \frac{1}{N} \sum_{k=1}^{N} \frac{1}{(2\pi\sigma^2)^{n/2}} \exp\left(-\frac{\|x - x_k\|_2^2}{2\sigma^2}\right) \tag{1.38}$$

especially in the context of density estimation, where a careful choice of the kernel width h should be made and no other parameters are to be determined.

RBF networks are inspired on (1.36) but are a *parameterized version* of it. One parameterizes a model into the form

$$f_{\mathrm{rbf}}(x; w_i, c_i) = \sum_{i=1}^{n_h} w_i \, G(\|x - c_i\|) \tag{1.39}$$

where one takes a number of n_h hidden neurons. The unknown parameters are $w_i \in \mathbb{R}, c_i \in \mathbb{R}^n$ for $i = 1, ..., n_h$. Note that the number of parameters to be estimated in this case is fixed and independent of the number of data which is not the case for (1.36), (1.37). The goal is then to minimize

$$\min_{w_i, c_i} J(w_i, c_i) = \sum_{k=1}^{N} \left(y_k - f_{\mathrm{rbf}}(x_k; w_i, c_i)\right)^2 + \nu \|P f_{\mathrm{rbf}}(x_k; w_i, c_i)\|^2. \tag{1.40}$$

In the case of fixed centers the optimal solution is given by

$$w = (G^T G + \nu R)^{-1} G^T y \qquad (1.41)$$

with $G_{ki} = G(x_k, c_i)$ and $R_{ij} = G(c_i, c_j)$. Often, in a first stage, one tries to find a good choice of the centers by means of an unsupervised learning algorithm or cluster algorithm such as K-means and compute then the solution (1.41) for the output weights once the centers have been fixed. An alternative is to solve the entire nonlinear optimization problem in all unknowns w_i, c_i in a brute force way. One can also work with a weighted norm and consider the elements of the weighting matrix as additional unknowns to the optimization problem. Finally, one should note that in the case the centers c_i for $i = 1, ..., n_h$ coincide with the training data points x_k for $k = 1, ..., N$ one has $R = G$ and $G = G^T$. One can then write $w = (GG + \nu G)^{-1} Gy = (G + \nu I)^{-1} y$ which equals (1.37). These insights clarify links and differences between the non-parametric (1.37) and parametric solution (1.41) and ridge regression.

1.7 Feedforward versus recurrent network models

When using neural networks in a dynamical systems context it is important to decide about the model structure. From a system and identification theory viewpoint [144; 216; 230] one has input/output (I/O) models and state space models. In the context of I/O models it is important to make a distinction between NARX (Nonlinear AutoRegressive with eXogenous input) and NOE (Nonlinear Output Error) models. In NARX models one has

$$\hat{y}_k = f(y_{k-1}, y_{k-2}, ..., y_{k-q}, u_{k-1}, u_{k-2}, ..., u_{k-q}) \qquad (1.42)$$

where y_k denotes the true output at discrete time instant k, u_k the input at time k and $\hat{y}_k$ the estimated output at time k. The number q corresponds to the order of the system. In NOE models one has

$$\hat{y}_k = f(\hat{y}_{k-1}, \hat{y}_{k-2}, ..., \hat{y}_{k-q}, u_{k-1}, u_{k-2}, ..., u_{k-q}). \qquad (1.43)$$

Note that one has a recursion now on the variable $\hat{y}_k$ in contrast with the NARX model. From a neural networks perspective, the NARX model may be considered as a feedforward model, while the NOE model is *recurrent*.

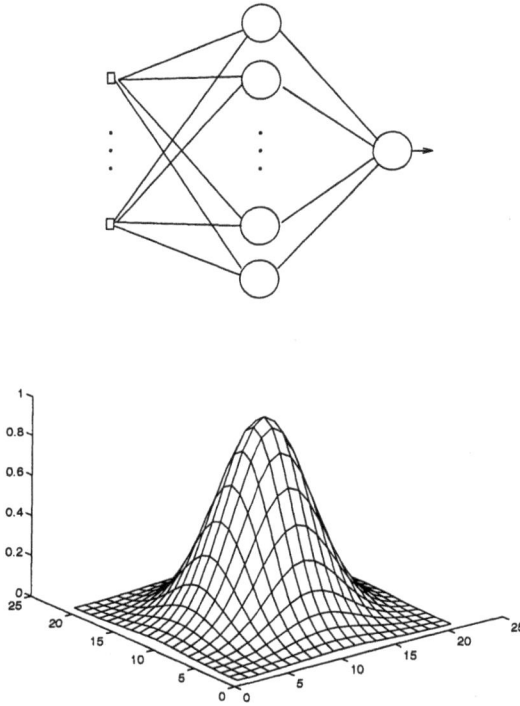

Fig. 1.12 *Radial basis function network with Gaussian activation function.*

The nonlinear function $f(\cdot)$ is parameterized by means of a multilayer perceptron or an RBF network. The unknown parameters of the neural network are determined then by minimizing a suitable cost function for given training data.

Models for time-series prediction are closely related to these models by omitting the input variable u. One obtains then

$$\hat{y}_{k+1} = f(y_k, y_{k-1}, ..., y_{k-q}) \tag{1.44}$$

which is parameterized by an MLP as

$$\hat{y}_{k+1} = w^T \tanh(V[y_k; y_{k-1}; ...; y_{k-q}] + \beta). \tag{1.45}$$

It is not necessary that the past values $y_k, y_{k-1}, ..., y_{k-q}$ are subsequent in time; certain values could be omitted or values at different time scales could be taken. In order to generate predictions, the true values y_k are replaced then by the estimated values $\hat{y}_k$ and the iterative prediction is generated by the recurrent network

$$\hat{y}_{k+1} = w^T \tanh(V[\hat{y}_k; \hat{y}_{k-1}; ...; \hat{y}_{k-q}] + \beta) \tag{1.46}$$

for a given initial condition.

Instead of I/O models one may also take discrete time nonlinear state space descriptions

$$\begin{cases} \hat{x}_{k+1} &= f(\hat{x}_k, u_k) \\ \hat{y}_k &= g(\hat{x}_k) \end{cases} \tag{1.47}$$

where $f(\cdot) : \mathbb{R}^n \times \mathbb{R}^{n_u} \to \mathbb{R}^n$ and $g(\cdot) : \mathbb{R}^n \to \mathbb{R}^{n_v}$ are nonlinear mappings. For the autonomous case (external inputs set to zero), (1.47) and (1.44) are related through Takens' embedding theorem. This states that by taking $q \geq 2n$ one can reconstruct the state space from a single time series $\{y_k\}$. When one parameterizes these nonlinear functions by means of a feedforward neural network (such as MLP or RBF) one obtains a recurrent neural network.

Recurrent models are e.g. used in control applications, where one first identifies a model and then designs a controller based upon the identified model and applies it to the real system, either in a non-adaptive or adaptive setting. In [230] a stability theory has been developed for such neural control systems. When using neural networks in a dynamical systems context, one should be aware that even very simple recurrent networks can lead to complex behaviour such as chaos. In this sense stability issues of multi-layer recurrent networks are important e.g. towards applications in signal processing and control. The training of recurrent networks is also more complicated than for feedforward networks. In the recurrent network case a cost function is defined on a dynamical system (iterative system) which leads to more complicated analytic expressions for the gradient of the cost function. In order to compute the gradient one can simulate a sensitivity model together with the recurrent network model.

Training mode

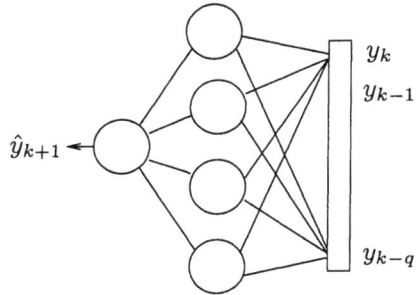

Iterative prediction mode

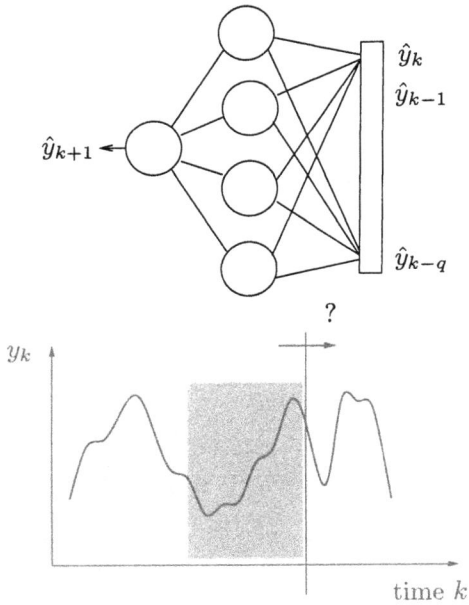

Fig. 1.13 *Time-series prediction with neural networks: (Top) identification with a NARX model structure; (Bottom) iterative prediction as a recurrent network by replacing the true values y_k with the estimated values $\hat{y}_k$ at the input of the network.*

Chapter 2

Support Vector Machines

In this Chapter we give a short overview on the formulations of standard Support Vector Machines as introduced by Vapnik. We discuss linear and nonlinear SVM classifiers with the separable and non-separable case as well as linear and nonlinear function estimation by SVMs based on the Vapnik ϵ-insensitive loss function. The extension from the linear to the nonlinear case is done by the so-called kernel trick. In a second part of this Chapter, we discuss a number of modifications and extensions related to this SVM methodology. For textbooks and overview material we refer to [35; 51; 73; 202; 206; 219; 223; 279; 281].

2.1 Maximal margin classification and linear SVMs

2.1.1 *Margin*

While the weight decay term is an important aspect for obtaining good generalization in the context of neural networks for regression, the margin plays a somewhat similar role in classification problems. This margin concept is a first important step towards understanding the formulation of support vector machines.

In Fig. 2.1 an illustrative example is given of a separable problem in a two-dimensional input space. One can see that there exist several separating hyperplanes that separate the data of the two classes (data depicted by 'x' and '+'). On the other hand, one can also define a unique separating hyperplane. In order to achieve this, Vapnik considered a rescaling of the problem such that the points closest to the hyperplane satisfy

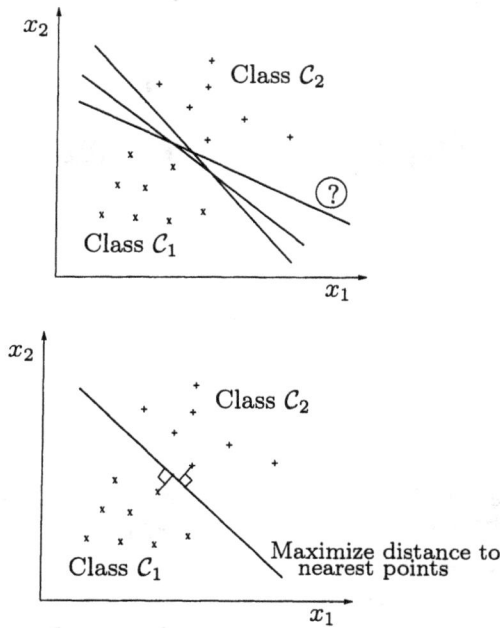

Fig. 2.1 *Linear classification: (Top) example of classification problem where the separating hyperplane is not unique; (Bottom) definition of a unique hyperplane which corresponds to a maximal distance between the nearest points of the two classes C_1 and C_2.*

$|w^T x_k + b| = 1$. One obtains a canonical form then for (w, b) of the hyperplane satisfying $y_k(w^T x_k + b) \geq 1$. In this case the margin equals $2/\|w\|_2$. Hence, the points closest to the hyperplane have a distance $1/\|w\|_2$ and in this way one maximizes the distance to the nearest points of the two classes. It is a desirable property then to maximize the margin which corresponds to minimizing $\|w\|_2$. This minimization of $w^T w$ is closely related to the use of a weight decay term in the training of neural networks.

2.1.2 *Linear SVM classifier: separable case*

Although the general nonlinear version of Support Vector Machines (SVM) is quite recent [279], the roots of the SVM approach about constructing an optimal separating hyperplane for pattern recognition dates back to the

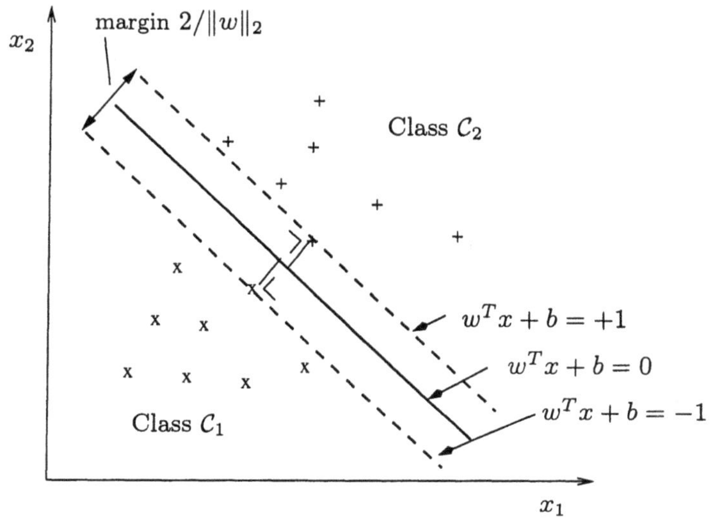

Fig. 2.2   *Linear classification: definition of a unique separating hyperplane illustrated in a two-dimensional input space. The margin is the distance between the dashed lines.*

work of Vapnik & Lerner in 1963 and Vapnik & Chervonenkis in 1964 [276; 277]. This original linear SVM formulation was made for separable data.

Consider a given training set $\{x_k, y_k\}_{k=1}^N$ with input data $x_k \in \mathbb{R}^n$ and output data $y_k \in \mathbb{R}$ with class labels $y_k \in \{-1, +1\}$ and the linear classifier

$$y(x) = \mathrm{sign}[w^T x + b]. \tag{2.1}$$

When the data of the two classes are separable one can say

$$\begin{cases} w^T x_k + b \geq +1, & \text{if } y_k = +1 \\ w^T x_k + b \leq -1, & \text{if } y_k = -1. \end{cases} \tag{2.2}$$

These two sets of inequalities can be combined into one single set as follows

$$y_k [w^T x_k + b] \geq 1, \quad k = 1, ..., N. \tag{2.3}$$

Support vector machine formulations are done within a context of convex optimization theory. The general methodology is to start formulating the problem in the primal weight space as a constrained optimization problem, next formulate the Lagrangian, then take the conditions for optimality and finally solve the problem in the dual space of Lagrange multipliers. The latter will be called support values.

One formulates an optimization problem which expresses that one should maximize the margin subject to the fact that all training data points need to be correctly classified. This gives the following primal problem in w:

$$\boxed{P}: \quad \min_{w,b} J_P(w) = \tfrac{1}{2}w^T w$$
$$\text{such that} \quad y_k[w^T x_k + b] \geq 1, \quad k = 1,...,N. \tag{2.4}$$

The Lagrangian for this problem is

$$\mathcal{L}(w,b;\alpha) = \frac{1}{2}w^T w - \sum_{k=1}^{N} \alpha_k \left(y_k[w^T x_k + b] - 1 \right) \tag{2.5}$$

with Lagrange multipliers $\alpha_k \geq 0$ for $k = 1,...,N$. The solution is characterized by the saddle point of the Lagrangian

$$\max_{\alpha} \min_{w,b} \mathcal{L}(w,b;\alpha). \tag{2.6}$$

One obtains

$$\begin{cases} \frac{\partial \mathcal{L}}{\partial w} = 0 & \rightarrow \quad w = \sum_{k=1}^{N} \alpha_k y_k x_k \\ \frac{\partial \mathcal{L}}{\partial b} = 0 & \rightarrow \quad \sum_{k=1}^{N} \alpha_k y_k = 0 \end{cases} \tag{2.7}$$

with resulting classifier

$$y(x) = \text{sign}[\sum_{k=1}^{N} \alpha_k y_k x_k^T x + b]. \tag{2.8}$$

By replacing the expression for w (2.7) in the Lagrangian (2.5) one obtains the following Quadratic Programming (QP) problem as the dual problem in the Lagrange multipliers α_k

$$\boxed{D}: \quad \max_{\alpha} J_D(\alpha) = -\frac{1}{2}\sum_{k,l=1}^{N} y_k y_l x_k^T x_l \, \alpha_k \alpha_l + \sum_{k=1}^{N} \alpha_k$$
$$\text{such that} \quad \sum_{k=1}^{N} \alpha_k y_k = 0, \; \alpha_k \geq 0, \; \forall k \tag{2.9}$$

Note that this problem is solved in $\alpha = [\alpha_1;...;\alpha_N]$, not in w. This QP problem has a number of interesting properties:

- *Global and unique solution*:
  The matrix related to the quadratic term in α in this quadratic form is positive definite or positive semidefinite. In the case that the matrix is positive definite (all eigenvalues strictly positive) the solution α to this QP problem is global and unique. When the matrix is positive semidefinite (all eigenvalues positive but zero eigenvalues possible) then the solution is global but not necessarily unique [82]. Some interesting discussions in relation to these aspects are given in [35]. It may happen that the solution of (w, b) is unique but α not. In terms of $w = \sum_k \alpha_k y_k x_k$ this means that there may exist equivalent expansions of w which require fewer support vectors.

- *Sparseness*:
  An interesting property is that many of the resulting α_k values are equal to zero. Hence the obtained solution vector is sparse. This means that in the resulting classifier the sum should be taken only over the non-zero α_k values instead of all training data points:

$$y(x) = \text{sign}[\sum_{k=1}^{\#\text{SV}} \alpha_k y_k x_k^T x + b]$$

  where the index k runs now over the number of support vectors. The training data points corresponding to non-zero α_k values are called *support vectors*.

- *Geometrical meaning of support vectors*:
  The support vectors obtained from the QP problem are located close to the decision boundary.

- *Non-parametric/parametric issues*:
  It is important to note that the size of the solution vector α grows with the number of data points N. Hence, the dual problem corresponds in fact to a non-parametric approach, while in the primal problem on the other hand the problem is parametric because the size of w is fixed independently of the number of data points. This means that for large data sets it might be advantageous to solve the primal problem but in higher dimensional input spaces it is better to solve the dual problem as the size of the solution vector α does

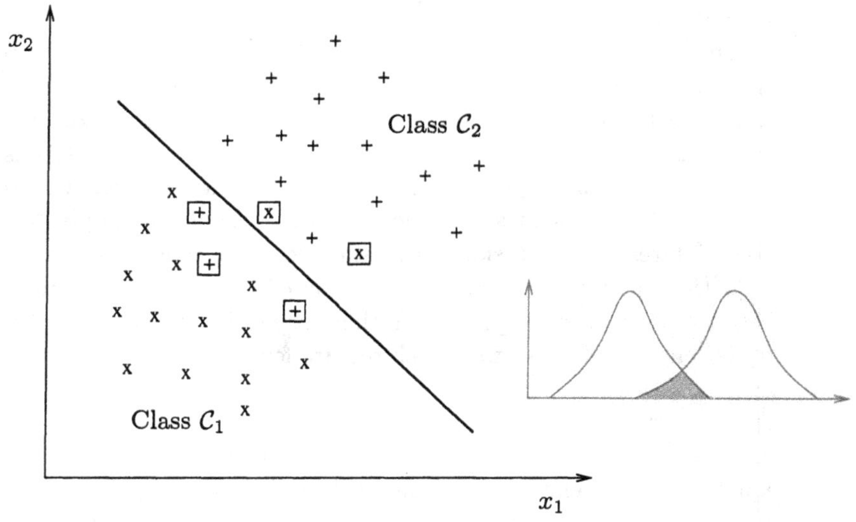

Fig. 2.3   *Problem of non-separable data, due to overlapping distributions.*

not depend on the dimension n of the input space.

2.1.3   *Linear SVM classifier: non-separable case*

In the previous subsection the SVM solution to a linearly separable classification problem was explained. On the other hand, one should note that most real-life problems that one encounters are *non-separable* cases either in a linear or nonlinear sense. Usually, one tries to find a set of inputs such that one can separate the classes as much as possible but often relevant inputs are missing, the data are incomplete, unreliable or noisy etc. Also in the previous Chapter it was shown that if for a binary classification problem the data are generated from a Gaussian density with the same covariance matrix for the two classes, the optimal decision boundary is a linear separating hyperplane according to Bayesian decision theory, no matter how small or large the overlap between these Gaussian densities. As a consequence this means that in the linearly non-separable case with overlapping distributions between the two classes, one should *tolerate misclassifications*.

The extension of linear SVMs to the non-separable case was made by

Cortes & Vapnik in 1995 [46]. Basically, it is done by taking additional slack variables in the problem formulation. In order to tolerate misclassifications, one modifies the set of inequalities into

$$y_k[w^T x_k + b] \geq 1 - \xi_k, \qquad k = 1, ..., N \tag{2.10}$$

with slack variables $\xi_k > 0$. When $\xi_k > 1$ the k-th inequality becomes violated in comparison with the corresponding inequality from the linearly separable case.

In the primal weight space the optimization problem becomes

$$
\left[
\begin{array}{l}
\boxed{\text{P}}: \quad \min_{w,b,\xi} J_{\text{P}}(w,\xi) = \frac{1}{2} w^T w + c \sum_{k=1}^{N} \xi_k \\[2mm]
\qquad \text{such that} \qquad y_k[w^T x_k + b] \geq 1 - \xi_k, \qquad k = 1, ..., N \\[2mm]
\qquad\qquad\qquad\quad\;\; \xi_k \geq 0, \qquad k = 1, ..., N
\end{array}
\right]
\tag{2.11}
$$

where c is a positive real constant. The following Lagrangian should be considered then

$$\mathcal{L}(w, b, \xi; \alpha, \nu) = J_{\text{P}}(w, \xi) - \sum_{k=1}^{N} \alpha_k \left(y_k[w^T x_k + b] - 1 + \xi_k \right) - \sum_{k=1}^{N} \nu_k \xi_k \tag{2.12}$$

with Lagrange multipliers $\alpha_k \geq 0$, $\nu_k \geq 0$ for $k = 1, ..., N$. The second set of Lagrange multipliers is needed due to the additional slack variables ξ_k. The solution is given by the saddle point of the Lagrangian:

$$\max_{\alpha, \nu} \min_{w, b, \xi} \mathcal{L}(w, b, \xi; \alpha, \nu). \tag{2.13}$$

One obtains

$$
\begin{cases}
\frac{\partial \mathcal{L}}{\partial w} = 0 & \rightarrow \quad w = \sum_{k=1}^{N} \alpha_k y_k x_k \\[3mm]
\frac{\partial \mathcal{L}}{\partial b} = 0 & \rightarrow \quad \sum_{k=1}^{N} \alpha_k y_k = 0 \\[3mm]
\frac{\partial \mathcal{L}}{\partial \xi_k} = 0 & \rightarrow \quad 0 \leq \alpha_k \leq c, \; k = 1, ..., N
\end{cases}
\tag{2.14}
$$

36   *Support Vector Machines*

which gives the following dual QP problem after replacing (2.14) in (2.12):

$$
\boxed{D}: \quad \max_{\alpha} J_D(\alpha) = -\frac{1}{2}\sum_{k,l=1}^{N} y_k y_l\, x_k^T x_l\, \alpha_k \alpha_l + \sum_{k=1}^{N}\alpha_k
$$

$$
\text{such that} \quad \sum_{k=1}^{N}\alpha_k y_k = 0
$$

$$
0 \le \alpha_k \le c, \; k = 1, ..., N. \tag{2.15}
$$

In comparison with the linearly separable case (2.9) this problem has additional box constraints.

2.2 Kernel trick and Mercer condition

Important progress in SVM theory has been made thanks to the fact that the linear method has been extended to a nonlinear technique by Vapnik in 1995 [279; 281]. In order to achieve this, one maps the input data into a high dimensional feature space which can be infinite dimensional. A construction of the linear separating hyperplane is done then in this high dimensional feature space*, after a nonlinear mapping $\varphi(x)$ of the input data to the feature space (Fig. 2.4).

Surprisingly, no explicit construction of the nonlinear mapping $\varphi(x)$ is needed. This is motivated by the following result. For any symmetric, continuous function $K(x,z)$ satisfying Mercer's condition [160], there exists a Hilbert space $\mathcal{H}$, a map $\varphi : \mathbb{R}^n \to \mathcal{H}$ and positive numbers λ_i such that one can write [48]:

$$
K(x,z) = \sum_{i=1}^{n_{\mathcal{H}}} \lambda_i \phi_i(x)\phi_i(z), \tag{2.16}
$$

where $x, z \in \mathbb{R}^n$ and $n_{\mathcal{H}}$ is the dimension of $\mathcal{H}$ (which can be infinite dimensional). Mercer's condition requires that

$$
\int K(x,z)g(x)g(z)dxdz \ge 0 \tag{2.17}
$$

---

*From the viewpoint of neural networks it would be better in fact to call this a high-dimensional *hidden layer*, because in pattern recognition one frequently uses the term feature space with another meaning of input space. Nevertheless we will use the term feature space in the sequel because it is always used in this area.

for any square integrable function $g(x)$. The integral is taken here over a compact subset of $\mathbb{R}^n$. One can write $K(x, z) = \sum_{i=1}^{n_H} \sqrt{\lambda_i}\phi_i(x) \sqrt{\lambda_i}\phi_i(z)$ and define $\varphi_i(x) = \sqrt{\lambda_i}\phi_i(x)$ and $\varphi_i(z) = \sqrt{\lambda_i}\phi_i(z)$ such that the kernel function can be expressed as the inner product (often called the dot product)

$$K(x, z) = \varphi(x)^T \varphi(z). \tag{2.18}$$

Hence, having a positive semidefinite[†] kernel is a condition to guarantee that one may write (2.18). This also implies that the kernel K is separable. Such separable kernels have been used for example also for solving integral equations [58].

The application of (2.18) is often called the *kernel trick*. It enables us to work in huge dimensional feature spaces without actually having to do explicit computations in this space. Computations are done in another space after applying this kernel trick. In the case of support vector machines, one starts from a formulation in the primal weight space with a high dimensional feature space by applying transformations $\varphi(\cdot)$. The problem is not solved in this primal weight space but in the dual space of Lagrange multipliers after applying the kernel trick. In this way one can implicitly work in high dimensional feature spaces (hidden layer space) without doing computations in that space (Fig. 2.5).

2.3   Nonlinear SVM classifiers

The extension from linear SVM classifiers to nonlinear SVM classifiers is straightforward. One can in fact formally replace x by $\varphi(x)$ and apply the kernel trick where possible. One should, however, be aware that $\varphi(x)$ can be infinite dimensional, and hence, also the w vector. While for linear SVMs one can in fact equally well solve the primal problem in w as the dual problem in the support values α, this is no longer the same for the nonlinear SVM case because in the primal problem the unknown w can be infinite dimensional.

In a similar way as for the linear SVM case, we can now write for the

---

[†]We use here the linear algebra terminology of *positive definite* and *positive semidefinite*. The corresponding terminology in functional analysis is often *strictly positive definite* and *positive definite*, respectively.

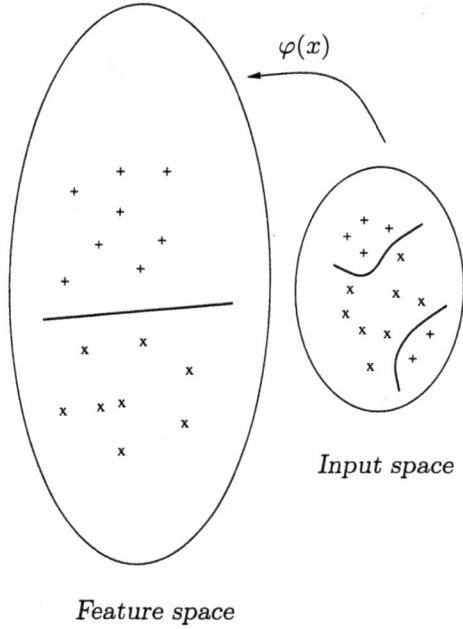

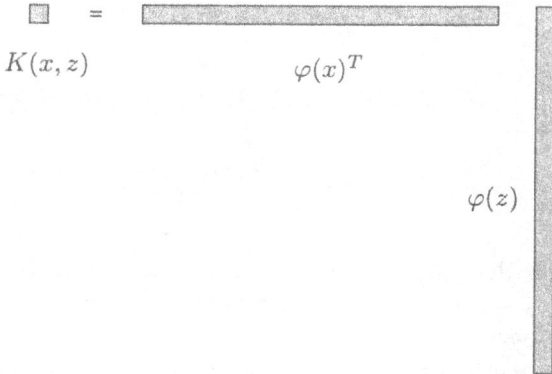

Fig. 2.4 *(Top) Mapping of the input space to a high dimensional feature space where a linear separation is made, which corresponds to a nonlinear separation in the original input space; (Bottom) Illustration of the kernel trick where $K(x,z)$ is chosen and the inner product representation exists provided $K(\cdot,\cdot)$ is a symmetric positive definite kernel.*

Primal problem $\boxed{P}$

Parametric: estimate $w \in \mathbb{R}^{n_h}$

$$y(x) = \text{sign}[w^T \varphi(x) + b]$$

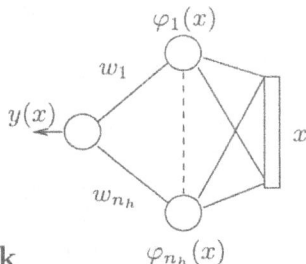

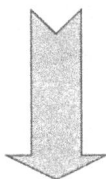

Kernel trick

$$K(x_k, x_l) = \varphi(x_k)^T \varphi(x_l)$$

Dual problem $\boxed{D}$

Non-parametric: estimate $\alpha \in \mathbb{R}^N$

$$y(x) = \text{sign}[\sum_{k=1}^{\#\text{sv}} \alpha_k y_k K(x, x_k) + b]$$

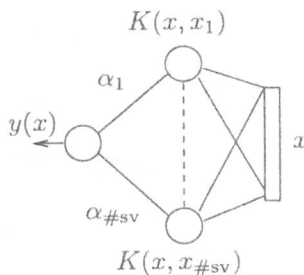

Fig. 2.5 *Primal-dual neural network interpretations of support vector machines.*

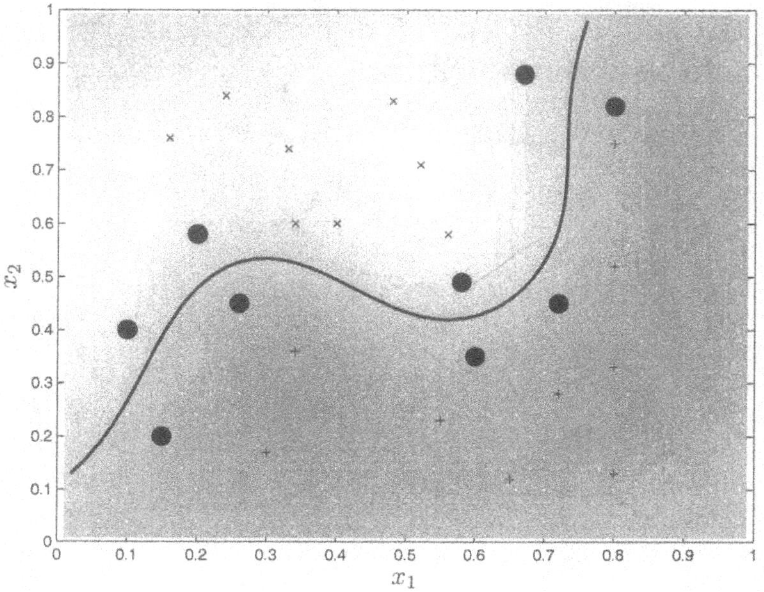

Fig. 2.6 *For nonlinear SVMs the support vectors (indicated by the black dots) are located close to the decision boundary as illustrated here by a simple nonlinearly separable binary class problem.*

nonlinear case

$$\begin{cases} w^T \varphi(x_k) + b \geq +1, & \text{if } y_k = +1 \\ w^T \varphi(x_k) + b \leq -1, & \text{if } y_k = -1 \end{cases} \qquad (2.19)$$

which is equivalent to

$$y_k[w^T \varphi(x_k) + b] \geq 1, \quad k = 1, ..., N \qquad (2.20)$$

in the case of separable data. At this point no explicit form or construction is made of $\varphi(\cdot) : \mathbb{R}^n \to \mathbb{R}^{n_h}$. The number n_h is the number of hidden units in the primal weight space representation of the nonlinear classifier (Fig. 2.5)

$$y(x) = \text{sign}[w^T \varphi(x) + b] \qquad (2.21)$$

where n_h corresponds to the dimension $n_{\mathcal{H}}$.

The optimization problem becomes

$$
\boxed{P}: \quad \min_{w,b,\xi} J_P(w,\xi) = \frac{1}{2}w^T w + c\sum_{k=1}^{N}\xi_k
$$
$$
\text{such that} \quad y_k[w^T\varphi(x_k)+b] \geq 1-\xi_k, \quad k=1,...,N
$$
$$
\xi_k \geq 0, \quad k=1,...,N.
$$
(2.22)

One constructs the Lagrangian:

$$
\mathcal{L}(w,b,\xi;\alpha,\nu) = J_P(w,\xi) - \sum_{k=1}^{N}\alpha_k\left(y_k[w^T\varphi(x_k)+b]-1+\xi_k\right) - \sum_{k=1}^{N}\nu_k\xi_k
$$
(2.23)

with Lagrange multipliers $\alpha_k \geq 0$, $\nu_k \geq 0$ for $k=1,...,N$. The solution is given by the saddle point of the Lagrangian:

$$
\max_{\alpha,\nu}\min_{w,b,\xi}\mathcal{L}(w,b,\xi;\alpha,\nu).
$$
(2.24)

One obtains

$$
\begin{cases}
\frac{\partial \mathcal{L}}{\partial w} = 0 & \rightarrow \quad w = \sum_{k=1}^{N}\alpha_k y_k\varphi(x_k) \\
\frac{\partial \mathcal{L}}{\partial b} = 0 & \rightarrow \quad \sum_{k=1}^{N}\alpha_k y_k = 0 \\
\frac{\partial \mathcal{L}}{\partial \xi_k} = 0 & \rightarrow \quad 0 \leq \alpha_k \leq c, \; k=1,...,N.
\end{cases}
$$
(2.25)

The quadratic programming problem (dual problem) becomes

$$
\boxed{D}: \quad \max_{\alpha} J_D(\alpha) = -\frac{1}{2}\sum_{k,l=1}^{N}y_k y_l\, K(x_k,x_l)\,\alpha_k\alpha_l + \sum_{k=1}^{N}\alpha_k
$$
$$
\text{such that} \quad \sum_{k=1}^{N}\alpha_k y_k = 0
$$
$$
0 \leq \alpha_k \leq c, \; k=1,...,N.
$$
(2.26)

In this quadratic form one makes use of the kernel trick

$$
K(x_k,x_l) = \varphi(x_k)^T\varphi(x_l)
$$
(2.27)

for $k, l = 1, ..., N$. Finally, the nonlinear SVM classifier takes the form

$$y(x) = \text{sign}[\sum_{k=1}^{N} \alpha_k \, y_k \, K(x, x_k) + b] \qquad (2.28)$$

with α_k positive real constants which are the solution to the QP problem. However, we still need to determine b. The complementarity conditions from the Karush-Kuhn-Tucker (KKT) conditions (see Appendix) for the above problem state that the product of the dual variables and the constraints should be zero at the optimal solution. The KKT conditions yield

$\boxed{\text{KKT}}$:

$$\begin{cases} \frac{\partial \mathcal{L}}{\partial w} = 0 \rightarrow \quad w = \sum_{k=1}^{N} \alpha_k y_k \varphi(x_k) \\[2mm] \frac{\partial \mathcal{L}}{\partial b} = 0 \rightarrow \quad \sum_{k=1}^{N} \alpha_k y_k = 0 \\[2mm] \frac{\partial \mathcal{L}}{\partial \xi_k} = 0 \rightarrow \quad c - \alpha_k - \nu_k = 0 \\[2mm] \alpha_k \{ y_k [w^T \varphi(x_k) + b] - 1 + \xi_k \} = 0, \quad k = 1, .., N \\[2mm] \nu_k \xi_k = 0, \qquad\qquad\qquad\qquad\qquad\quad k = 1, .., N \\[2mm] \alpha_k \geq 0, \qquad\qquad\qquad\qquad\qquad\quad\; k = 1, .., N \\[2mm] \nu_k \geq 0, \qquad\qquad\qquad\qquad\qquad\quad\; k = 1, .., N. \end{cases} \qquad (2.29)$$

From $\nu_k \xi_k = 0$ we have for the solution $w^\star, b^\star, \xi^\star, \alpha^\star, \nu^\star$ to this problem that $\xi_k^\star = 0$ for $\alpha_k^\star \in (0, c)$. Hence

$$y_k [w^T \varphi(x_k) + b] - 1 = 0 \text{ for } \alpha_k \in (0, c), \qquad (2.30)$$

which means that one can take any training data point for which $0 < \alpha_k < c$ and use that equation to compute the bias term b (numerically it might be better to take an average over these training data points).

Some properties of the nonlinear SVM classifier and its solution are the following:

- *Choice of kernel function:*
  Several choices are possible for the kernel $K(\cdot, \cdot)$. Some typical choices are

$$K(x, x_k) = x_k^T x \text{ (linear SVM)}$$
$$K(x, x_k) = (\tau + x_k^T x)^d \text{ (polynomial SVM of degree } d)$$
$$K(x, x_k) = \exp(-\|x - x_k\|_2^2/\sigma^2) \text{ (RBF kernel)}$$
$$K(x, x_k) = \tanh(\kappa_1 x_k^T x + \kappa_2) \text{ (MLP kernel)} .$$

The Mercer condition holds for all σ values in the RBF kernel case and positive τ values in the polynomial case but not for all possible choices of κ_1, κ_2 in the MLP kernel case. A further discussion about kernel functions will be given in the sequel of this Chapter.

- *Global and unique solution:*
  As in the linear SVM case the solution to the convex QP problem is again global and unique provided that one chooses a positive definite kernel for $K(\cdot, \cdot)$. This choice guarantees that the matrix involved in the QP problem is positive definite as well, and that one can apply the kernel trick. For a positive semidefinite kernel the solution to the QP problem is global but not necessarily unique.

- *Sparseness:*
  As in the linear SVM classifier case many α_k values are equal to zero in the solution vector. In the dual space the nonlinear SVM classifier takes the form

$$y(x) = \text{sign}[\sum_{k=1}^{\#SV} \alpha_k y_k K(x, x_k) + b]$$

where the sum is taken over the non-zero α_k values which correspond to support vectors x_k of the training data set. In Fig. 2.5 a neural network interpretation is given of this. The number of hidden units equals the number of support vectors SV, that one finds from a convex QP problem. In the case of an RBF kernel, one has

$$
\begin{aligned}
y(x) &= \text{sign}[\sum_{k=1}^{N} \alpha_k y_k \exp\left(-\|x - x_k\|_2^2/\sigma^2\right) + b] \\
&= \text{sign}[\sum_{k=1}^{\#SV} \alpha_k y_k \exp\left(-\|x - x_k\|_2^2/\sigma^2\right) + b]
\end{aligned}
\tag{2.31}
$$

where $\#$ SV denotes the number of support vectors.

- *Geometrical meaning of support vectors*:
  The support vectors obtained from the QP problem are located close to the nonlinear decision boundary illustrated on (Fig. 2.6) and (Fig. 2.7) for the synthetic Ripley data set [193].

- *Non-parametric/parametric issues*:
  Both the primal and the dual problem have neural network representation interpretations (Fig. 2.5). The problem in the primal weight space is parametric, while the dual problem is non-parametric. Note that in the dual problem the size of the QP problem is not influenced by the dimension n of the input space.

In comparison with traditional multilayer perceptron neural networks that suffer from the existence of multiple local minima solutions, convexity is an important and interesting property of nonlinear SVM classifiers. However, strictly speaking, this is less surprising if one notes that w and α are in fact interconnection weights of the output layer. Also in MLPs one would have a convex problem if one would fix a hidden layer interconnection matrix and one would compute the output layer (with linear characteristic at the output) from a sum squared error cost function. The nice aspect, however, of SVMs is the powerful representation of the network. If one takes an RBF kernel, almost all unknowns follow as the solution to a convex problem, which is not the case for traditional MLP networks. Furthermore, in classical MLPs one has to fix the number of hidden units beforehand while in nonlinear SVMs this number of hidden units follows from the QP problem as the number of support vectors.

2.4   VC theory and structural risk minimization

2.4.1   *Empirical risk versus generalization error*

From the linear and nonlinear SVM formulations it is clear so far that the margin plays an important role in the formulations. This importance can be put within the broader perspective of statistical learning theory (VC theory) [279; 281]. While in classical neural networks practice one selects models based upon specific validation sets, a major goal of VC (Vapnik-Chervonenkis) theory is to characterize the generalization error instead of the error on specific data sets.

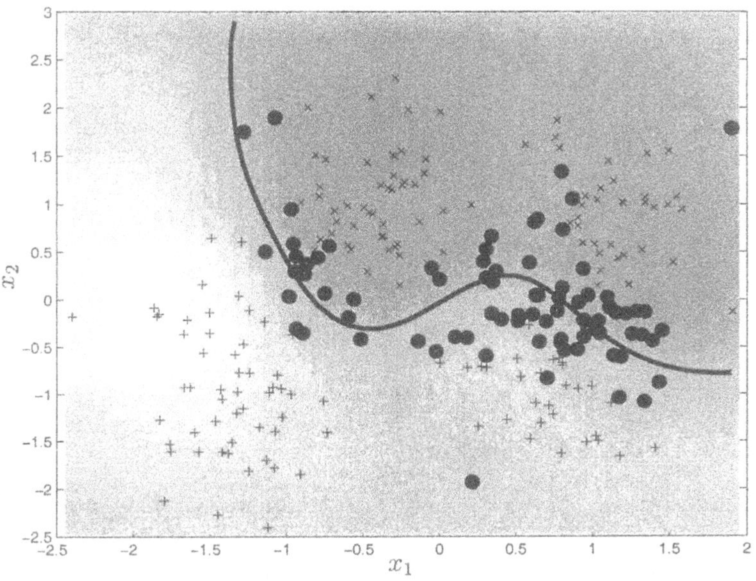

Fig. 2.7   *Nonlinear SVM classifier with RBF kernel on the Ripley data set binary classification problem with '+' and 'x' given training data points of the two classes. The black dots correspond to the support vectors, which have non-zero α_k values.*

Let us consider a binary classification problem and a set of functions with adjustable parameter vector θ such that $\{f(x;\theta) : \mathbb{R}^n \rightarrow \{-1,+1\}\}$ together with a training set $\{(x_k, y_k)\}_{k=1}^N$ with input patterns $x_k \in \mathbb{R}^n$ and output $y_k \in \{-1,+1\}$. The following error defined on the training data set is usually called empirical risk

$$R_{\text{emp}}(\theta) = \frac{1}{2N} \sum_{k=1}^{N} |y_k - f(x_k;\theta)|. \tag{2.32}$$

The generalization error (or risk) on the other hand is defined as

$$R(\theta) = \int \frac{1}{2}|y - f(x;\theta)|\, p(x,y)\, dxdy \tag{2.33}$$

which measures the error for all input/output patterns that are generated from the underlying generator of the data characterized by the probability

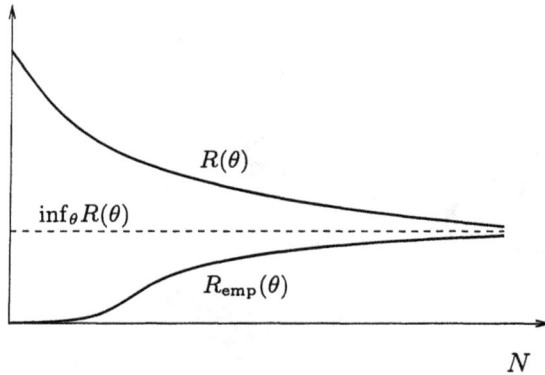

Fig. 2.8 *If the expected risk $R(\theta)$ and the empirical risk $R_{emp}(\theta)$ both converge to the value $\inf_\theta R(\theta)$ when the number of data N goes to infinity, then the learning process is consistent.*

distribution $p(x, y)$ (which is unfortunately not known in practice). However, one can derive bounds on this generalization error.

An important result by Vapnik (1979) states that one can give an upper bound to this generalization error in a probabilistic sense [279; 281]. Vapnik showed that for any parameter vector θ of the set of function $\{f(x; \theta)|\theta \in \Theta\}$, i.i.d. data and $N > h$, the bound

$$R(\theta) \leq R_{\text{emp}}(\theta) + \sqrt{\frac{h(\ln(2N/h) + 1) - \ln(\eta/4)}{N}} \qquad (2.34)$$

holds with probability $1 - \eta$, where the second term is a confidence term which depends on the VC dimension h that characterizes the capacity of the set of functions and it is a combinatorial measure for the model complexity. An important aspect of this risk bound is that it assumes no specific form for the underlying density $p(x, y)$, only that i.i.d. data are drawn from this density $p(x, y)$. In addition to this risk upper bound also a lower bound on $R(\theta)$ can be derived. An important insight of VC theory is that the worst case over all functions that a learning machine can implement determines consistency (Fig. 2.8) of the empirical risk minimization. Without restricting the set of admissible functions, empirical risk minimization is not consistent. For the empirical risk minimization principle to be consistent, it is necessary and sufficient that the empirical risk $R_{\text{emp}}(\theta)$ converges uniformly to the risk $R(\theta)$ in the following sense: $\lim_{N \to \infty} P\{\sup_{\theta \in \Theta}(R(\theta) - R_{\text{emp}}(\theta)) > \epsilon\} = 0, \forall \epsilon > 0$ [279].

Fig. 2.9 *Example of $N = 4$ points in an $n = 2$ dimensional input space. The points can be labelled then in $2^4 = 16$ possible ways. At most 3 points can be separated by straight lines, because it's well known that the last two cases are XOR problems which cannot be separated by a straight line. This results in a VC dimension equal to 3. For nonlinear classifiers this VC dimension can be larger.*

The VC dimension can be understood as follows. If a given set of N points can be labelled in all possible 2^N ways and for each labelling a member of the set $\{f(x;\theta)\}$ can be found which correctly assigns those labels, we can say that that set of points is *shattered* (separated) by that set of functions. The VC dimension for the set of functions $\{f(x;\theta)\}$ is defined as the maximum number of training points that can be shattered by members of $\{f(x;\theta)\}$. For linear separating hyperplanes in an input space $\mathbb{R}^n$, one can show that the VC dimension equals $h = n + 1$. Consider for example $N = 4$ points in an $n = 2$ dimensional input space. The points can be labelled then in $2^4 = 16$ possible ways. Not all of these 16 cases are linearly separable, e.g., it is well known that XOR cases are not linearly separable (Fig. 2.9). At most 3 points can be separated or said in terms of the existing connection between VC theory and Popper's work in philosophy: four points can falsify any linear law for this example. In larger dimensions one can relate this problem of linear separability also to Cover's theorem [23].

Intuitively one would expect that models with many parameters lead to a high VC dimension and models with few parameters to a low VC dimension. However, this intuition is not entirely true as it is easy to construct counterexamples of models having only one parameter that can possess an infinite VC dimension [35].

In the previous Section we have discussed the prominent role played by convex optimization within SVM methodology. On the other hand one should be aware of the fact that if one takes e.g. an RBF kernel, then the choice of σ for this RBF kernel does not follow as the solution to the QP problem. It should be fixed beforehand. In fact one has to try several possible values for this tuning parameter, in combination also with the other constant c in the algorithm. This VC upper bound (2.34) can then be used in order to select a good value of σ, c which guarantees a good generalization of the model. Although one exactly knows the VC dimension for linear classifiers, for nonlinear SVM classifiers no exact expressions are available for the VC dimension h. One often works then with an upper bound on h.

2.4.2 *Structural risk minimization*

In order to apply the VC bound to SVM classifiers, Vapnik employs the concept of so-called structural risk minimization. One considers a structure

$$\mathcal{S}_i = \{w^T \varphi(x) + b : \|w\|_2^2 \leq c_i\}, \quad c_1 < c_2 < ... < c_i < c_{i+1} < ... \quad (2.35)$$

which consists of nested sets of functions of increasing complexity. This is illustrated in Fig. 2.10 [206]. The concept is quite similar to previously discussed complexity criteria and bias-variance trade-off curves. In this case of VC theory one may consider sets of functions with increasing VC dimension. The larger this VC dimension the smaller the training set error can become (empirical risk) but the confidence term (second term in (2.34)) will grow. The minimum of the sum of these two terms is then a good compromise solution as the trade-off between these two curves.

For SVM classifiers one can find an upper bound on the VC dimension as follows. Vapnik has shown that hyperplanes satisfying $\|w\|_2 < a$ have a

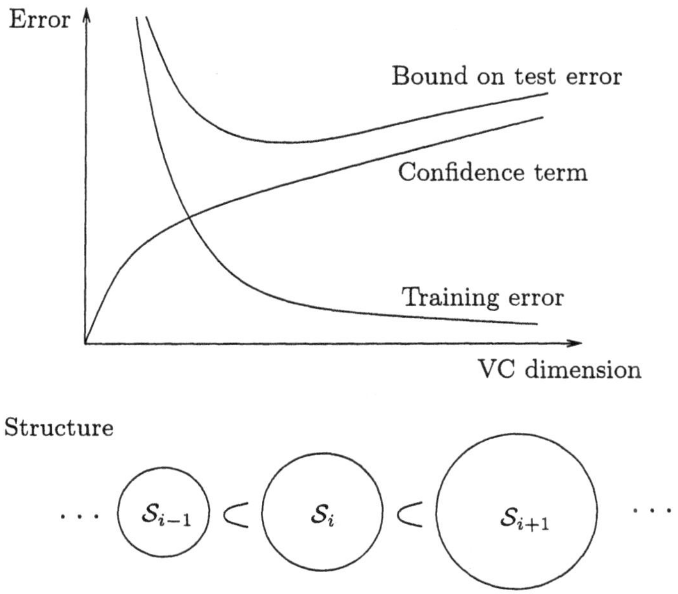

Fig. 2.10   *Principle of structural risk minimization and illustration of the generalization bound depending on the VC dimension.*

VC-dimension h which is upper bounded by

$$h \le \min([r^2 a^2], n) + 1 \qquad (2.36)$$

where $[\cdot]$ denotes the integer part and r is the radius of the smallest ball containing the points $\varphi(x_1), ..., \varphi(x_N)$ in the high dimensional feature space (Fig. 2.11).

SVM parameters which do not result from the QP problem such as the σ value of the RBF kernel are often selected then in such a way that this upper bound (2.36) is minimal. A nice property is that this upper bound can again be computed by solving a convex QP problem. Hence one may proceed then by solving in an alternating way QP problems in α and QP problems related to (2.36), both by convex optimization. The value $w^T w$ can be computed by applying the kernel trick. One has $w^T w = \alpha^T \Omega \alpha$ with $\Omega_{kl} = y_k y_l K(x_k, x_l)$ where $K(x_k, x_l) = \varphi(x_k)^T \varphi(x_l)$ for $k, l = 1, ..., N$.

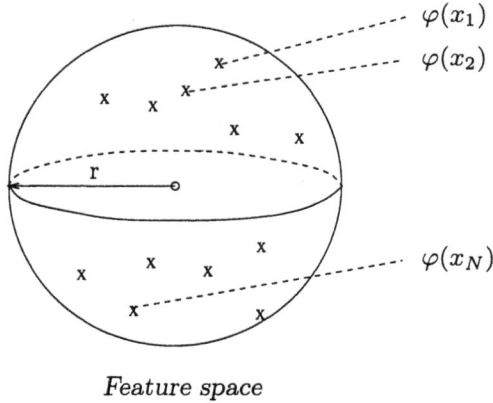

Feature space

Fig. 2.11   *Smallest ball containing the points $\varphi(x_1), ..., \varphi(x_N)$ in the feature space. The radius r follows from a QP problem. Together with the margin this number r plays a role in the bound on the VC dimension for SVM classifiers.*

The radius r of the ball in the feature space can be computed as follows. Consider as optimization problem in the primal space

$$
\left[\begin{array}{ll} \boxed{P}: & \displaystyle\min_{r,q} \quad r \\ & \text{such that} \quad \|\varphi(x_k) - q\|_2 < r, \ k = 1, 2, ..., N \end{array} \right] \tag{2.37}
$$

where q is a point inside the ball to be determined. The optimization problem expresses that all points should be contained within the ball with radius r in the feature space. The Lagrangian is

$$
\mathcal{L}(r, q, \lambda) = r^2 - \sum_{k=1}^{N} \lambda_k (r^2 - \|\varphi(x_k) - q\|_2^2) \tag{2.38}
$$

with positive Lagrange multipliers λ_k. After taking the conditions for optimality one obtains the following dual QP problem:

$$
\boxed{\text{D}} : \quad \max_{\lambda} \quad -\sum_{k,l=1}^{N} K(x_k, x_l) \lambda_k \lambda_l + \sum_{k=1}^{N} \lambda_k K(x_k, x_k)
$$

$$
\text{such that} \quad \sum_{k=1}^{N} \lambda_k = 1 \tag{2.39}
$$

$$
\lambda_k \geq 0 \ , k = 1, ..., N
$$

where the kernel trick $K(x_k, x_l) = \varphi(x_k)^T \varphi(x_l)$ has been applied again. Furthermore one also obtains $q = \sum_{k=1}^{N} \lambda_k \varphi(x_k)$.

A remark that should be made concerning the VC bound is that one has experienced that the bound (2.36) is often conservative. Nevertheless it can also be sufficiently indicative. Sharper bounds have been derived in terms of covering numbers and entropy numbers and can be expressed in terms of the eigenvalues of the Gram matrix as shown in [206; 207] and Rademacher complexity [15; 159]. A recent overview is given by Herbrich in [111] with results in the VC, PAC and luckiness framework, algorithmic stability and compression bounds and a PAC-Bayesian framework. For further reading on the problem of learning and generalization and VC theory, see also [11; 51; 285].

2.5 SVMs for function estimation

2.5.1 *SVM for linear function estimation*

In addition to classification, the support vector methodology has also been introduced for linear and nonlinear function estimation problems [279]. Let us discuss first the linear function estimation case. Consider regression in the set of linear functions

$$
f(x) = w^T x + b \tag{2.40}
$$

with N given training data with input values $x_k \in \mathbb{R}^n$ and output values $y_k \in \mathbb{R}$. For empirical risk minimization in the sense of Vapnik one employs the cost function

$$
R_{\text{emp}} = \frac{1}{N} \sum_{k=1}^{N} |y_k - w^T x_k - b|_\epsilon \tag{2.41}
$$

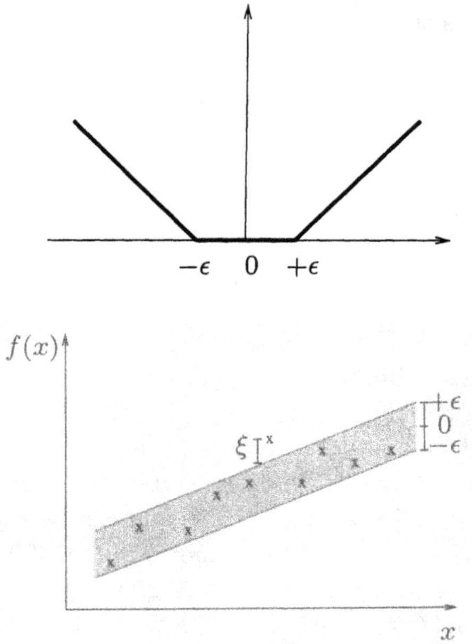

Fig. 2.12 *(Top) Vapnik ε-insensitive loss function for function estimation; (Bottom) Tube of ε-accuracy and points which cannot meet this accuracy, motivating the use of slack variables.*

with the so-called Vapnik's ε-insensitive loss function defined as

$$|y - f(x)|_\epsilon = \begin{cases} 0, & \text{if } |y - f(x)| \le \epsilon \\ |y - f(x)| - \epsilon, & \text{otherwise} \end{cases} \tag{2.42}$$

which is shown in Fig. 2.12.

Estimation of a linear function is done then by formulating the following primal problem in w, b:

$$\left[\begin{array}{l} \boxed{\text{P}}: \quad \min_{w,b} J_P(w) = \tfrac{1}{2} w^T w \\[2mm] \qquad \text{such that} \quad y_k - w^T x_k - b \le \epsilon, \quad k = 1, ..., N \\[1mm] \qquad\qquad\qquad\quad w^T x_k + b - y_k \le \epsilon, \quad k = 1, ..., N. \end{array}\right] \tag{2.43}$$

The value ϵ in the Vapnik ϵ-insensitive loss function is the accuracy that one requires for the approximation. The formulation (2.43) corresponds to a case in which all training data points would belong to an ϵ-tube of accuracy. Of course when one chooses ϵ small, certain points will be outside this ϵ-tube and the problem (2.43) will become infeasible. Therefore additional slack variables ξ_k, ξ_k^* for $k = 1, ..., N$ are introduced. This is somewhat similar to the classifier case with overlapping distributions where also slack variables were introduced.

The problem (2.43) is modified into

$$
\boxed{\text{P}} : \quad \min_{w,b,\xi,\xi^*} \; J_\text{P}(w, \xi, \xi^*) = \frac{1}{2} w^T w + c \sum_{k=1}^{N} (\xi_k + \xi_k^*)
$$

$$
\begin{aligned}
\text{such that} \quad & y_k - w^T x_k - b \leq \epsilon + \xi_k \;, \; k = 1, ..., N \\
& w^T x_k + b - y_k \leq \epsilon + \xi_k^* \;, \; k = 1, ..., N \\
& \xi_k, \xi_k^* \geq 0 \;, \; k = 1, ..., N.
\end{aligned}
$$

(2.44)

The constant $c > 0$ determines the amount up to which deviations from the desired ϵ accuracy are tolerated.

The Lagrangian for this problem is

$$
\mathcal{L}(w, b, \xi, \xi^*; \alpha, \alpha^*, \eta, \eta^*) =
$$

$$
\frac{1}{2} w^T w + c \sum_{k=1}^{N} (\xi_k + \xi_k^*) - \sum_{k=1}^{N} \alpha_k (\epsilon + \xi_k - y_k + w^T x_k + b)
$$

$$
- \sum_{k=1}^{N} \alpha_k^* (\epsilon + \xi_k^* + y_k - w^T x_k - b) - \sum_{k=1}^{N} (\eta_k \xi_k + \eta_k^* \xi_k^*)
$$

(2.45)

with positive Lagrange multipliers $\alpha_k, \alpha_k^*, \eta_k, \eta_k^* \geq 0$. The saddle point of the Lagrangian is characterized by

$$
\max_{\alpha, \alpha^*, \eta, \eta^*} \min_{w,b,\xi,\xi^*} \mathcal{L}(w, b, \xi, \xi^*; \alpha, \alpha^*, \eta, \eta^*)
$$

(2.46)

with conditions for optimality:

$$
\begin{cases}
\frac{\partial \mathcal{L}}{\partial w} = 0 & \rightarrow \quad w = \sum_{k=1}^{N} (\alpha_k - \alpha_k^*) x_k \\
\frac{\partial \mathcal{L}}{\partial b} = 0 & \rightarrow \quad \sum_{k=1}^{N} (\alpha_k - \alpha_k^*) = 0 \\
\frac{\partial \mathcal{L}}{\partial \xi_k} = 0 & \rightarrow \quad c - \alpha_k - \eta_k = 0 \\
\frac{\partial \mathcal{L}}{\partial \xi_k^*} = 0 & \rightarrow \quad c - \alpha_k^* - \eta_k^* = 0.
\end{cases}
$$

(2.47)

The dual problem is again a QP problem:

$$
\boxed{\text{D}}: \quad \max_{\alpha,\alpha^*} J_D(\alpha,\alpha^*) = -\frac{1}{2} \sum_{k,l=1}^{N} (\alpha_k - \alpha_k^*)(\alpha_l - \alpha_l^*) x_k^T x_l
$$
$$
-\epsilon \sum_{k=1}^{N} (\alpha_k + \alpha_k^*) + \sum_{k=1}^{N} y_k(\alpha_k - \alpha_k^*) \tag{2.48}
$$
$$
\text{such that} \quad \sum_{k=1}^{N} (\alpha_k - \alpha_k^*) = 0
$$
$$
\alpha_k, \alpha_k^* \in [0, c].
$$

The SVM in primal weight space for linear function estimation is $f(x) = w^T x + b$. With $w = \sum_{k=1}^{N}(\alpha_k - \alpha_k^*)x_k$ this becomes in the dual space:

$$
f(x) = \sum_{k=1}^{N} (\alpha_k - \alpha_k^*) x_k^T x + b \tag{2.49}
$$

where α_k, α_k^* are the solution to the QP problem (2.48). The bias term b follows from the complementarity KKT conditions.

The properties of this solution are comparable with the results on classification. The solution is global and unique. Many of the elements in the solution vector will be equal to zero, which gives us again a sparseness property. In this linear case one can equally well solve the primal problem as the dual problem. The former is parametric while the latter is non-parametric. The size of the dual problem is independent of the dimension of the input space, but depends on the number of training data points.

2.5.2 *SVM for nonlinear function estimation*

Linear support vector regression can now be extended from the linear to the nonlinear case, again by application of the kernel trick.

In the primal weight space the model takes the form

$$
f(x) = w^T \varphi(x) + b, \tag{2.50}
$$

with given training data $\{x_k, y_k\}_{k=1}^{N}$ and $\varphi(\cdot) : \mathbb{R}^n \to \mathbb{R}^{n_h}$ a mapping to a high dimensional feature space which can be infinite dimensional and is only implicitly defined. Note that in this nonlinear case the vector w can also become infinite dimensional. The optimization problem in the primal weight space becomes

$$
\boxed{\text{P}}: \quad \min_{w,b,\xi,\xi^*} \ J_\text{P}(w,\xi,\xi^*) = \frac{1}{2} w^T w + c \sum_{k=1}^{N}(\xi_k + \xi_k^*)
$$

such that
$$
y_k - w^T \varphi(x_k) - b \le \epsilon + \xi_k \ , \ k = 1, ..., N
$$
$$
w^T \varphi(x_k) + b - y_k \le \epsilon + \xi_k^* \ , \ k = 1, ..., N
$$
$$
\xi_k, \xi_k^* \ge 0 \ , \ k = 1, ..., N.
$$
(2.51)

In fact one can formally replace here x from the linear function estimation case by $\varphi(x)$.

After taking the Lagrangian and conditions for optimality one obtains the following dual problem:

$$
\boxed{\text{D}}: \quad \max_{\alpha,\alpha^*} \ J_\text{D}(\alpha,\alpha^*) = \ -\frac{1}{2} \sum_{k,l=1}^{N}(\alpha_k - \alpha_k^*)(\alpha_l - \alpha_l^*)\, K(x_k, x_l)
$$
$$
-\epsilon \sum_{k=1}^{N}(\alpha_k + \alpha_k^*) + \sum_{k=1}^{N} y_k(\alpha_k - \alpha_k^*)
$$

such that
$$
\sum_{k=1}^{N}(\alpha_k - \alpha_k^*) = 0
$$
$$
\alpha_k, \alpha_k^* \in [0, c].
$$
(2.52)

Here the kernel trick has been applied with $K(x_k, x_l) = \varphi(x_k)^T \varphi(x_l)$ for $k, l = 1, ..., N$. The dual representation of the model becomes

$$
f(x) = \sum_{k=1}^{N}(\alpha_k - \alpha_k^*) K(x, x_k) + b \tag{2.53}
$$

where α_k, α_k^* are the solution to the QP problem (2.74) and b follows from the complementarity KKT conditions.

The solution to the QP problem is global and unique provided that the chosen kernel function is positive definite. As in the classifier case, the solution is sparse. The size of the QP problem does not depend on the dimension of the input space. While in the linear case one might equally well solve the primal problem, this is no longer the case for the nonlinear function estimation problem as the w vector might become infinite dimensional and often $\varphi(\cdot)$ is not explicitly known. In comparison with the nonlinear SVM classifier case, the nonlinear SVM regression formulation contains more tuning parameters. In the case of an RBF kernel σ, c, ϵ are to be considered as additional tuning parameters which do not follow as the solution to the QP problem but should be determined in another way, e.g. based on cross-validation or by applying VC bounds for the regression case. In Fig. 2.13 and Fig. 2.14 some illustrations are given on the effect of changing ϵ for a noisy sinc function and the corresponding support vectors using the SVM Matlab toolbox by Steve Gunn.

In order to have a better control upon the number of support vectors one has proposed other formulations such as the *ν-tube support vector regression* [205]. In this method the primal objective function is modified as follows:

$$\frac{1}{2} w^T w + c \left(\nu \epsilon + \frac{1}{N} \sum_{k=1}^{N} (\xi_k + \xi_k^*) \right). \tag{2.54}$$

In this approach one can control by ν the fraction of support vectors that is allowed to lie outside the tube, which is directly proportional to the number of support vectors in an asymptotic sense.

2.5.3   *VC bound on generalization error*

For the nonlinear function estimation case VC bounds have also been derived [41; 42; 206; 281]. For the empirical risk with squared loss function

$$R_{\text{emp}}(\theta) = \frac{1}{N} \sum_{k=1}^{N} (y_k - f(x_k; \theta))^2 \tag{2.55}$$

and the predicted risk

$$R(\theta) = \int (y - f(x; \theta))^2 p(x, y) dx dy \tag{2.56}$$

one has the following VC bound

$$R(\theta) \leq R_{\text{emp}}(\theta) \left(1 - c \sqrt{\frac{h(\ln(aN/h) + 1) - \ln \eta}{N}} \right)_+^{-1} \tag{2.57}$$

where h denotes the VC dimension of the set of approximating functions. In [42] $a = c = 1$ has been chosen for the constants. This bound holds

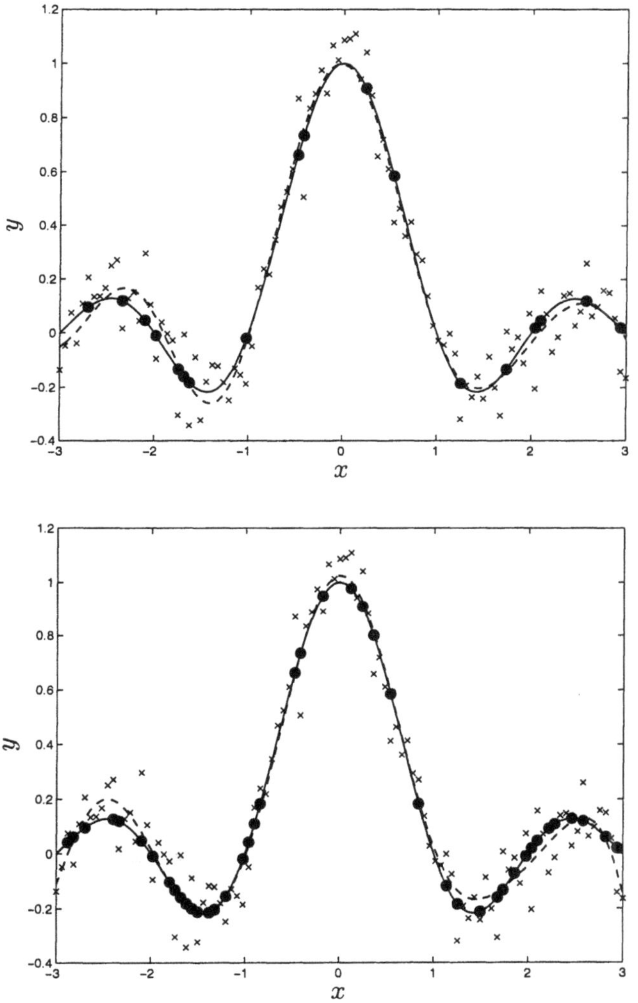

Fig. 2.13   Illustration of SVM with Vapnik ϵ-insensitive loss function on a noisy sinc function (zero mean Gaussian noise with standard dev. equal to 0.01) with (Top) $c = 100$ and $\epsilon = 0.15$ and RBF kernel with $\sigma = 0.8$ giving 18 support vectors and (Bottom) $\epsilon = 0.15/2$ giving 43 support vectors. True sinc function (solid line); SVM output (dashed line).

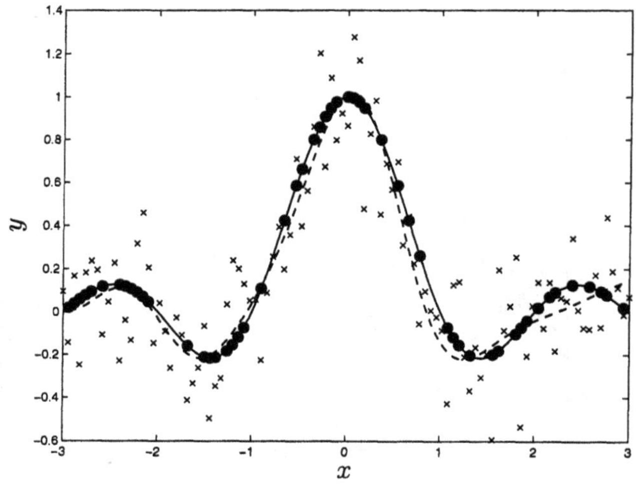

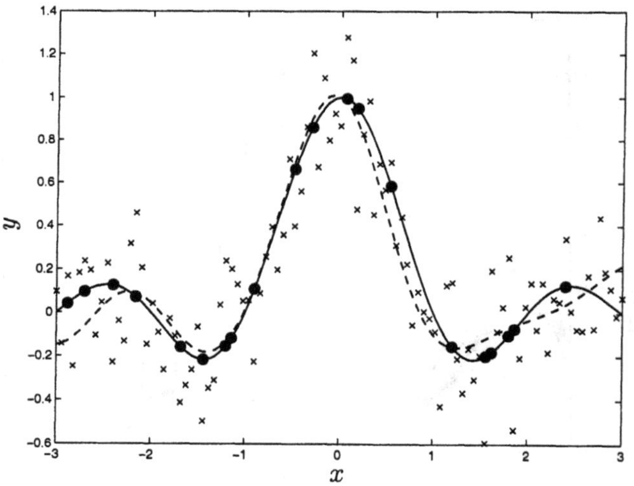

Fig. 2.14   Illustration of SVM with Vapnik ϵ-insensitive loss function on a noisy sinc function (zero mean Gaussian noise with standard dev. equal to 0.05) with (Top) $c = 100$ and $\epsilon = 0.15$ and RBF kernel with $\sigma = 0.8$ resulting into 54 support vectors and (Bottom) $\epsilon = 0.3$ giving 20 support vectors. True sinc function (solid line); SVM output (dashed line).

with probability $1 - \eta$. The notation $(x)_+$ means $(x)_+ = x$ if $x > 0$ and 0 otherwise.

The estimated risk can be expressed as

$$R_{\text{est}} = g(h, N)\frac{1}{N}\sum_{k=1}^{N}(y_k - f(x_k; \theta))^2 \qquad (2.58)$$

where $g(h, N)$ is a correcting function.

After bringing the formula in this form it is straightforward to establish the link with well-known classical criteria in statistics (see Vapnik [281]) such as

- *Finite prediction error* (FPE) (Akaike) [5]

$$g(d, N) = \frac{1 + d/N}{1 - d/N} \qquad (2.59)$$

- *Generalized cross-validation* (GCV) (Craven & Wahba) [49]

$$g(d, N) = \frac{1}{\left(1 - \frac{d}{N}\right)^2} \qquad (2.60)$$

- *Shibata's model selector* (SMS) [214]

$$g(d, N) = 1 + 2\frac{d}{N} \qquad (2.61)$$

- *Schwarz criteria* (MDL criteria) [209]

$$g(d, N) = 1 + \frac{\frac{d}{N}\log N}{2\left(1 - \frac{d}{N}\right)} \qquad (2.62)$$

with d the number of free parameters for a model which is linear in the parameters.

Recent work on the topic of generalization and the mathematical foundations of learning theory have been given by Cucker & Smale in [54]. In this work bounds on the generalization error are derived with a careful analysis of the approximation error and sample error, with results specified to e.g. Sobolev spaces and reproducing kernel Hilbert spaces.

2.6   Modifications and extensions

2.6.1   *Kernels*

Kernels from kernels

In the previous Sections some typical choices of positive definite kernels were discussed such as the linear, polynomial, RBF and MLP kernel. In classification problems one frequently employs the linear, polynomial and RBF kernel. In nonlinear function estimation and nonlinear modelling problems one often uses the RBF kernel.

However, there are many other opportunities to go beyond such basic choices. A first extension is that it is allowed to do many operations on positive definite kernels such that they still remain positive definite. For symmetric positive definite kernels the following operations are allowed as explained for example by Cristianini & Shawe-Taylor in [51]:

- $K(x, z) = aK_1(x, z) \quad (a > 0)$
- $K(x, z) = K_1(x, z) + b \quad (b > 0)$
- $K(x, z) = x^T P z \quad (P = P^T > 0)$
- $K(x, z) = K_1(x, z) + K_2(x, z)$
- $K(x, z) = K_1(x, z) K_2(x, z)$
- $K(x, z) = f(x) f(z)$
- $K(x, z) = K_3(\phi(x), \phi(z))$
- $K(x, z) = p_+(K_4(x, z))$
- $K(x, z) = \exp(K_4(x, z))$

where $a, b \in \mathbb{R}^+$ and K_1, K_2, K_3, K_4 are symmetric positive definite kernel functions, $f(\cdot) : \mathbb{R}^n \to \mathbb{R}$, $\phi(\cdot) : \mathbb{R}^n \to \mathbb{R}^{n_h}$ and $p_+(\cdot)$ is a polynomial with positive coefficients.

One may also normalize kernels. For the linear kernel one can take:

$$K(x, z) = \frac{x^T z}{\|x\|_2 \|z\|_2} = \cos(\theta_{\{x,z\}}) \qquad (2.63)$$

where $\theta_{\{x,z\}}$ denotes the angle between the vectors x, z in the input space, or in general

$$
\begin{aligned}
\tilde{K}(x, z) &= \frac{K(x, z)}{\sqrt{K(x, x)}\sqrt{K(z, z)}} \\
&= \cos(\theta_{\{\varphi(x), \varphi(z)\}})
\end{aligned}
\tag{2.64}
$$

where $\tilde{K}$ is the normalized kernel [108] and $\theta_{\{\varphi(x), \varphi(z)\}}$ is the angle between $\varphi(x), \varphi(z)$ in the feature space. Also note that one can express $d(\varphi(x), \varphi(z))$ as the Euclidean distance between $\varphi(x), \varphi(z)$ in the feature space as

$$
d(\varphi(x), \varphi(z))^2 = \|\varphi(x) - \varphi(z)\|_2^2 = K(x, x) + K(z, z) - 2K(x, z). \tag{2.65}
$$

Hence instead of working with $d(x, z)$ which is the Euclidean distance between x, z one may take the RBF kernel $K(x, z) = \exp(-\|x - z\|_2^2/\sigma^2) = \exp(-d(x, z)^2/\sigma^2)$ such that:

$$
\begin{aligned}
\tilde{K}(x, z) &= \exp\left(-\frac{K(x, x) + K(z, z) - 2K(x, z)}{\sigma^2}\right) \\
&= \exp\left(-\frac{d(\varphi(x), \varphi(z))^2}{\sigma^2}\right).
\end{aligned}
\tag{2.66}
$$

The normalization of the kernel (2.64) may be considered as a special case of a conformal transformation of the kernel

$$
\tilde{K}(x, z) = c(x)K(x, z)c(z) \tag{2.67}
$$

with factor $c(x)$. This was successfully applied by Amari [7] for modifying the kernel in a data dependent way in order to increase the margin by enlarging the spatial resolution around the decision boundary. Also differential geometrical interpretations have been given at this point by showing that the metric can be directly derived from the kernel. The link between the choice of the kernel and differential geometry has also been investigated by Schölkopf *et al.* [203] and Burges in [202]. From this study one finds e.g. for which values of κ_1, κ_2 the MLP kernel $\tanh(\kappa_1 x^T z + \kappa_2)$ is positive definite.

Compactly supported kernels

Towards computational methods it can be an advantage to work with a compactly supported kernel, e.g. in the case of an RBF kernel one may try to cut-off the kernel in order to create zeros (sparseness) in the kernel matrix.

However, this will often result in a kernel which is no longer positive definite and should be avoided. Therefore, in [93] a general way has been discussed to obtain a compactly supported kernel from a given kernel without loosing positive definiteness and has been applied in [104].

Training a classical neural network by SVM methodology

While SVM methodology largely makes use of the fact that one chooses a kernel function without worrying about the underling feature space map $\varphi(\cdot)$, unless showing that it exists by taking a positive definite kernel, one may also explicitly choose $\varphi(\cdot)$ and compute $K(x, z) := \varphi(x)^T \varphi(z)$. In [234] this idea was applied to show that one can also train classical MLP neural networks by SVM methodology. A classical MLP classifier of the form

$$y(x) = \text{sign}[w^T \tanh(Vx + \beta)] \tag{2.68}$$

was interpreted as an SVM classifier in the primal weight space by defining $\varphi(x) = \tanh(Vx + \beta)$ where the feature space explicitly corresponds to the hidden layer space of the MLP classifier. The kernel then becomes $K(x, z) = \tanh(Vx + \beta)^T \tanh(Vz + \beta)$. In this case all the elements of the hidden layer matrix V and the bias vector β are to be considered as tuning parameters, which shows the computational advantage and powerful representation of choosing e.g. an RBF kernel in SVM which has only one single tuning parameter.

String kernels and textmining

Instead of letting kernels operate on real numbers one can also go far beyond such limitations. Special type of kernels have been considered for example that can operate on strings as shown by Cristianini [51] and Haussler [108]. Furthermore, in applications of text categorization for classification of documents the following choice of a kernel is meaningful [51]

$$\varphi_i(x) = a \, \text{TF}_i \log(\text{IDF}_i) \tag{2.69}$$

and in accordance with the information retrieval literature. Here TF_i denotes the number of occurrences of a term i in the document x, IDF_i the number of documents containing the term and a a normalization constant which is chosen such that $\|\varphi(x)\|_2 = 1$. The vector $\varphi(x)$ is typically very large and sparse. The kernel is chosen then as $K(x,z) := \varphi(x)^T \varphi(z)$ and again positive definite by construction.

Bioinformatics applications

Special type of kernels have also been developed towards certain bioinformatics applications [206; 303]. For example in order to recognize translation initiation sites from which coding starts and to determine which parts of a sequence will be translated one may employ the kernel

$$K(x,z) = \left(\sum_{p=1}^{l} \mathrm{win}_p(x,z) \right)^{d_2}$$

$$\text{with} \qquad \mathrm{win}_p(x,z) = \left(\sum_{j=-l}^{+l} v_j \mathrm{match}_{p+j}(x,z) \right)^{d_1} \qquad (2.70)$$

where $\mathrm{match}_{p+j}(x,z)$ is defined to be 1 for matching nucleotides at position $p+j$ and zero otherwise, for the DNA four letter alphabet of nucleotides $\{A, C, G, T\}$. This kernel is so-called *locality improved* as it emphasizes local correlations. The window scores computed with win_p are summed over the whole length of the sequence. Correlations between windows of order up to d_2 are taking into account. At each sequence position one compares two sequences locally, within a small window length $2l+1$ around that position. Matching nucleotides are summed and multiplied with weights v_j having increasing values towards the center of the window.

For microarray data classification the use of traditional linear, polynomial and RBF kernels has been investigated [31; 89; 297].

Optimal kernels

In general the problem of which is the most suitable kernel for a particular application or problem is still an open problem up till now. Attempts in this direction are [121; 52] using Fisher kernels and optimizing kernel alignment and kernel-target alignment.

2.6.2   *Extension to other convex cost functions*

In the previous Sections commonly used cost functions were discussed as originally proposed by Vapnik. It is possible to generalize the results of

SVM regression to any convex cost function.

According to Vapnik [282] and Smola & Schölkopf [219], consider a loss function

$$L_\epsilon(y - f(x)) = \begin{cases} 0, & \text{if } |y - f(x)| \le \epsilon \\ L(y - f(x)) - \epsilon, & \text{otherwise} \end{cases} \tag{2.71}$$

where $L(\cdot)$ is convex. One can formulate then the primal problem

$$\boxed{P}: \quad \min_{w,b,\xi,\xi^*} \quad \frac{1}{2}w^T w + c\sum_{k=1}^{N}(L(\xi_k) + L(\xi_k^*))$$

$$\text{such that} \quad y_k - w^T\varphi(x_k) - b \le \epsilon + \xi_k$$
$$w^T\varphi(x_k) + b - y_k \le \epsilon + \xi_k^*$$
$$\xi_k, \xi_k^* \ge 0 \tag{2.72}$$

where ξ, ξ^* are slack variables. The Lagrangian for this problem is

$$\mathcal{L}(w,b,\xi,\xi^*;\alpha,\alpha^*,\eta,\eta^*) =$$
$$\frac{1}{2}w^T w + c\sum_{k=1}^{N}(L(\xi_k) + L(\xi_k^*)) - \sum_{k=1}^{N}\alpha_k(\epsilon + \xi_k - y_k + w^T\varphi(x_k) + b)$$
$$- \sum_{k=1}^{N}\alpha_k^*(\epsilon + \xi_k^* + y_k - w^T\varphi(x_k) - b) - \sum_{k=1}^{N}(\eta_k\xi_k + \eta_k^*\xi_k^*) \tag{2.73}$$

with Lagrange multipliers $\alpha_k, \alpha_k^*, \eta_k, \eta_k^* \ge 0$ for $k = 1, ..., N$.

One obtains the following dual problem

$$\boxed{D}: \quad \max_{\alpha,\alpha^*,\eta,\eta^*} \quad J_D(\alpha,\alpha^*,\eta,\eta^*)$$

$$\text{such that} \quad \sum_{k=1}^{N}(\alpha_k - \alpha_k^*) = 0$$
$$cL'(\xi_k) - \alpha_k - \eta_k = 0, \quad k = 1,...,N$$
$$cL'(\xi_k^*) - \alpha_k^* - \eta_k^* = 0, \quad k = 1,...,N$$
$$\alpha_k, \alpha_k^*, \eta_k, \eta_k^* \ge 0, \quad k = 1,...,N \tag{2.74}$$

where $J_D(\alpha,\alpha^*,\eta,\eta^*)$ is the objective function for this dual problem obtained after elimination of the primal variables. In these expressions one further needs to eliminate $L'(\xi_k)$. This is possible for example for $L(\xi) = \xi$ (Vapnik ϵ-insensitive loss function), $L(\xi) = \xi^2$ (quadratic loss function with

ϵ-zone) and Huber loss function with ϵ-zone as explained in [282]. The Huber loss function is well-known in the area of robust statistics. For these problems the dual variables η, η^* can be eliminated from the problem. In [219] the results have been further generalized to any convex cost function.

From these equations one can see that the obtained *sparseness* for the model is due to the ϵ-zone around the origin of the loss function. In this region one has $L'(\cdot) = 0$ such that one obtains $\alpha_k = -\eta$, $\alpha_k^* = -\eta^*$ and because $\alpha_k, \alpha_k^*, \eta_k, \eta_k^* \geq 0$, one obtains then $\alpha_k = 0$, $\alpha_k^* = 0$.

2.6.3 *Algorithms*

Interior point algorithms

The dual problems in SVM formulations can usually be cast in the form

$$\begin{aligned} \min \quad & \frac{1}{2}q(\alpha) + c^T\alpha \\ \text{such that} \quad & A\alpha = b \\ & l \leq \alpha \leq u \end{aligned} \tag{2.75}$$

where α denotes here the vector of all dual variables, $c, \alpha, l, u \in \mathbb{R}^n$, $A \in \mathbb{R}^{m \times n}$, $b \in \mathbb{R}^m$ and $q(\alpha)$ a convex function in α.

An important class of methods for solving convex programs are *interior point algorithms* [86; 173; 174; 190; 261]. Since the seminal work of Karmarkar in linear programming, this approach has been successfully extended to nonlinear programs. Several techniques have been developed for interior point algorithms including e.g. central path methods, path following methods and potential reduction methods.

In [219] the implementation of a path-following method has been investigated for SVMs by Smola. In (2.75) one first translates the inequality constraints into equality constraints by adding slack variables g, t as follows

$$\begin{aligned} \min \quad & \frac{1}{2}q(\alpha) + c^T\alpha \\ \text{such that} \quad & A\alpha = b \\ & \alpha - g = l \\ & \alpha + t = u \\ & g, t \geq 0. \end{aligned} \tag{2.76}$$

The Wolfe dual to (2.76) is

$$\max \quad \tfrac{1}{2}(q(\alpha) - \nabla q(\alpha)^T \alpha) + b^T y + l^T z - u^T s$$

$$\text{such that} \quad \tfrac{1}{2}\nabla q(\alpha) + c - (Ay)^T + s = z \tag{2.77}$$

$$s, z \geq 0$$

together with

$$\begin{aligned} g_i z_i &= 0 \\ s_i t_i &= 0, \ i = 1, ..., n \end{aligned} \tag{2.78}$$

from the KKT conditions. For the interior point algorithm the latter is modified then into

$$\begin{aligned} g_i z_i &= \mu \\ s_i t_i &= \mu, \ i = 1, ..., n \end{aligned} \tag{2.79}$$

and the problem is iteratively solved by gradually decreasing the value of μ. The above system of equations is linearized and the resulting equations are solved by a predictor-corrector approach until the duality gap is small enough. This linearization gives the following after neglecting terms in Δ^2:

$$\begin{aligned} A\Delta\alpha & & &= b - A\alpha & &=: \rho \\ \Delta\alpha - \Delta g & & &= l - \alpha + g & &=: \nu \\ \Delta\alpha + \Delta t & & &= u - \alpha - t & &=: \tau \\ (A\Delta y)^T + \Delta z - \Delta s - \tfrac{1}{2}\nabla^2 q(\alpha)\Delta\alpha & & &= c - (Ay)^T + \tfrac{1}{2}\nabla q(\alpha) + s - z & &=: \sigma \\ g_{\text{inv}_i} z_i \Delta g_i + \Delta z_i & & &= \mu g_{\text{inv}_i} - z_i - g_{\text{inv}_i}\Delta g_i \Delta z_i & &=: \gamma_{z_i} \\ t_{\text{inv}_i} s_i \Delta t_i + \Delta s_i & & &= \mu t_{\text{inv}_i} - s_i - t_{\text{inv}_i}\Delta t_i \Delta s_i & &=: \gamma_{s_i} \end{aligned} \tag{2.80}$$

where $g_{\text{inv}} = [1/g_1; ...; 1/g_n]$ and similarly for t_{inv}, z_{inv}, s_{inv}. Solving for $\Delta g, \Delta t, \Delta z, \Delta s$ one gets

$$\begin{aligned} \Delta g_i &= z_{\text{inv}_i} g_i (\gamma_{z_i} - \Delta z_i) \\ \Delta t_i &= s_{\text{inv}_i} t_i (\gamma_{s_i} - \Delta s_i) \\ \Delta z_i &= g_{\text{inv}_i} z_i (\hat{\nu}_i - \Delta\alpha_i) \\ \Delta s_i &= t_{\text{inv}_i} s_i (\Delta\alpha_i - \hat{\tau}_i) \end{aligned} \tag{2.81}$$

where $\hat{\nu}_i = \nu_i - z_{\text{inv}_i} g_i \gamma_{z_i}$, $\hat{\tau}_i = \tau_i - s_{\text{inv}_i} t_i \gamma_{s_i}$ for $i = 1, ..., n$. One also obtains the *reduced KKT system*

$$\begin{bmatrix} -(\tfrac{1}{2}\nabla^2 q(\alpha) + D) & A^T \\ A & 0 \end{bmatrix} \begin{bmatrix} \Delta\alpha \\ \Delta y \end{bmatrix} = \begin{bmatrix} \sigma - r \\ \rho \end{bmatrix} \tag{2.82}$$

where $D = \mathrm{diag}[g_{\mathrm{inv}_1}z_1 + t_{\mathrm{inv}_1 s_1}; ...; g_{\mathrm{inv}_n}z_n + t_{\mathrm{inv}_n s_n}]$ and $r_i = g_{\mathrm{inv}_i}z_i\hat{\nu}_i - t_{\mathrm{inv}_i}s_i\hat{\tau}_i$.

In the predictor step one solves the system (2.81) and (2.82) with $\nu = 0$ and all Δ terms on the right hand side set to zero, i.e. $\gamma_z = z$, $\gamma_s = s$. The values in Δ are substituted back into the definitions for γ_z and γ_s and (2.81) and (2.82) are solved again in the corrector step. The matrix needs to be inverted then only once in a given iteration step. To ensure that the variables meet the positivity constraints, a careful choice of the steplength ξ is needed. The value of μ is often chosen as

$$\mu = \frac{g^T z + s^T t}{2n}\left(\frac{\xi - 1}{\xi + 10}\right)^2. \tag{2.83}$$

The reduced KKT system is often additionally regularized in order to further improve the behaviour of the algorithm [262]. Also a good initialization of the algorithm will improve its performance. For further details on this interior point algorithm the reader may consult [219; 262].

Large data sets: chunking, decomposition and others

The interior point method as described here is only suitable for typically about 2000 data points (depending on the computer memory) assuming that the matrices are stored. A possible way for applying SVMs to larger data sets is to start with a first *chunk* of the data set that fits into memory. One trains an SVM on this chunk, keeps the support vectors and deletes the non-support vector data. This removed part can be replaced then by a new part of the large training set, where one selects data on which the current SVM makes an error. The system is retrained and one iterates until the KKT conditions are satisfied for all training data.

In *decomposition* methods [142] one works with a fixed size subset while freezing the other variables, which is often more convenient. Results on chunking and decomposition were made by Vapnik [279], Osuna *et al.* [175], Joachims [202], Lin [142] and others. An extreme form of decomposition is sequential minimal optimization (SMO) [182] where very small subsets of size 2 are selected. In this case one can find an analytic expression for the solution to the two-datapoint sub-problem.

On the other hand other optimization algorithms have also been used such as SOR (Successive OverRelaxation) on massive data sets with millions of data points by Mangasarian & Musicant in [155]. Within the class of iterative methods conjugate gradient methods perform well, see e.g. Kaufman in [202]. These methods are suitable for processing large data sets as well.

2.6.4 *Parametric versus non-parametric approaches*

The original Vapnik formulation for SVMs emphasizes primal-dual interpretations and makes use of the kernel trick. In recent years one has also recognized that this kernel trick can be applied in a more general fashion. Often existing linear techniques can be kernelized into nonlinear version. However, sometimes there might arise confusion with respect to parametric versus non-parametric issues at this point.

As explained in the previous Chapter the link between parametric models of radial basis function (RBF) networks and their parametric version could be understood by taking the centers of the Gaussians at the data points itself. However, in this case one should keep in mind that the unknown parameter vector size is fixed and the centers accidently coincide with the training data points. One could take a parametric kernel model as follows:

$$f(x; w) = \sum_{i=1}^{N} w_i K(x, x_i) \qquad (2.84)$$

with $w \in \mathbb{R}^N$ a fixed size parameter vector and estimate the parameter vector as if it were a non-kernel based model (which traditionally belongs to the class of non-parametric methods). The notation w is chosen here instead of α in order to avoid confusion with SVM models in the dual space of Lagrange multipliers. Also note that there is no need at all at this point that the kernel function should be positive definite and satisfy the Mercer condition, because in (2.84) one may choose in principle *any* parameterization that one would like.

As explained in the previous chapter one could proceed then as with the training of classical MLPs where one considers a weight decay term and find w from

$$\min_{w} J(w) = \|w\|_2^2 + \sum_{i=1}^{N} \left(y_i - \sum_{i=1}^{N} w_i K(x, x_i) \right)^2 . \qquad (2.85)$$

The technique of automatic relevance determination (ARD) has then been used within a Bayesian inference context in order to assess the importance of the weights and hence on the data points, because by construction of the model each weight w_i corresponds to a data point $x_i \in \mathbb{R}^n$ (which is not the case e.g. in MLP neural networks). This method is related to what has been called the *relevance vector machine* in [253]. From a neural networks perspective such a model should rather be considered as a radial basis function network than as a support vector machine.

One can achieve sparseness for this model also by taking another regularization term

$$\min_{w} J(w) = \|w\|_1 + \sum_{i=1}^{N} \left(y_i - \sum_{i=1}^{N} w_i K(x, x_i) \right)^2 \qquad (2.86)$$

where the 1-norm $\|w\|_1 = \sum_{i=1}^{N} |w_i|$ is used. This regularization term corresponds to a Laplacian prior when interpreted within a Bayesian inference context. Such methods are related to methods of sparse approximation [38; 95] and the Lasso and shrinkage estimators [107] in statistics.

Presently many methods are called kernel machines. However, not all of these possess the same primal-dual interpretations of SVMs as proposed by Vapnik or make use of the kernel trick. LP kernel machine [206; 225] formulations are based on the Vapnik ϵ-insensitive loss function. One optimizes

$$R_{\text{reg}}(w) = \frac{1}{N} \sum_{i=1}^{N} |w_i| + c \frac{1}{N} \sum_{i=1}^{N} |y_i - f(x_i)|_\epsilon \qquad (2.87)$$

which leads to the following LP problem

$$\min \quad \frac{1}{N} \sum_{i=1}^{N} (w_i + w_i^*) + \frac{c}{N} \sum_{i=1}^{N} (\xi_i + \xi_i^*)$$

$$\text{such that} \quad \sum_{i=1}^{N} (w_i - w_i^*) K(x_i, x_j) + b - y_i \le \epsilon + \xi_i \qquad (2.88)$$

$$y_i - \sum_{i=1}^{N} (w_i - w_i^*) K(x_i, x_j) - b \le \epsilon + \xi_i^*$$

$$w_i, w_i^*, \xi_i, \xi_i^* \ge 0$$

for $i, j = 1, ..., N$. Here w_i has been replaced by the positive variables w_i, w_i^* in order to avoid $|w_i|$ in the cost function. According to Smola in [225] there is no computational advantage for this LP problem to compute the Wolfe dual. Methods for large scale LP problems are available in the optimization literature including simplex methods and interior point algorithms.

Chapter 3

Basic Methods of Least Squares
Support Vector Machines

In this Chapter we discuss basic methods of Least Squares Support Vector Machines (LS-SVMs) for classification and nonlinear function estimation. For LS-SVM classifiers we discuss the link with a kernel version of Fisher discriminant analysis in high dimensional feature spaces. LS-SVM regression is closely related to regularization networks, Gaussian processes and reproducing kernel Hilbert spaces but emphasizes primal-dual interpretations in the context of constrained optimization problems. LS-SVM models for classification and nonlinear regression are characterized by linear Karush-Kuhn-Tucker systems. A simple method for imposing sparseness to LS-SVM models is explained by applying pruning techniques as known in neural networks.

3.1 Least Squares Support Vector Machines for classification

A nice aspect of nonlinear SVMs is that one solves nonlinear classification and regression problems by means of convex quadratic programs. Moreover, one also obtains sparseness as a result of this QP problem. A natural question that one may ask is *how much one may simplify the SVM formulation without losing any of its advantages?*

We will propose here a modification to the Vapnik SVM classifier formulation which leads to solving a set of linear equations, which is for many practitioners in different areas, easier to use than QP solvers. The following

SVM modification was originally proposed by Suykens in [235]:

$$
\left[\begin{array}{l}
\boxed{P}: \quad \min_{w,b,e} J_P(w,e) = \quad \frac{1}{2} w^T w + \gamma \frac{1}{2} \sum_{k=1}^{N} e_k^2 \\[2em]
\text{such that} \quad y_k \left[w^T \varphi(x_k) + b \right] = 1 - e_k, \quad k = 1, ..., N
\end{array}\right]
$$

(3.1)

for a classifier in the primal space that takes the form

$$
y(x) = \text{sign}[w^T \varphi(x) + b]
$$

(3.2)

where $\varphi(\cdot) : \mathbb{R}^n \rightarrow \mathbb{R}^{n_h}$ is the mapping to the high dimensional feature space as in the standard SVM case. The Vapnik formulation is modified here at two points. First, instead of inequality constraints one takes equality constraints where the value 1 at the right hand side is rather considered as a target value than a threshold value. Upon this target value an error variable e_k is allowed such that misclassifications can be tolerated in the case of overlapping distributions. These error variables play a similar role as the slack variables ξ_k in SVM formulations. Second, a squared loss function is taken for this error variable. As we will see now these modifications will greatly simplify the problem.

In the case of a linear classifier one could easily solve the primal problem, but in general w might become infinite dimensional. Therefore let us derive the dual problem for this LS-SVM nonlinear classifier formulation. The Lagrangian for the problem is

$$
\mathcal{L}(w,b,e;\alpha) = J_P(w,e) - \sum_{k=1}^{N} \alpha_k \{ y_k [w^T \varphi(x_k) + b] - 1 + e_k \}
$$

(3.3)

where the α_k values are the Lagrange multipliers, which can be positive or negative now due to the equality constraints.

The conditions for optimality yield

$$
\begin{cases}
\frac{\partial \mathcal{L}}{\partial w} = 0 & \rightarrow \quad w = \sum_{k=1}^{N} \alpha_k y_k \varphi(x_k) \\[2ex]
\frac{\partial \mathcal{L}}{\partial b} = 0 & \rightarrow \quad \sum_{k=1}^{N} \alpha_k y_k = 0 \\[2ex]
\frac{\partial \mathcal{L}}{\partial e_k} = 0 & \rightarrow \quad \alpha_k = \gamma e_k, \qquad\qquad\qquad k = 1, ..., N \\[2ex]
\frac{\partial \mathcal{L}}{\partial \alpha_k} = 0 & \rightarrow \quad y_k[w^T \varphi(x_k) + b] - 1 + e_k = 0, \quad k = 1, ..., N.
\end{cases}
\tag{3.4}
$$

Defining $Z^T = [\varphi(x_1)^T y_1; ...; \varphi(x_N)^T y_N]$, $y = [y_1; ...; y_N]$, $1_v = [1; ...; 1]$, $e = [e_1; ...; e_N]$, $\alpha = [\alpha_1; ...; \alpha_N]$ and eliminating w, e, one obtains the following linear Karush-Kuhn-Tucker (KKT) system [82; 174]

$$
\left[\boxed{D} : \quad \text{solve in } \alpha, b : \\[2ex]
\left[\begin{array}{c|c} 0 & y^T \\ \hline y & \Omega + I/\gamma \end{array} \right] \left[\begin{array}{c} b \\ \alpha \end{array} \right] = \left[\begin{array}{c} 0 \\ 1_v \end{array} \right] \right]
\tag{3.5}
$$

where $\Omega = Z^T Z$ and the kernel trick can be applied within the Ω matrix

$$
\begin{aligned}
\Omega_{kl} &= y_k y_l \, \varphi(x_k)^T \varphi(x_l) \\
&= y_k y_l \, K(x_k, x_l) \, , \quad k, l = 1, ..., N.
\end{aligned}
\tag{3.6}
$$

The classifier in the dual space takes the form

$$
y(x) = \text{sign} \left[\sum_{k=1}^{N} \alpha_k y_k K(x, x_k) + b \right]
\tag{3.7}
$$

similar to the standard SVM case. Note that one could equally well express the solution in terms of the unknown error variables e by eliminating α instead of e.

This LS-SVM classifier has the following properties:

- *Choice of kernel function:*
  The chosen kernel function should be positive definite and satisfy the Mercer condition. All the comments on kernel functions equally well apply to the use of kernels in the LS-SVM context. For the

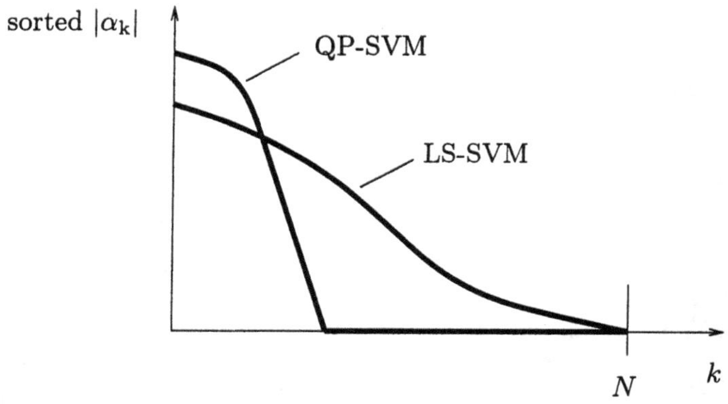

Fig. 3.1   *Qualitative comparison of the sorted $|\alpha|$ solution vectors between QP-SVMs (including the standard Vapnik SVM) and LS-SVMs. In the LS-SVM case each data point is contributing to the model and sparseness is lost.*

sequel of this book we will focus however on the use of linear, polynomial and RBF kernels.

- *Global and unique solution:*
  The dual problem for linear and nonlinear LS-SVMs corresponds to solving a linear KKT system which is a square system with a unique solution when the matrix has full rank.

- *KKT system as a core problem:*
  Solving linear KKT systems is a fundamental issue in constrained nonlinear optimization problems in general. At every iteration step KKT systems of a similar form as (3.5) are solved.
  Also we have seen in the previous Chapter that when one uses interior point methods for solving SVMs with any convex cost function, one finally obtains reduced KKT systems (2.82). Therefore, solving SVM classifiers can be considered as iteratively solving KKT systems where each takes a similar form as one single LS-SVM classifier [171].

- *Lack of sparseness and interpretation of support vectors:*
  A drawback of the simplified formulation is the lack of sparseness.

This is clear from the condition for optimality

$$\alpha_k = \gamma e_k.$$

In the LS-SVM classifier case every data point is a support vector as, normally, no α_k values will be exactly equal to zero. In Fig. 3.1 an illustrative comparison is shown between sorted $|\alpha_k|$ values in the Vapnik SVM case versus the LS-SVM classifier case after taking the absolute value and sorting the values from large to small. In the LS-SVM case this is rather a decaying spectrum without a black/white decision between non-zero and zero α_k values. Hence, one could heuristically say that in the LS-SVM case all training data points will contribute to the model, and certain data points are more important than others. The points with large $|\alpha_k|$ are located close and far from the decision boundary.

- *Non-parametric/parametric issues*:
  LS-SVM classifiers have the same primal-dual neural network interpretations as shown for SVMs in Fig. 2.5. In the primal weight space the problem is parametric with fixed size vector $w \in \mathbb{R}^{n_h}$ where n_h is the number of hidden units in the network interpretation for this space. In the dual space the problem is non-parametric as the size of the solution vector $\alpha \in \mathbb{R}^N$ grows with the number of training data N. Note that the size of the KKT system is not influenced by the dimension of the input space n, but is only determined by N. Hence, typically for large dimensional input spaces one will benefit by solving the dual problem instead of the primal one and often it is only possible to solve the dual (when the dimension of the feature space is very large).

- *Tuning parameters*:
  If one takes for example an RBF kernel $K(x, x_k) = \exp(-\|x - x_k\|_2^2/\sigma^2)$ then only α, b result from solving the linear KKT system (5.12), not the parameters (γ, σ). There are several ways then to proceed in order to determine these parameters. A simple way is to work with a training set, validation set and test set, explore a meaningful grid of possible (γ, σ) combinations and select the values in such a way that they give the best performance on this

validation set. As we discussed in the introductory chapter, the results in this case might be too sensitive with respect to the chosen validation set. In a statistical sense it is therefore better (but computationally heavier) to do n_{CV}-fold cross-validation. Further alternatives are bootstrap techniques and determination of hyper-parameters at higher levels by Bayesian inference. VC bounds as discussed in the previous chapter can be applied as well. The advantage of VC bounds is that they do not take any assumptions about the underlying density of the data, except that the data are i.i.d.

In Fig. 3.2 an example is shown of an LS-SVM classifier with RBF kernel for a two spiral classification problem. This problem is known to be difficult for classical MLP classifiers as explained e.g. in [192] due to the fact that a highly nonlinear decision boundary has to be realized. The LS-SVM classifier obtains a zero error on the training set and excellent generalization with a smooth decision boundary as shown in the Figure. This problem is difficult in the sense that a complicated nonlinear decision boundary should be determined, but on the other hand is simple in the sense that the problem is separable (non-overlapping class distributions). A second illustrative example is given in Fig. 3.3 for the Ripley data set. This figure shows the decision boundary and the support values. The good performance on many UCI benchmark data sets will be discussed further in this Chapter.

3.2   Multi-class formulations

The binary LS-SVM classifier can be easily extended to multi-class problems. A traditional approach in neural networks is to take additional output variables as was illustrated e.g. on the alphabet recognition problem in the introductory chapter.

Instead of a single output value y, let us take now multiple output values $y^{(i)}$ with $i = 1, ..., n_y$ where n_y denotes the number of output values. With respect to the choice of the number of outputs the coding is important. In principle it is possible to encode 2^{n_y} classes by means of a number of n_y outputs, but usually this will certainly not be the best possible choice. A better choice is normally to take as many outputs as the number of classes. Let us postpone the discussion here on the best coding/decoding issues

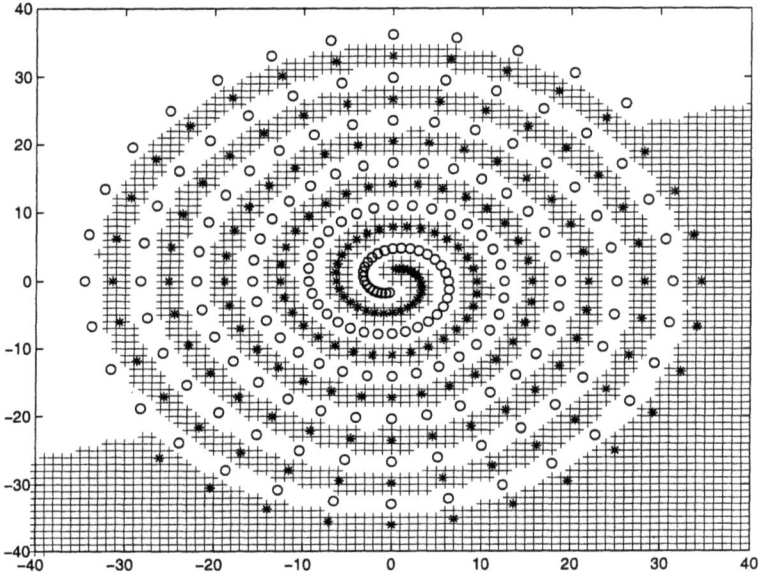

Fig. 3.2 *Excellent classification and generalization performance obtained by an LS-SVM classifier with RBF kernel on a two spiral classification problem. The given training data points for the two classes are shown as (o) and (*). The black/white regions give a visualization of the generalization performance with classification as class 1 (white) or class 2 (black) for given inputs in $\mathbb{R}^2$ with inputs in the range [-40,40] which are the variables given at the two axes.*

and focus now on a simple LS-SVM classifier formulation extension to the multi-class case [237].

In the primal weight space the multi-class classification system is based on the outputs of the following binary classifiers

$$
\left\{
\begin{array}{rcl}
y^{(1)}(x) & = & \text{sign}[w^{(1)^T}\varphi^{(1)}(x) + b^{(1)}] \\
y^{(2)}(x) & = & \text{sign}[w^{(2)^T}\varphi^{(2)}(x) + b^{(2)}] \\
& \vdots & \\
y^{(n_y)}(x) & = & \text{sign}[w^{(n_y)^T}\varphi^{(n_y)}(x) + b^{(n_y)}]
\end{array}
\right.
\tag{3.8}
$$

with mappings to high dimensional feature spaces $\varphi^{(i)}(\cdot) : \mathbb{R}^n \to \mathbb{R}^{n_{h_i}}$ (for $i = 1, 2, ..., n_y$) with dimensions $n_{h_1}, n_{h_2}, ..., n_{h_{n_y}}$. One has corresponding vectors $w^{(i)} \in \mathbb{R}^{n_{h_i}}$ and bias terms $b^{(i)} \in \mathbb{R}$.

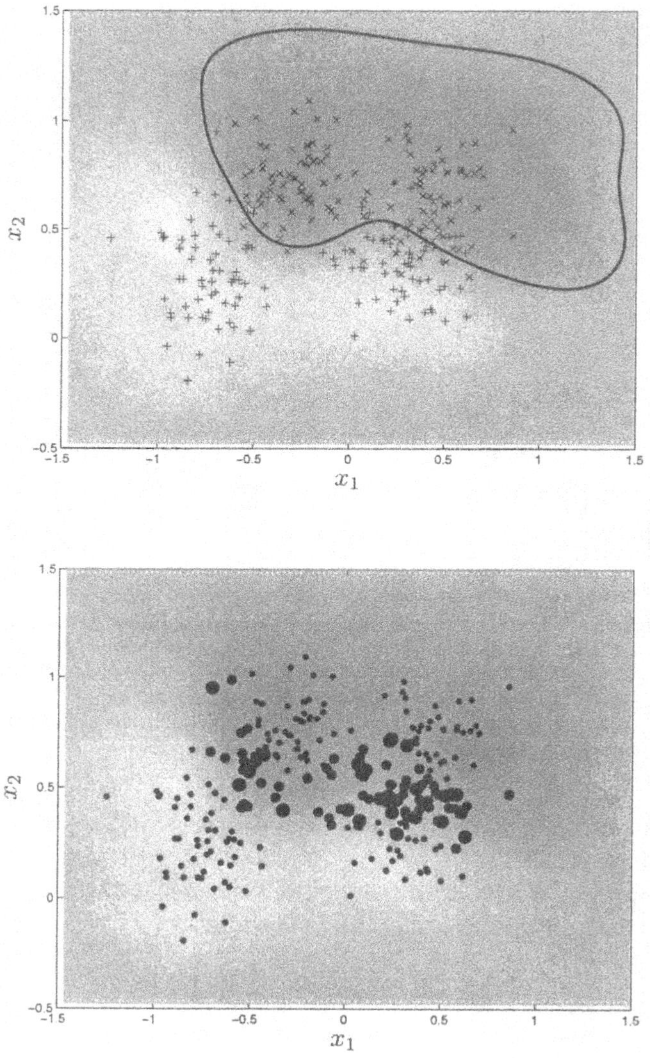

Fig. 3.3   *Illustration of the LS-SVM classifier on the Ripley binary classification data sets: (Top) Decision boundary for LS-SVM with a well-tuned RBF kernel; (Bottom) illustration of the support values α_k with size of the black dots chosen proportional to the α_k values.*

The primal optimization problem for multi-class LS-SVMs becomes then

$$\boxed{\text{P}}: \quad \min_{w^{(i)}, b^{(i)}, e_k^{(i)}} J_{\text{P}}(w^{(i)}, e_k^{(i)}) =$$

$$\frac{1}{2} \sum_{i=1}^{n_y} w^{(i)^T} w^{(i)} + \frac{1}{2} \sum_{i=1}^{n_y} \gamma_i \sum_{k=1}^{N} \left(e_k^{(i)} \right)^2$$

such that

$$y_k^{(1)} [w^{(1)^T} \varphi^{(1)}(x_k) + b^{(1)}] = 1 - e_k^{(1)}, \ k = 1, ..., N$$

$$y_k^{(2)} [w^{(2)^T} \varphi^{(2)}(x_k) + b^{(2)}] = 1 - e_k^{(2)}, \ k = 1, ..., N$$

$$\vdots$$

$$y_k^{(n_y)} [w^{(n_y)^T} \varphi^{(n_y)}(x_k) + b^{(n_y)}] = 1 - e_k^{(n_y)}, \ k = 1, ..., N.$$

(3.9)

The Lagrangian for this problem is

$$\mathcal{L}(w^{(i)}, b^{(i)}, e_k^{(i)}; \alpha_k^{(i)}) = J_{\text{P}}(w^{(i)}, e_k^{(i)}) -$$

$$\sum_{i=1}^{n_y} \sum_{k=1}^{N} \alpha_k^{(i)} \left(y_k^{(i)} [w^{(i)^T} \varphi^{(i)}(x_k) + b^{(i)}] - 1 + e_k^{(i)} \right)$$

(3.10)

with conditions for optimality

$$\begin{cases} \dfrac{\partial \mathcal{L}}{\partial w^{(i)}} = 0 \quad \rightarrow \quad w^{(i)} = \displaystyle\sum_{k=1}^{N} \alpha_k^{(i)} y_k^{(i)} \varphi^{(i)}(x_k) \\[4mm] \dfrac{\partial \mathcal{L}}{\partial b^{(i)}} = 0 \quad \rightarrow \quad \displaystyle\sum_{k=1}^{N} \alpha_k^{(i)} y_k^{(i)} = 0 \\[4mm] \dfrac{\partial \mathcal{L}}{\partial e_k^{(i)}} = 0 \quad \rightarrow \quad \alpha_k^{(i)} = \gamma e_k^{(i)} \\[4mm] \dfrac{\partial \mathcal{L}}{\partial \alpha_k^{(i)}} = 0 \quad \rightarrow \quad y_k^{(i)} [w^{(i)^T} \varphi^{(i)}(x_k) + b^{(i)}] = 1 - e_k^{(i)} \end{cases}$$

(3.11)

for $k = 1, ..., N$ and $i = 1, ..., n_y$.

Finally, after elimination of the variables $w^{(i)}$ and $e_k^{(i)}$ we obtain then the following KKT system as the dual problem

$$
\left[\boxed{\text{D}} : \quad \text{solve in } \alpha_k^{(i)}, b^{(i)} : \\
\left[\begin{array}{c|c} 0 & Y_M^T \\ \hline Y_M & \Omega_M + D_M \end{array} \right] \left[\begin{array}{c} b_M \\ \alpha_M \end{array} \right] = \left[\begin{array}{c} 0 \\ 1_v \end{array} \right] \right] \tag{3.12}
$$

where

$$
Y_M = \text{blockdiag}\left\{ \left[\begin{array}{c} y_1^{(1)} \\ \vdots \\ y_N^{(1)} \end{array} \right], ..., \left[\begin{array}{c} y_1^{(n_y)} \\ \vdots \\ y_N^{(n_y)} \end{array} \right] \right\}
$$

$$
\Omega_M = \text{blockdiag}\{\Omega^{(1)}, ..., \Omega^{(n_y)}\}, \quad \Omega_{kl}^{(i)} = y_k^{(i)} y_l^{(i)} K^{(i)}(x_k, x_l)
$$

$$
D_M = \text{blockdiag}\{D^{(1)}, ..., D^{(n_y)}\}, \quad D_{kl}^{(i)} = \delta_{kl}/\gamma_i
$$

for $k, l = 1, ..., N$, $i = 1, ..., n_y$ and $b_M = [b^{(1)}; ...; b^{(n_y)}]$, $\alpha_M = [\alpha_1^{(1)}; ...;$ $\alpha_N^{(1)}; ....; \alpha_1^{(n_y)}; ...; \alpha_N^{(n_y)}]$ and δ_{kl} denotes the Kronecker delta ($\delta_{kl} = 1$ if $k = l$ and 0 otherwise). The kernel trick is applied as follows

$$
\begin{aligned}
K^{(i)}(x_k, x_l) &= \varphi^{(i)}(x_k)^T \varphi^{(i)}(x_l) \\
&= \exp\left(-\|x_k - x_l\|_2^2/\sigma_i^2\right), \quad i = 1, ..., n_y
\end{aligned} \tag{3.13}
$$

illustrated for RBF kernels. Note that for each individual (binary) subclassifier it is important to take different optimal values for (γ_i, σ_i) in the case of an RBF kernel. The resulting classifiers in the dual space are

$$
y^{(i)}(x) = \text{sign}\left[\sum_{k=1}^{N} \alpha_k^{(i)} y_k^{(i)} K^{(i)}(x, x_k) + b^{(i)}\right] \tag{3.14}
$$

for $i = 1, ..., n_y$.

The KKT system (3.12) also has a clear block-diagonal structure. Therefore it is clear that it is not needed to solve the system (3.12) as a whole. One can decompose it into n_y smaller subproblems thanks to the block-diagonal structure. In fact this comes as no surprise because it is well known that multiclass problems can be solved as a set of binary class problems. In general it is important then to find an optimal coding/decoding for the problem [66; 260].

Usually, multiclass categorization problems are solved by reformulating the multiclass problem with n_C classes into a set of n_y binary classification problems. To each class C_i a unique codeword $[y_i^{(1)}; y_i^{(2)}; \ldots; y_i^{(n_y)}] \in \{-1, +1\}^{n_y}$ for $i = 1, \ldots, n_C$ is assigned. There are several ways then to construct the set of binary classifiers. If one uses n_y outputs to encode up to 2^{n_y} classes, one applies a minimum output coding (MOC). In one-versus-all (1vsA) coding one takes the number of outputs equal to the number of classes or $n_C = n_y$ and makes binary decisions between each class and all other classes. The use of error correcting output codes (ECOC) [66] is motivated by information theory. One introduces redundancy by taking more binary classifiers than the number of classes ($n_y > n_C$) in the output coding. In one-versus-one (1vs1) output coding, one uses $n_C(n_C - 1)/2$ binary plug-in classifiers, where each binary classifier discriminates between two opposite classes in a pairwise way. The 1vs1 output coding can also be represented by n_y length codewords belonging to $\{-1, 0, +1\}^{n_y}$ where one uses 0 for the classes that are not considered. For example, three classes can be represented using the codewords $[-1; +1; 0]$, $[-1; 0; +1]$ and $[0; -1; +1]$. While MOC uses only 2 outputs to encode 3 classes, the use of three outputs in the 1vs1 coding typically results into simpler decision boundaries. In the decoding process, class labels are usually assigned according to the codeword with minimal Hamming distance to the output, but other options are possible as well. When one considers the 1vs1 output coding scheme as a voting between each pair of classes, the codeword with minimal Hamming distance corresponds to the class with the maximal number of votes.

3.3 Link with Fisher discriminant analysis in feature space

In the previous Chapter we have explained that the standard SVM formulation conceptually starts from the margin concept according to which a unique separating hyperplane is defined, while in the separable case in fact several separating hyperplanes exist in general.

We will argue now that a natural link exists between the LS-SVM classifier formulation and Fisher discriminant analysis which is an old and well-known method in pattern recognition, but done now in the high dimensional feature space.

In order to establish and explain this link let us first review the method of linear Fisher discriminant analysis and then explain its extension to the

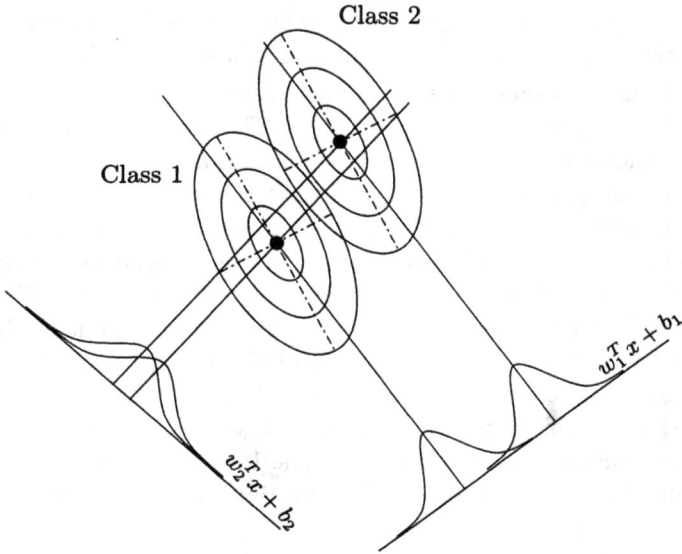

Fig. 3.4   In Fisher discriminant analysis one aims at maximizing the between-class scatter and minimizing the within-class scatter. One finds a projection of data $x \in \mathbb{R}^n$ onto a one-dimensional space with variable $z \in \mathbb{R}$ such that a Rayleigh quotient is maximized. This figure shows that projection on the line $z = w_1^T x + b_1$ gives a much better discriminatory power than on $z = w_2^T x + b_2$.

high dimensional feature space.

3.3.1   *Linear Fisher discriminant analysis*

A major goal of Fisher discriminant analysis [80; 81] is to project data $x_k \in \mathbb{R}^n$ from the original input space to a one-dimensional variable $z_k \in \mathbb{R}$ (in a statistical sense this means projecting multivariate data to univariate data) and make a discrimination based on this projected variable. In this one-dimensional space (in fact a straight line) one tries to achieve a high discriminatory power. One tries to maximize the between-class variances and minimize the within-class variances for the two classes (Fig. 3.4).

The data are projected as follows

$$z = f(x) = w^T x + b \tag{3.15}$$

with $f(\cdot) : \mathbb{R}^n \rightarrow \mathbb{R}$. We are interested then in finding a line such that the following objective of a *Rayleigh quotient* is maximized:

$$
\begin{aligned}
\max_{w,b} J_{\text{FD}}(w,b) &= \frac{[\mathcal{E}[z^{(1)}] - \mathcal{E}[z^{(2)}]]^2}{\mathcal{E}\{[z^{(1)} - \mathcal{E}[z^{(1)}]]^2\} + \mathcal{E}\{[z^{(2)} - \mathcal{E}[z^{(2)}]]^2\}} \\
&= \frac{[w^T(\mu^{(1)} - \mu^{(2)})]^2}{w^T \mathcal{E}\{[x - \mu^{(1)}][x - \mu^{(1)}]^T\}w + w^T \mathcal{E}\{[x - \mu^{(2)}][x - \mu^{(2)}]^T\}w} \\
&= \frac{w^T \Sigma_B w}{w^T \Sigma_W w}
\end{aligned}
$$

(3.16)

where $z_k^{(1)} = w^T x_k^{(1)} + b$, $z_k^{(2)} = w^T x_k^{(2)} + b$ denote the transformed variables belonging to class 1 and class 2, respectively. The means of the input variables for class 1 and class 2 are $\mathcal{E}[x^{(1)}] = \mu^{(1)}, \mathcal{E}[x^{(2)}] = \mu^{(2)}$. The between and within covariance matrices related to class 1 and class 2 are $\Sigma_B = [\mu^{(1)} - \mu^{(2)}][\mu^{(1)} - \mu^{(2)}]^T$, $\Sigma_W = \mathcal{E}\{[x - \mu^{(1)}][x - \mu^{(1)}]^T\} + \mathcal{E}\{[x - \mu^{(2)}][x - \mu^{(2)}]^T\}$ where the latter is the sum of the two covariance matrices $\Sigma_{W_1}, \Sigma_{W_2}$ for the two classes. Note that the Rayleigh quotient is independent of the bias term b.

It is well-known that a Rayleigh quotient characterizes the optimality for eigenvalue problems [114; 290]. By taking $\partial J_{\text{FD}}(w)/\partial w = 0$ we indeed obtain the generalized eigenvalue problem $\Sigma_W w = \Sigma_B w \, (w^T \Sigma_W w / w^T \Sigma_B w)$. By using the expression for Σ_B one obtains then the following w_{FD} direction for the optimal line

$$
w_{\text{FD}} \propto \Sigma_W^{-1}[\mu^{(1)} - \mu^{(2)}].
$$

(3.17)

The projections of the means $\mu^{(1)}, \mu^{(2)}$ to the one-dimensional space give

$$
\begin{aligned}
f(\mu^{(1)}) &= w_{\text{FD}}^T \mu^{(1)} + b \propto [\mu^{(1)} - \mu^{(2)}]^T \Sigma_W^{-1} \mu^{(1)} \\
f(\mu^{(2)}) &= w_{\text{FD}}^T \mu^{(2)} + b \propto [\mu^{(1)} - \mu^{(2)}]^T \Sigma_W^{-1} \mu^{(2)}.
\end{aligned}
$$

(3.18)

In practice one works with the sample means

$$
\hat{\mu}^{(1)} = \frac{1}{N_1} \sum_{k=1}^{N_1} x_k^{(1)}, \quad \hat{\mu}^{(2)} = \frac{1}{N_2} \sum_{k=1}^{N_2} x_k^{(2)}
$$

(3.19)

for $\mu^{(1)}, \mu^{(2)}$ of class 1 and 2, respectively, and the sample covariance ma-

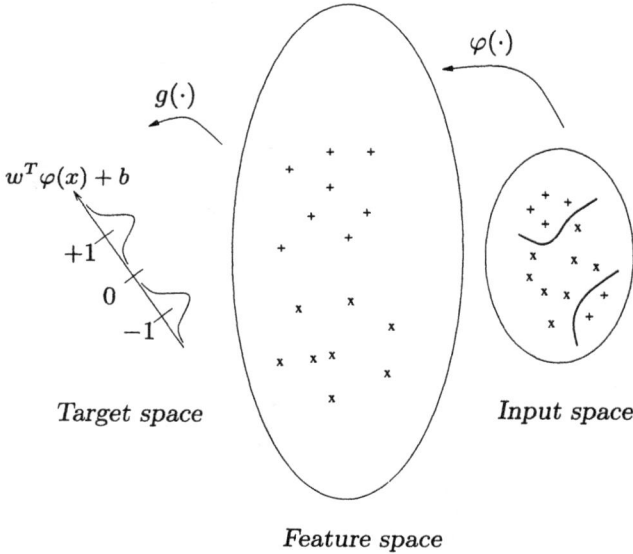

Fig. 3.5 *Fisher discriminant analysis in the feature space is closely related to LS-SVM classification. In LS-SVM classification the constraints $y_k[w^T\varphi(x_k)+b] = 1 - e_k$ can be interpreted as having target values +1 and −1 on a line for which one aims at minimizing the within-class scattering, which is characterized by the term $\sum_k e_k^2$ in the objective function of the LS-SVM classifier.*

trices

$$S_{W_1} = \frac{1}{N_1 - 1}\sum_{i=1}^{N_1}[x_k^{(1)} - \hat{\mu}^{(1)}][x_k^{(1)} - \hat{\mu}^{(1)}]^T$$

$$S_{W_2} = \frac{1}{N_2 - 1}\sum_{i=1}^{N_2}[x_k^{(2)} - \hat{\mu}^{(2)}][x_k^{(2)} - \hat{\mu}^{(2)}]^T$$

(3.20)

as estimates for the covariance matrices. Here N_1, N_2 denote the number of data points for class 1 and 2, respectively.

As a classification rule one often computes then the midpoint of the line and checks at which side a projected data point is located in order to make a decision.

3.3.2 Fisher discriminant analysis in feature space

Let us now formulate the Fisher discriminant algorithm in a high dimensional feature space instead of the original input space and point out the link with LS-SVM classifiers.

The data are projected as follows

$$z = f(x) = g(\varphi(x)) = w^T \varphi(x) + b \qquad (3.21)$$

with $f(\cdot) : \mathbb{R}^n \to \mathbb{R}$ the mapping from the input space to the one-dimensional target space, $g(\cdot) : \mathbb{R}^{n_h} \to \mathbb{R}$ the mapping from the high dimensional feature space to the target space and $\varphi(\cdot) : \mathbb{R}^n \to \mathbb{R}^{n_h}$ the mapping from the input space to the high dimensional feature space, which can be infinite dimensional.

As for linear Fisher discriminant analysis we aim now at maximizing the between-class scatter and minimize the within-class scatter for the projected one-dimensional variable z. Note that the LS-SVM classifier constraints

$$y_k[w^T \varphi(x_k) + b] = 1 - e_k, \quad k = 1, ..., N \qquad (3.22)$$

can be interpreted as having target values $+1$ and -1 upon which one wants to minimize the errors e_k (Fig. 3.5). This is done by taking a squared error in the cost function which corresponds to minimizing the within-class scatter for classes 1 and 2. Hence, in the LS-SVM classifier formulation one fixes in fact the between-class scatter by taking target values $+1$ and -1 for the z target variable on the line $z = w^T \varphi(x) + b$.

Fisher discriminant analysis in the feature space optimizes then the same Rayleigh quotient as in (3.16), where $z_k^{(1)} = w^T \varphi(x_k^{(1)}) + b$, $z_k^{(2)} = w^T \varphi(x_k^{(2)}) + b$ denote now the projected variables belonging to class 1 and class 2, respectively. The means of the input variables for class 1 and class 2 are $\mathcal{E}[\varphi(x^{(1)})] = \mu^{(1)}, \mathcal{E}[\varphi(x^{(2)})] = \mu^{(2)}$. The between and within covariance matrices related to class 1 and class 2 are $\Sigma_B = [\mu^{(1)} - \mu^{(2)}][\mu^{(1)} - \mu^{(2)}]^T$, $\Sigma_W = \mathcal{E}\{[\varphi(x) - \mu^{(1)}][\varphi(x) - \mu^{(1)}]^T\} + \mathcal{E}\{[\varphi(x) - \mu^{(2)}][\varphi(x) - \mu^{(2)}]^T\}$ where the latter is the sum of the two covariance matrices $\Sigma_{W_1}, \Sigma_{W_2}$ for the two classes.

Note that these matrices can become *infinite dimensional* now, as well as the solution vector w_{FD}. Therefore, for the nonlinear case one may better solve the dual problem and apply the kernel trick. This dual problem corresponds then to solving the LS-SVM classifier problem (5.12). Of course instead of fixing the targets at $+1$ and -1, one could generalize this to other values and allow more freedom in the numerator of the Rayleigh quotient in this way. This would lead to a LS-SVM classifier formulation in the primal weight space as follows

$$
\left[\begin{array}{l}
\boxed{\text{P}}: \quad \min_{w,b,e} J_\text{P}(w,e) = \frac{1}{2}w^T w + \gamma\frac{1}{2}\left(\sum_{k=1}^{N_1}(e_k^{(1)})^2 + \sum_{l=1}^{N_2}(e_l^{(2)})^2\right) \\
\text{such that} \quad y_k[w^T\varphi(x_k) + b] = t_1 - e_k^{(1)}, k = 1,\dots,N_1 \\
\qquad\qquad\quad y_l[w^T\varphi(x_l) + b] = t_2 - e_l^{(2)}, l = 1,\dots,N_2
\end{array}\right]
$$
(3.23)

where t_1, t_2 are fixed and positive constant target values for class 1 and 2 and N_1, N_2 are the number of data points belonging to class 1 and 2 respectively. The indices k, l run over the elements of class 1 and 2, respectively. In a regression interpretation context one has targets $z_1 = t_1$ and $z_2 = -t_2$. It is an easy exercise to work out the dual problem of this. In [264] it was shown, however, that it is usually better to work with $t_1 = t_2 = 1$ as in the original LS-SVM classifier formulation.

As we explained, the link between LS-SVM classification and Fisher discriminant analysis is very natural. From this analysis we can state that the goal of LS-SVM classifiers is basically twofold:

(1) Maximizing the soft margin by minimizing $\|w\|_2$ for separable or non-separable data either in the linear or nonlinear case.

(2) Minimizing the within-class scatter for fixed targets on a line with a similar objective as Fisher discriminant analysis but in a high dimensional feature spaces.

Other formulations and links with kernel based methods have been discussed also in [16; 163; 206].

3.4   Solving the LS-SVM KKT system

3.4.1   *Conjugate gradient iterative methods*

The solution to linear and nonlinear LS-SVM classifiers is characterized by a square linear system of equations. For many decades several methods have been developed in numerical analysis for reliably solving sets of linear equations and many specialized algorithms have been developed that maximally try to exploit structure of the problem [98; 100; 256; 301].

The size of the matrix grows with the number of data points. There-fore, direct elimination methods are usually restricted to data sets of size of about $N = 2000$, depending on the available computer memory and as-suming that one stores the full matrix into memory. For larger data sets the use of iterative methods is recommended. In principle, various methods can be used at this point including SOR (Successive Over-Relaxation), CG (Conjugate Gradient), GMRES (Generalized Minimal Residual) etc. How-ever, not all of these iterative methods can be applied to any kind of linear system. For example, in order to apply CG the matrix should be positive definite. Due to the presence of the b bias term in the LS-SVM model the resulting matrix is not positive definite (in fact it is well known that KKT systems as arising in constrained optimization problems are indefinite). So before we can apply such methods we have to transform the linear system into a positive definite system. According to [236] this can be done in a simple way as follows.

The LS-SVM KKT system is of the form

$$\begin{bmatrix} 0 & y^T \\ y & H \end{bmatrix} \begin{bmatrix} \xi_1 \\ \xi_2 \end{bmatrix} = \begin{bmatrix} d_1 \\ d_2 \end{bmatrix} \tag{3.24}$$

more specifically with $H = \Omega + I/\gamma$, $\xi_1 = b$, $\xi_2 = \alpha$, $d_1 = 0$, $d_2 = 1_v$. This can be transformed into

$$\begin{bmatrix} s & 0 \\ 0 & H \end{bmatrix} \begin{bmatrix} \xi_1 \\ \xi_2 + H^{-1}y\xi_1 \end{bmatrix} = \begin{bmatrix} -d_1 + y^T H^{-1}d_2 \\ d_2 \end{bmatrix} \tag{3.25}$$

with $s = y^T H^{-1}y > 0$ $(H = H^T > 0)$. Because s is positive and H positive definite the overall matrix is positive definite. This form is very suitable because different kinds of iterative methods can be applied to problems involving positive definite matrices. This leads to the following algorithm for solving LS-SVM classifiers with application of a conjugate gradient al-gorithm.

LS-SVM CG large scale algorithm:

(1) Solve η, ν from $H\eta = y$ and $H\nu = 1_v$.

(2) Compute $s = y^T \eta$.

(3) Find solution: $b = \eta^T 1_v / s$ and $\alpha = \nu - \eta b$.

The system (3.25) is of the form $\mathcal{A}x = \mathcal{B}$ with $\mathcal{A} = \mathcal{A}^T > 0$. A basic *Hestenes-Stiefel CG algorithm* [98] works as follows

$$
\left[
\begin{array}{l}
i = 0; x_0 = 0; r_0 = \mathcal{B}; \\
\text{while } r_i \neq 0 \\
\quad i = i + 1 \\
\quad \text{if } i = 1 \\
\quad\quad p_1 = r_0 \\
\quad \text{else} \\
\quad\quad \beta_i = r_{i-1}^T r_{i-1} / r_{i-2}^T r_{i-2} \\
\quad\quad p_i = r_{i-1} + \beta_i p_{i-1} \\
\quad \text{end} \\
\quad \lambda_i = r_{i-1}^T r_{i-1} / p_i^T \mathcal{A} p_i \\
\quad x_i = x_{i-1} + \lambda_i p_i \\
\quad r_i = r_{i-1} - \lambda_i \mathcal{A} p_i \\
\text{end} \\
x = x_i
\end{array}
\right.
$$

Some comments on the application of this conjugate gradient method:

- The underlying cost function optimized by the CG algorithm is of the quadratic form $V(x) = \frac{1}{2}x^T \mathcal{A}x - x^T \mathcal{B}$. In the application of this algorithm it has been experienced that especially the decrease of $V(x)$ is important towards a suitable stopping criterion in this LS-SVM context. The main reason for using a measure based on the difference $V(x_i) - V(x_{i-1})$ between two iteration steps i and $i-1$, is that $V(x)$ monotonically decreases. This is not the case for the norm on the residual $\mathcal{A}x - \mathcal{B}$.

- The complexity of CG methods can be understood as follows [25; 98; 217]. Let x_i be the i-th iteration by the CG method when solving $\mathcal{A}x = \mathcal{B}$ with $\mathcal{A}$ real symmetric and positive definite, then one has

$$
\|x_i - x_*\|_{\mathcal{A}} \leq \left(\frac{\sqrt{\kappa} - 1}{\sqrt{\kappa} + 1} \right)^i \|x_0 - x_*\|_{\mathcal{A}}
\tag{3.26}
$$

where the $\mathcal{A}$-norm of a vector v is defined as $\|v\|_{\mathcal{A}} = (v^T \mathcal{A} v)^{1/2}$ and $\kappa = \|\mathcal{A}\| \|\mathcal{A}^{-1}\|$ denotes the condition number. One can conclude

[217] that if

$$i \geq \frac{\log \frac{2}{\epsilon}}{\log \left(\frac{\sqrt{\kappa}-1}{\sqrt{\kappa}+1} \right)} = \mathcal{O} \left(\frac{\sqrt{\kappa}}{2} \log \frac{2}{\epsilon} \right) \tag{3.27}$$

then $\|x_i - x_*\|_{\mathcal{A}} \leq \epsilon \|x_0 - x_*\|_{\mathcal{A}}$ with accuracy ϵ. Hence, the time needed to solve the problem depends on the requested precision $\log \frac{1}{\epsilon}$ and the condition number of matrix $\mathcal{A}$. In contrast with direct algorithms such as Gauss elimination the computational complexity does not just depend on the size of the matrix $\mathcal{A}$. It is important to note that when using e.g. an LS-SVM classifier with RBF kernel, the matrix $\mathcal{A}$ will depend on the tuning parameters (γ, σ) and as a result also the condition number κ and the speed of convergence. Also in [293] quick convergence has been reported of conjugate gradient algorithms in the context of kernel based learning.

- The basic CG algorithm shown here might be further improved by using preconditioners. On the other hand this basic CG algorithm has shown to perform sufficiently well and fast on a large variety of problems. For the implementation of the algorithm the matrix $\mathcal{A}$ is usually not stored. The elements of the matrix are re-calculated every iteration step. Clever tricks may however be applied to optimally use the computer memory.

The CG method has also been compared with SOR methods in [103] where CG and block CG methods performed much better on several tests. Recently in [132] also squential minimal optimization (SMO) has been successfully applied to solving LS-SVM systems.

3.4.2 *LS-SVM classifiers: UCI benchmarking results*

Tuning parameter selection

Let us take a look now at some benchmarking results on UCI binary and multiclass data sets [24] as reported by Van Gestel *et al.* in [264]. The LS-SVM classifiers are tested for linear, polynomial and RBF kernels. A simple method of tuning parameter selection based on 10-fold cross-validation with several randomizations for the data sets is applied. In the case of an RBF kernel, the hyperparameter γ, the kernel parameter σ and the test set performance of the binary LS-SVM classifier are estimated by applying the following algorithm of a shrinking grid search.

LS-SVM tuning - grid search algorithm:

(1) Divide data set into 2/3 for training and validation and 1/3 for a test set.

(2) Start $i = 0$. Apply 10-fold cross-validation on the training/validation data for each (σ, γ) combination from the initial candidate tuning sets at iteration 0:

$$\begin{aligned} \Sigma_0 &= \{0.5, 5, 10, 15, 25, 50, 100, 250, 500\} \text{ (elements} \times \sqrt{n}) \\ \Gamma_0 &= \{0.01, 0.05, 0.1, 0.5, 1, 5, 10, 50, 100, 500, 1000\}. \end{aligned}$$

The square root $\sqrt{n}$ of the number of inputs n is considered in the grid since the value of $||x - x_k||_2^2$ in the RBF kernel grows with the value of n.

(3) Choose optimal (σ, γ) from the tuning sets Σ_i and Γ_i by looking at the best cross-validation performance.

(4) If $i = i_{max}$, go to step 5; else $i := i + 1$, construct a locally refined grid $\Sigma_i \times \Gamma_i$ around the optimal hyperparameters (σ, γ) and go to step 3.

(5) Construct the LS-SVM classifier using the total training/validation set for the optimal choice of the tuned hyperparameters (σ, γ).

(6) Assess the test set accuracy by means of the independent test set.

In this benchmark study, i_{max} was chosen as $i_{max} = 3$. For the polynomial kernels (γ, τ) was tuned following a similar procedure, while the γ parameter for the linear kernel was selected from a refined set Γ based upon the cross-validation performance. For multiclass problems, this cross-validation procedure is applied to each binary subproblem and the different coding/decoding schemes.

Benchmark data sets

The UCI datasets can be obtained from the UCI benchmark repository [24] at `http://kdd.ics.uci.edu/`. The US postal service is available from `http://www.kernel-machines.org/`. These datasets have been referred numerous times in the literature, which makes them very suitable for benchmarking purposes. As a preprocessing step, all records containing unknown values are removed and all data sets were normalized for the benchmarking process. The following binary datasets were retrieved from [24]: Statlog Australian credit (acr), Bupa liver disorders (bld), Statlog German credit (gcr), Statlog heart disease (hea), Johns Hopkins university ionosphere (ion), Pima Indians diabetes (pid), sonar (snr), tic-tac-toe endgame (ttt), Wisconsin breast cancer (wbc) and adult (adu) dataset. The main characteristics of these datasets are summarized in Table 3.1. The following multiclass datasets were used: balance scale (bal), contraceptive method choice (cmc), image segmentation (ims), iris (iri), LED display (led), thyroid disease (thy), US postal service (usp), Statlog vehicle silhouette (veh), waveform (wav) and wine recognition (win) dataset. The main characteristics of the multiclass datasets are summarized in Table 3.2.

Compared methods

According to [264] the following methods are compared in Tables 3.3 & 3.4: linear, polynomial and RBF kernels for LS-SVM classifiers, standard SVM classifiers with linear and RBF kernel; the decision tree algorithm C4.5 [188], Holte's one-rule classifier (oneR) [113]; linear discriminant analysis (LDA), quadratic discriminant analysis (QDA), logistic regression (Logit) [23; 71; 193]; instance based learners (IB) [3] and Naive Bayes [126]. The Matlab SVM toolbox `http://theoval.sys.uea.ac.uk/~gcc/svm/tool-box` with SMO solver was used to train and evaluate the Vapnik SVM classifier. The C4.5, IB1, IB10, Naive Bayes and oneR algorithms were implemented using the Weka workbench [295], while the Discriminant Analysis Toolbox (M. Kiefte) for Matlab was applied for LDA, QDA and Logit. The oneR, LDA, QDA, Logit, NBk and NBn require no parameter tuning. Both standard Naive Bayes with the normal approximation (NB$_n$) [71] and the kernel approximation (NB$_k$) for numerical attributes [126] were used. The default classifier or majority rule (Maj. Rule) was included as a baseline in the comparison tables. All comparisons were made on the same randomizations. For a further comparison study among 22 decision tree, 9 statistical and 2 neural network algorithms, see [141]. For each algorithm, the average test set performance and sample standard deviation on 10 randomizations in each domain is reported [18; 62; 68; 141]. Averaging over all domains, the Average Accuracy (AA) and Average Rank (AR) are shown for each algorithm [141]. A Wilcoxon signed rank test of equality of medians is used on both AA and AR to check whether the performance of an algorithm is significantly different from the algorithm with the highest accuracy. A Probability of a Sign Test (P_{ST}) is also reported comparing each algorithm to the algorithm with best accuracy.

LS-SVM classifier performance on binary class problems

The benchmarking results for the binary class problem are shown in Table 3.3. For the kernel types, RBF kernels, linear (Lin) and polynomial (Pol) kernels (with degree $d = 2, \dots, 10$) were tried. Both the performance of LS-SVM targets $\{-1, +1\}$ and Regularized Kernel Fisher Discriminant Analysis (LS-SVM$_F$) targets $\{-N/N_2, +N/N_1\}$ are reported.

The experimental results indicate that the RBF kernel yields the best validation and test set performance, while also polynomial kernels yield good performances (if one allows τ values non equal to 1). LS-SVMs with polynomial kernels ($\tau = 1$) of degrees $d = 2$ and $d = 10$ yielded on all domains average test set performances of 84.3% and 65.9%, respectively. Comparing this performance with the average test set performance of 85.6% and 85.5% obtained when using scaling ($\tau \neq 1$), this clearly motivates the use of bandwidth or kernel parameters. This is especially important for polynomial kernels with degree $d \geq 5$. The regularization parameter c and kernel parameter σ of the SVM classifiers with linear and RBF kernels were selected in a similar way as for the LS-SVM classifier using the 10-fold cross-validation procedure.

The LS-SVM classifier with RBF kernel achieves the best average test set performance on 3 of the 10 benchmark domains, while its accuracy is not significantly worse than the best algorithm in 3 other domains. LS-SVM classifiers with polynomial and linear kernel yield the best performance on two and one datasets, respectively. Also RBF SVM, IB1, NB$_k$ and C4.5 achieve the best performance on one dataset each. Comparison of the accuracy achieved by the nonlinear polynomial and RBF kernel with the accuracy of the linear kernel illustrates that most domains are only weakly nonlinear. The LS-SVM formulation with targets $\{-1, +1\}$ yields a better performance than the LS-SVM$_F$ regression formulation related to regularized kernel Fisher's discriminant with targets $\{-N/N_2, +N/N_1\}$, although not all tests report a significant difference. Noticing that the LS-SVM with linear kernel without regularization ($\gamma \to \infty$) corresponds to the LDA classifier, we also remark that a comparison of both accuracies indicates that the use of regularization slightly improves the generalization behaviour. Considering the Average Accuracy (AA) and Average Ranking (AR) over all domains [18; 62; 68], the RBF SVM gets the best average accuracy and the RBF LS-SVM yields the best average rank. There is no significant difference between the performance of both classifiers. The average performance of Pol LS-SVM and Pol LS-SVM$_F$ is not significantly different with respect to the best algorithms. The performances of many other advanced SVM algorithms are in line with the above results [30; 163; 202]. The significance tests on the average performances of the other classifiers do not always yield the same results. Generally speaking, the performance of Lin LS-SVM, Lin SVM, Logit, NB$_k$ and IB1 are not significantly different at the 1% level. Also the performances of LS-SVMs with Fisher targets (LS-SVM$_F$) are not significantly different at the 1%. From these results we may conclude that both the SVM and LS-SVM classifiers achieve very good test set performances in comparison with the other reference classification algorithms.

LS-SVM classifier performance on multiclass problems

The benchmarking results for the binary class problem are shown in Table 3.4. Each multiclass problem is decomposed into a set of binary classification problems using minimum output coding (MOC) and one-versus-one (1vs1) output coding. The same kernel types as for the binary domain were considered: RBF kernels, linear (Lin) and polynomial (Pol) kernels with degrees $d = 2, \ldots, 10$. Both the performance of LS-SVM and LS-SVM$_F$ classifiers are reported. The MOC and 1vs1 output coding were also applied to SVM classifiers with linear and RBF kernels.

The average test set accuracies of the different LS-SVM and LS-SVM$_F$ classifiers, with RBF, Lin and Pol kernel ($d = 2, \ldots, 10$) and using MOC and 1vs1 output coding, are reported in Table 3.4. The use of QDA yields the best average test set accuracy on two domains, while LS-SVMs with 1vs1 coding using a RBF and Lin kernel and LS-SVM$_F$ with Lin kernel each yield the best performance on one domain. SVMs with RBF kernel with MOC and 1vs1 coding yield the best performance on one domain each. Also C4.5, Logit and IB1 each achieve one time the best performance. The use of 1vs1 coding generally results in a better classification accuracy. Averaging over all 10 multiclass domains, the LS-SVM classifier with RBF kernel and 1vs1 output coding achieves the best average accuracy (AA) and average ranking, while its performance is only on three domains significantly worse at 1% than the best algorithm. This performance is not significantly different from the SVM with RBF kernel and 1vs1 output coding. The results illustrate that the SVM and LS-SVM classifier with RBF kernel using 1vs1 output coding consistently yield very good test set accuracies on the multiclass domains.

	acr	bld	gcr	hea	ion	pid	snr	ttt	wbc	adu
N_{CV}	460	230	666	180	234	512	138	638	455	33000
N_{test}	230	115	334	90	117	256	70	320	228	12222
N	690	345	1000	270	351	768	208	958	683	45222
n_{num}	6	6	7	7	33	8	60	0	9	6
n_{cat}	8	0	13	6	0	0	0	9	0	8
n	14	6	20	13	33	8	60	9	9	14

Table 3.1 *Characteristics of binary classification UCI datasets, where N_{CV} stands for the number of data points used in the cross-validation based tuning procedure, N_{test} for the number of observations in the test set and N for the total dataset size. The number of numerical and categorical attributes is denoted by n_{num} and n_{cat} respectively, n is the total number of attributes.*

	bal	cmc	ims	iri	led	thy	usp	veh	wav	win
N_{CV}	416	982	1540	100	2000	4800	6000	564	2400	118
N_{test}	209	491	770	50	1000	2400	3298	282	1200	60
N	625	1473	2310	150	3000	7200	9298	846	3600	178
n_{num}	4	2	18	4	0	6	256	18	19	13
n_{cat}	0	7	0	0	7	15	0	0	0	0
n	4	9	18	4	7	21	256	18	19	13
M	3	3	7	3	10	3	10	4	3	3
$n_{y,MOC}$	2	2	3	2	4	2	4	2	2	2
$n_{y,1vs1}$	3	3	21	3	45	3	45	6	2	3

Table 3.2 *Characteristics of the multiclass datasets, where N_{CV} stands for the number of data points used in the cross-validation based tuning procedure, N_{test} for the number of data in the test set and N for the total amount of data. The number of numerical and categorical attributes is denoted by n_{num} and n_{cat} respectively, n is the total number of attributes. The M row denotes the number of classes for each dataset, encoded by $n_{y,MOC}$ and $n_{y,1vs1}$ bits for MOC and 1vs1 output coding, respectively.*

	acr	bld	gcr	hea	ion	pid	snr	ttt	wbc	adu	AA	AR	P_{ST}
N_{test}	230	115	334	90	117	256	70	320	228	12222			
n	14	6	20	13	33	8	60	9	9	14			
RBF LS-SVM	**87.0**(2.1)	70.2(4.1)	76.3(1.4)	84.7(4.8)	**96.0**(2.1)	76.8(1.7)	73.1(4.2)	99.0(0.3)	96.4(1.0)	84.7(0.3)	84.4	3.5	**0.727**
RBF LS-SVM$_F$	86.4(1.9)	65.1(2.9)	70.8(2.4)	83.2(5.0)	93.4(2.7)	72.9(2.0)	73.6(4.6)	97.9(0.7)	96.8(0.7)	77.6(1.3)	81.8	8.8	0.109
Lin LS-SVM	86.8(2.2)	65.6(3.2)	75.4(2.3)	**84.9**(4.5)	87.9(2.0)	76.8(1.8)	72.6(3.7)	66.8(3.9)	95.8(1.0)	81.8(0.3)	79.4	7.7	0.109
Lin LS-SVM$_F$	86.5(2.1)	61.8(3.3)	68.6(2.3)	82.8(4.4)	85.0(3.5)	73.1(1.7)	73.3(3.4)	57.6(1.9)	**96.9**(0.7)	71.3(0.3)	75.7	12.1	0.109
Pol LS-SVM	86.5(2.2)	**70.4**(3.7)	**76.3**(1.4)	83.7(3.9)	91.0(2.5)	**77.0**(1.8)	76.9(4.7)	**99.5**(0.5)	96.4(0.9)	84.6(0.3)	84.2	4.1	**0.727**
Pol LS-SVM$_F$	86.6(2.2)	65.3(2.9)	70.3(2.3)	82.4(4.6)	91.7(2.6)	79.0(1.8)	77.3(2.6)	98.1(0.8)	**96.9**(0.7)	77.9(0.2)	82.0	8.2	0.344
RBF SVM	86.3(1.8)	**70.4**(3.2)	75.9(1.4)	84.7(4.8)	95.4(1.7)	**77.3**(2.2)	75.0(6.6)	98.6(0.5)	96.4(1.0)	84.4(0.3)	**84.4**	4.0	**1.000**
Lin SVM	86.7(2.4)	67.7(2.6)	75.4(1.7)	83.2(4.2)	87.1(3.4)	77.0(2.4)	74.1(4.2)	66.2(3.6)	96.3(1.0)	83.9(0.2)	79.8	7.5	0.021
LDA	85.9(2.2)	65.4(3.2)	75.9(2.0)	**83.9**(4.3)	87.1(2.3)	76.7(2.0)	67.9(4.9)	68.0(3.0)	95.6(1.1)	82.2(0.3)	78.9	9.6	0.004
QDA	80.1(1.9)	62.2(3.6)	72.5(1.4)	78.4(4.0)	90.6(2.2)	74.2(3.3)	53.6(7.4)	75.1(4.0)	94.5(0.6)	80.7(0.3)	76.2	12.6	0.002
Logit	**86.8**(2.4)	66.3(3.1)	**76.3**(2.1)	82.9(4.0)	86.2(3.5)	**77.1**(1.8)	68.4(5.2)	68.3(2.9)	96.1(1.0)	83.7(0.2)	79.2	7.8	0.109
C4.5	85.5(2.1)	63.1(3.8)	71.4(2.0)	78.0(4.2)	90.6(2.2)	73.5(3.0)	72.1(2.5)	84.2(1.6)	94.7(1.0)	**85.6**(0.3)	79.9	10.2	0.021
oneR	85.4(2.1)	56.3(4.4)	66.0(3.0)	71.7(3.6)	83.6(4.8)	71.3(2.7)	62.6(5.5)	70.7(1.5)	91.8(1.4)	80.4(0.3)	74.0	15.5	0.002
IB1	81.1(1.9)	61.3(6.2)	69.3(2.6)	74.3(4.2)	87.2(2.8)	69.6(2.4)	**77.7**(4.4)	82.3(3.3)	95.3(1.1)	78.9(0.2)	77.7	12.5	0.021
IB10	86.4(1.3)	60.5(4.4)	72.6(1.7)	80.0(4.3)	85.9(2.5)	73.6(2.4)	69.4(4.3)	94.8(2.0)	96.4(1.2)	82.7(0.3)	80.2	10.4	0.039
NB$_k$	81.4(1.9)	63.7(4.5)	74.7(2.1)	83.9(4.5)	92.1(2.5)	75.5(1.7)	71.6(3.5)	71.7(3.1)	**97.1**(0.9)	84.8(0.2)	79.7	**7.3**	0.109
NB$_n$	76.9(1.7)	56.0(6.9)	74.6(2.8)	83.8(4.5)	82.8(3.8)	75.1(2.1)	66.6(3.2)	71.7(3.1)	95.5(0.5)	82.7(0.2)	76.6	12.3	0.002
Maj. Rule	56.2(2.0)	56.5(3.1)	69.7(2.3)	56.3(3.8)	64.4(2.9)	66.8(2.1)	54.4(4.7)	66.2(3.6)	66.2(2.4)	75.3(0.3)	63.2	17.1	0.002

Table 3.3 Comparison of the 10 times randomized test set performance of LS-SVM and LS-SVM$_F$ (linear, polynomial and Radial Basis Function kernel) with the performance of LDA, QDA, Logit, C4.5, oneR, IB1, IB10, NB$_k$, NB$_n$ and the Majority Rule classifier on 10 binary domains. The Average Accuracy (AA), Average Rank (AR) and Probability of equal medians using the Sign Test (P_{ST}) taken over all domains are reported in the last three columns. Best performances are underlined and denoted in bold face, performances not significantly different at the 5% level are denoted in bold face, performances significantly different at the 1% level are emphasized. LS-SVMs with RBF kernels are performing very well in this comparative study.

	bal	cmc	ims	iri	led	thy	usp	veh	wav	win	AA	AR	P_{ST}
N_{test}	209	491	770	50	1000	2400	3298	282	1200	60			
n	4	9	18	4	7	21	256	18	19	13			
RBF LS-SVM (MOC)	92.7(1.0)	**54.1(1.8)**	95.5(0.6)	96.6(2.8)	70.8(1.4)	96.6(0.4)	95.3(0.5)	81.9(2.6)	**99.8(0.2)**	**98.7(1.3)**	**88.2**	**7.1**	0.344
RBF LS-SVM$_F$ (MOC)	86.8(2.4)	43.5(2.6)	69.6(3.2)	98.4(2.1)	36.1(2.4)	22.0(4.7)	86.5(1.0)	66.5(6.1)	99.5(0.2)	93.2(3.4)	70.2	17.8	0.109
Lin LS-SVM (MOC)	90.4(0.8)	46.9(3.0)	72.1(1.2)	89.6(5.6)	52.1(2.2)	93.2(0.6)	76.5(0.6)	69.4(2.3)	90.4(1.1)	97.3(2.0)	77.8	17.8	0.002
Lin LS-SVM$_F$ (MOC)	86.6(1.7)	42.7(2.0)	69.8(1.2)	77.0(3.8)	35.1(2.6)	54.1(1.3)	58.2(0.9)	69.1(2.0)	55.7(1.3)	85.5(5.1)	63.4	22.4	0.002
Pol LS-SVM (MOC)	94.0(0.8)	53.5(2.3)	87.2(2.6)	96.4(3.7)	70.9(1.5)	94.7(0.2)	95.0(0.8)	81.8(1.2)	99.6(0.3)	97.8(1.9)	87.1	9.8	0.109
Pol LS-SVM$_F$ (MOC)	93.2(1.9)	47.4(1.6)	86.2(3.2)	96.0(3.7)	67.7(0.8)	69.9(2.8)	87.2(0.9)	81.9(1.3)	96.1(0.7)	98.2(3.2)	81.8	15.7	0.002
RBF LS-SVM (1vs1)	94.2(2.2)	**55.7(2.2)**	96.5(0.5)	97.6(2.3)	74.1(1.3)	96.8(0.3)	94.8(2.5)	83.6(1.3)	99.3(0.4)	98.2(1.8)	<u>89.1</u>	<u>5.9</u>	**1.000**
RBF LS-SVM$_F$ (1vs1)	71.4(15.5)	42.7(3.7)	46.2(6.5)	79.8(10.3)	58.9(8.5)	92.6(0.2)	30.7(2.4)	84.9(2.5)	97.3(1.7)	67.3(14.6)	61.2	22.3	0.002
Lin LS-SVM (1vs1)	87.8(2.2)	50.8(2.4)	93.4(1.0)	98.4(1.8)	74.5(1.0)	93.2(0.3)	95.4(0.3)	79.8(2.1)	97.6(0.9)	98.3(2.5)	86.9	9.7	0.754
Lin LS-SVM$_F$ (1vs1)	87.7(1.8)	49.6(1.8)	93.4(0.9)	**98.8(1.3)**	**74.5(1.0)**	74.9(0.8)	95.3(0.3)	79.8(2.2)	98.2(0.6)	97.7(1.8)	85.0	11.1	0.344
Pol LS-SVM (1vs1)	95.4(1.0)	53.2(2.2)	95.2(0.6)	96.8(2.3)	72.8(2.6)	88.8(14.6)	96.0(0.2)	82.8(1.8)	99.0(0.1)	99.0(1.4)	87.9	8.9	0.344
Pol LS-SVM$_F$ (1vs1)	56.5(16.7)	41.8(1.8)	30.1(3.8)	71.4(12.4)	32.6(10.9)	92.6(0.7)	95.8(1.7)	20.3(6.7)	77.5(4.9)	82.9(12.2)	60.1	21.9	0.021
RBF SVM (MOC)	**99.2(0.5)**	51.0(1.4)	94.9(0.9)	96.6(3.4)	69.9(1.0)	96.6(0.2)	95.5(0.4)	77.6(1.7)	99.7(0.1)	97.8(2.1)	87.9	8.6	0.344
Lin SVM (MOC)	98.3(1.2)	45.8(1.6)	74.1(1.4)	95.0(10.5)	50.9(3.2)	92.5(0.3)	81.9(0.3)	70.3(2.5)	99.8(0.2)	97.3(2.6)	80.5	16.1	0.021
RBF SVM (1vs1)	98.3(1.2)	**54.7(2.4)**	96.0(0.4)	97.0(3.0)	64.6(5.6)	98.3(0.3)	74.7(0.7)	83.8(1.6)	99.6(0.2)	96.8(5.7)	88.6	6.5	**1.000**
Lin SVM (1vs1)	91.0(2.3)	50.8(1.6)	95.8(0.7)	98.0(1.9)	74.4(1.2)	97.1(0.3)	95.1(0.3)	78.1(2.4)	99.6(0.2)	98.3(3.1)	87.8	7.3	0.754
LDA	86.9(2.1)	51.8(2.2)	91.2(1.1)	98.6(1.0)	73.7(0.8)	93.7(0.3)	91.5(0.5)	77.4(2.7)	94.6(1.2)	98.7(1.5)	85.8	11.0	0.109
QDA	90.5(1.1)	50.6(2.1)	81.8(9.6)	98.2(1.8)	73.6(1.1)	93.4(0.3)	74.7(0.7)	84.8(1.5)	60.9(9.5)	99.2(1.2)	80.8	11.8	0.344
Logit	88.5(2.0)	51.6(2.4)	95.4(0.6)	97.0(3.9)	73.9(1.0)	95.8(0.5)	91.5(0.5)	78.3(2.3)	99.9(0.1)	95.0(3.2)	86.7	9.8	0.021
C4.5	66.0(3.6)	50.9(1.7)	96.1(0.7)	96.0(3.1)	73.6(1.3)	99.7(0.1)	88.7(0.3)	71.1(2.6)	89.8(0.1)	87.0(5.0)	82.9	11.8	0.109
oneR	59.5(3.1)	43.2(3.5)	62.9(2.4)	95.9(2.5)	17.8(0.8)	96.3(0.5)	92.9(1.1)	58.9(1.9)	67.4(1.1)	76.2(4.6)	60.4	21.6	0.002
IB1	81.5(2.7)	43.9(1.1)	**96.8(0.6)**	95.6(3.6)	74.0(1.3)	92.8(0.4)	97.0(0.2)	70.1(2.9)	99.7(0.1)	95.2(2.0)	84.5	12.9	0.344
IB10	83.6(2.3)	44.3(2.4)	94.3(0.7)	97.2(1.9)	74.2(1.3)	93.7(0.3)	96.1(0.3)	67.1(2.1)	99.4(0.1)	96.2(1.9)	84.6	12.4	0.344
NB$_k$	89.9(2.0)	51.2(2.3)	84.9(1.4)	97.0(2.5)	74.0(1.2)	96.4(0.2)	79.3(0.9)	60.0(2.3)	99.5(0.1)	97.7(1.6)	83.0	12.2	0.021
NB$_n$	89.9(2.0)	48.9(1.8)	80.1(1.0)	97.2(2.7)	74.0(1.2)	55.5(0.4)	44.9(2.8)	44.9(2.8)	99.5(0.1)	97.5(1.8)	80.6	13.6	0.021
Maj. Rule	48.7(2.3)	43.2(1.8)	15.5(0.6)	38.6(2.8)	11.4(0.0)	92.5(0.3)	16.8(0.4)	27.7(1.5)	34.2(0.8)	39.7(2.8)	36.8	24.8	0.002

Table 3.4 Comparison of the 10 times randomized **test** set performance of LS-SVM and LS-SVM$_F$ (linear, polynomial and Radial Basis Function kernel) with the performance of LDA, QDA, Logit, C4.5, oneR, IB1, IB10, NB$_k$, NB$_n$ and the Majority Rule classifier on 10 binary domains. The Average Accuracy (AA), Average Rank (AR) and Probability of equal medians using the Sign Test (P_{ST}) taken over all domains are reported in the last three columns. Best performances are underlined and denoted in bold face, performances not significantly different at the 5% level are denoted in bold face, performances significantly different at the 1% level are emphasised. Good results are obtained by LS-SVM 1vs1 with RBF kernel.

3.5 Least Squares Support Vector Machines for function estimation

So far we have discussed LS-SVMs for classification together with its link to Fisher discriminant analysis in high dimensional feature spaces. Let us now derive the LS-SVM formulation for the case of nonlinear function estimation. The derivation is similar to the LS-SVM classifier case which can also be interpreted as a regression problem with targets $+1$ and -1 with squared loss function taken on these targets.

Consider first a model in the primal weight space of the following form:

$$y(x) = w^T \varphi(x) + b \qquad (3.28)$$

where $x \in \mathbb{R}^n, y \in \mathbb{R}$ and $\varphi(\cdot) : \mathbb{R}^n \to \mathbb{R}^{n_h}$ is the mapping to the high dimensional and potentially infinite dimensional feature space. Given a training set $\{x_k, y_k\}_{k=1}^N$ we can formulate then the following optimization problem in the primal weight space

$$\left[\boxed{\text{P}} : \quad \min_{w,b,e} J_{\text{P}}(w,e) = \frac{1}{2}w^T w + \gamma \frac{1}{2} \sum_{k=1}^N e_k^2 \right.$$
$$\left. \text{such that} \quad y_k = w^T \varphi(x_k) + b + e_k, \quad k = 1, ..., N. \right]$$
$$(3.29)$$

Note that this is in fact nothing else but a ridge regression [98] cost function formulated in the feature space. However, one should be aware that when w becomes infinite dimensional, one cannot solve this primal problem. Therefore, let us proceed by constructing the Lagrangian and derive the dual problem. This problem was studied in [198] without the use of a bias term.

One constructs the Lagrangian

$$\mathcal{L}(w,b,e;\alpha) = J_{\text{P}}(w,e) - \sum_{k=1}^N \alpha_k \{w^T \varphi(x_k) + b + e_k - y_k\} \qquad (3.30)$$

where α_k are Lagrange multipliers. The conditions for optimality are given by

$$
\begin{cases}
\frac{\partial \mathcal{L}}{\partial w} = 0 & \rightarrow & w = \sum_{k=1}^{N} \alpha_k \varphi(x_k) \\[2mm]
\frac{\partial \mathcal{L}}{\partial b} = 0 & \rightarrow & \sum_{k=1}^{N} \alpha_k = 0 \\[2mm]
\frac{\partial \mathcal{L}}{\partial e_k} = 0 & \rightarrow & \alpha_k = \gamma e_k, & k = 1, ..., N \\[2mm]
\frac{\partial \mathcal{L}}{\partial \alpha_k} = 0 & \rightarrow & w^T \varphi(x_k) + b + e_k - y_k = 0, & k = 1, ..., N.
\end{cases}
\tag{3.31}
$$

After elimination of the variables w and e one gets the following solution

$$
\left[
\boxed{D} : \quad \text{solve in } \alpha, b :
\right.
$$
$$
\left[\begin{array}{c|c} 0 & 1_v^T \\ \hline 1_v & \Omega + I/\gamma \end{array}\right]
\left[\begin{array}{c} b \\ \hline \alpha \end{array}\right]
= \left[\begin{array}{c} 0 \\ \hline y \end{array}\right]
\left. \right]
\tag{3.32}
$$

where $y = [y_1; ...; y_N]$, $1_v = [1; ...; 1]$ and $\alpha = [\alpha_1; ...; \alpha_N]$. The kernel trick is applied here as follows

$$
\begin{aligned}
\Omega_{kl} &= \varphi(x_k)^T \varphi(x_l) \\
&= K(x_k, x_l) \quad k, l = 1, ..., N.
\end{aligned}
\tag{3.33}
$$

The resulting LS-SVM model for function estimation becomes then

$$
y(x) = \sum_{k=1}^{N} \alpha_k K(x, x_k) + b
\tag{3.34}
$$

where α_k, b are the solution to the linear system (3.32). The large scale algorithm as outlined for LS-SVM classifiers can be applied in a similar way to the function estimation case in order to handle large data sets. Note that in the case of RBF kernels, one has only two additional tuning parameters (γ, σ), which is less than for standard SVMs.

In Fig. 3.6 an illustration is given on a problem of time-series prediction of the Santa Fe chaotic laser data set [288]. An LS-SVM model with RBF kernel was taken to train an NARX model of the form $\hat{y}_{k+1} = f(y_k, y_{k-1}, ..., y_{k-q})$ with $q = 50$. Note that the value of q determines the number of inputs of the LS-SVM and that the size of the linear system to be solved is independent of the choice of q (while for MLPs the number of interconnections weights would grow with the value of q). The (γ, σ)

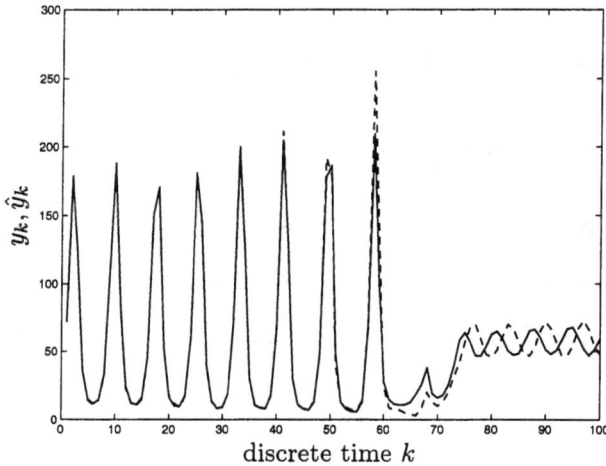

Fig. 3.6   *Time-series prediction by an LS-SVM with RBF kernel for the Santa Fe chaotic laser data. The figure shows the iterative prediction of an NARX LS-SVM model over a time horizon of 100 points, that was trained on the previous 1000 given data points; true data y_k (solid line), iterative prediction $\hat{y}_k$ (dashed line).*

parameters were tuned by 10-fold cross-validation. More sophisticated hyperparameter tuning methods may be further applied at this point, such as special cross-validation techniques for time-series.

3.6   Links with regularization networks and Gaussian processes

The LS-SVM solution (3.32) for the nonlinear function estimation case that we derived is closely related to results on regularization networks, reproducing kernel Hilbert spaces (RKHS), Gaussian processes, kriging and kernel ridge regression [1; 73; 153; 183; 286; 291]. However, a bias term b is included in the formulation as in the standard SVM case. Except for regularization networks, inclusion of a bias term is often not considered or treated in a different manner by including a constant in the kernel. The emphasis in LS-SVMs is on primal-dual optimization theory and neural networks interpretations (Fig. 3.7). A recent overview on the links between regularization networks and SVMs has been presented by Evgeniou, Pontil & Poggio in [73].

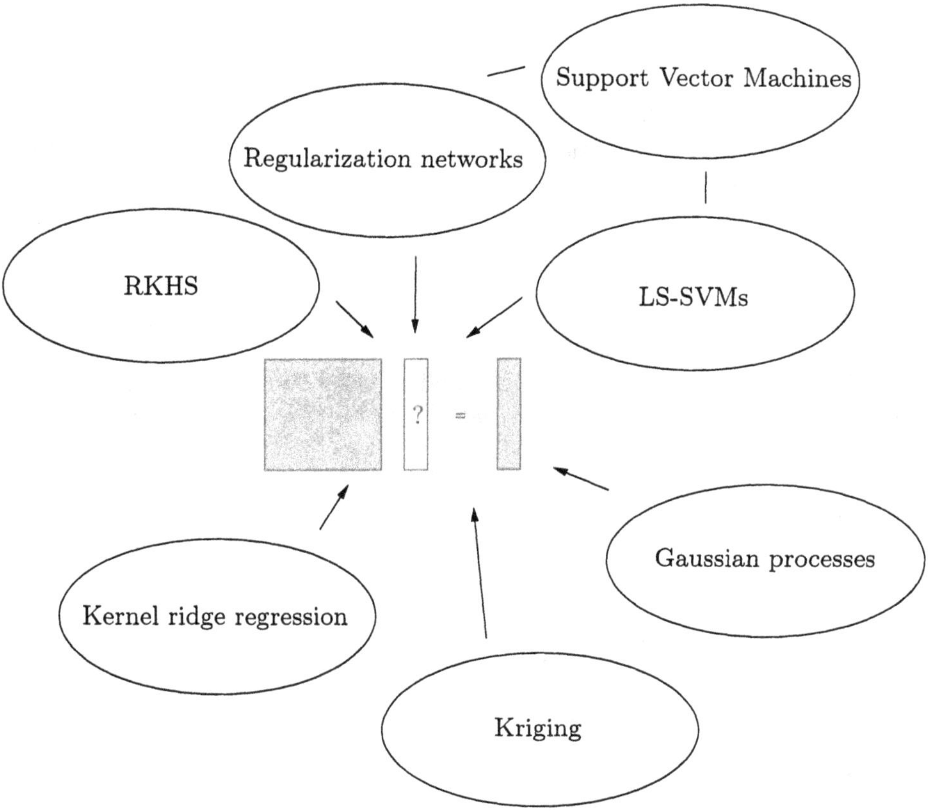

Fig. 3.7   *LS-SVMs are closely related to (and in certain cases lead to equivalent solutions as) regularization networks, estimation in RKHS, Gaussian processes, kriging and kernel ridge regression.   The emphasis in LS-SVMs is mainly on primal-dual insights from optimization theory and neural networks interpretations and aims at bridging many different areas in an interdisciplinary manner.*

3.6.1   *RKHS and reproducing property*

According to Kailath in [129] the theory of Reproducing Kernel Hilbert Spaces (RKHS) was first applied to detection and estimation problems by Parzen and first studied in 1910-1920 by Moore [168], in connection with a general theory of integral equations. Krein used them in his fundamental studies on the extension of positive definite functions. It was Aronszajn [13] who made a systematic abstract development of the theory in the 1940's, though it was discovered that many results were independently obtained in the USSR by Povzner [185]. Loeve [145] was the first to note that RKHS could be used to provide a representation for second-order random processes. On the other hand it was the work of Parzen that brought the RKHS to the fore in statistical problems. In particular in [177] the RKHS was used to clarify some relationships between time series, control theory and approximation theory. For an early review paper on links with linear filtering theory, see e.g. Kailath in [130].

A Reproducing Kernel Hilbert Space (RKHS) is a Hilbert space $\mathcal{H}$ of functions defined over a bounded domain $X \subset \mathbb{R}^n$ with the property that for each $x \in X$ the evaluation functionals $\mathcal{F}_x$ defined as $\mathcal{F}_x[f] = f(x)$ for all $f \in \mathcal{H}$ are linear bounded functionals. The boundedness means that there exists a $U = U_x \in \mathbb{R}^+$ such that $|\mathcal{F}_x[f]| = |f(x)| \leq U\|f\|$ for all f in the RKHS.

It can be proved that to every RKHS $\mathcal{H}$ there corresponds a unique positive definite function $K(x, z)$ of two variables in X (Moore-Aronszajn Theorem [13]), called the reproducing kernel of $\mathcal{H}$, that has the following reproducing property:

$$f(x) = \langle f(z), K(z, x) \rangle_{\mathcal{H}} , \ \forall f \in \mathcal{H} \tag{3.35}$$

where $\langle ., . \rangle_{\mathcal{H}}$ denotes the inner product in $\mathcal{H}$ and K is called a reproducing kernel for $\mathcal{H}$. The function K behaves in $\mathcal{H}$ as the delta function does in L_2. In particular one has

$$\langle K(x, \cdot), K(\cdot, z) \rangle_{\mathcal{H}} = K(x, z). \tag{3.36}$$

Consider now the following set of functions

$$f(x) = \sum_{i=1}^{\infty} c_i \phi_i(x). \tag{3.37}$$

This Hilbert space is a RKHS because

$$\langle f(z), K(z, x) \rangle_{\mathcal{H}} = \sum_{i=1}^{\infty} \frac{c_i \lambda_i \phi_i(x)}{\lambda_i} = f(x) \tag{3.38}$$

where the scalar product between two functions $f(x) = \sum_{i=1}^{\infty} c_i \phi_i(x)$ and

$g(x) = \sum_{i=1}^{\infty} d_i\phi_i(x)$ is defined as

$$\langle f(x), g(x) \rangle_{\mathcal{H}} = \langle \sum_{i=1}^{\infty} c_i\phi_i(x), \sum_{i=1}^{\infty} d_i\phi_i(x) \rangle_{\mathcal{H}} = \sum_{i=1}^{\infty} \frac{c_i d_i}{\lambda_i} \qquad (3.39)$$

with kernel

$$K(x, z) = \sum_{i=1}^{\infty} \lambda_i\phi_i(x)\phi_i(z) \qquad (3.40)$$

where a sequence of positive numbers λ_i and linearly independent basis functions $\phi_i(x)$ are assumed and the series converges. As an orthogonal basis one has the eigenfunctions $\{\phi_i(x)\}_{i=1}^{\infty}$ of the integral operator $\Gamma_K :$ $L_2(X) \to L_2(X)$ with $(\Gamma_K f)(x) = \int K(x, z)f(z)dz, \forall f \in L_2(X)$ and $\lambda_i > 0$ are the corresponding eigenvalues.

One can verify then that this kernel is positive definite. Note that when working with complex functions $\phi_i(x)$ one has $K(x, z) = \sum_{i=1}^{\infty} \lambda_i\phi_i(x)\phi_i^*(z)$, where $\phi_i^*(z)$ denotes the complex conjugate of $\phi_i(z)$. In relation to the LS-SVM formulations when having a function representation $f(x) = w^T\varphi(x)$ in the primal space and kernel trick $K(z, x) = \varphi(z)^T\varphi(x)$ with $K(\cdot, \cdot)$ symmetric and positive definite, the reproducing property means

$$\langle f(z), K(z, x) \rangle_{\mathcal{H}} = \langle w^T\varphi(z), \varphi(z)^T\varphi(x) \rangle_{\mathcal{H}} = w^T\varphi(x) = f(x). \qquad (3.41)$$

For functions $f(x) = w^T\varphi(x)$, $g(x) = v^T\varphi(x)$ we have $\langle f(x), g(x) \rangle_{\mathcal{H}} =$ $\langle w^T\varphi(x), v^T\varphi(x) \rangle_{\mathcal{H}} = w^Tv$. The norm is

$$\|f\|_K^2 = \langle f(x), f(x) \rangle_{\mathcal{H}} = \langle w^T\varphi(x), w^T\varphi(x) \rangle_{\mathcal{H}} = w^Tw. \qquad (3.42)$$

Note that for a Gaussian RBF kernel the RKHS is infinite dimensional, while for a polynomial kernel with a certain degree the RKHS is finite. The RBF kernel is translation invariant with $K(x, z) = K(x - z)$ and radial with $K(x, z) = K(\|x - z\|)$. A radial positive definite K defines a RKHS in which the basis functions $\{\phi_s(x)\}_{s=0}^{\infty}$ in the high dimensional feature space are Fourier components [73]:

$$K(x, z) = \sum_{s=0}^{\infty} \lambda_s\phi_s(x)\phi_s(z) = \sum_{s=0}^{\infty} \lambda_s \exp(j2\pi sx) \exp(-j2\pi sz) \qquad (3.43)$$

with $j = \sqrt{-1}$. Hence any positive definite radial kernel defines a RKHS with a scalar product equal to

$$\langle f(x), g(x) \rangle_{\mathcal{H}} = \sum_{s=0}^{\infty} \frac{\tilde{f}(s)\tilde{g}^*(s)}{\lambda_s} \tag{3.44}$$

where $\tilde{f}$ denotes the Fourier transform of f. The RKHS then becomes a subspace of $L_2([0,1]^n)$ of the functions such that

$$\|f\|_K^2 = \sum_{s=0}^{\infty} \frac{|\tilde{f}(s)|^2}{\lambda_s} < \infty. \tag{3.45}$$

Functionals of this form are known to be smoothness functionals.

3.6.2   *Representer theorem and regularization networks*

In order to see the link now with LS-SVM regression, let us consider the following variational problem as explained by Girosi in [95]:

$$\min_{f \in \mathcal{H}} H[f] = \gamma \sum_{k=1}^{N} L\left(y_k - f(x_k)\right) + \frac{1}{2}\|f\|_K^2. \tag{3.46}$$

Assume that $\mathcal{H}$ is a RKHS with kernel K. This is equivalent to assuming that the functions in $\mathcal{H}$ have a unique expansion of the form $f(x) = \sum_{i=1}^{\infty} c_i \phi_i(x)$ with norm $\|f\|_K^2 = \sum_{i=1}^{\infty} c_i^2/\lambda_i$. The solution to the problem follows from $\partial H[f]/\partial c_i = 0$ which gives

$$\gamma \sum_{k=1}^{N} L'\left(y_k - f(x_k)\right) \phi_i(x_k) + \frac{c_i}{\lambda_i} = 0. \tag{3.47}$$

One defines then new unknowns $a_k = \gamma L'(y_k - f(x_k))$. Expressing the coefficients c_i in terms of a_k one gets $c_i = \lambda_i \sum_{k=1}^{N} a_k \phi_i(x_k)$ such that the solution to the variational problem becomes

$$f(x) = \sum_{i=1}^{\infty} c_i \phi_i(x) = \sum_{i=1}^{\infty} \sum_{k=1}^{N} a_k \lambda_i \phi_i(x_k) \phi_i(x) = \sum_{k=1}^{N} a_k K(x, x_k). \tag{3.48}$$

Hence, independently of the form of L, the solution to the problem is always a linear superposition of kernel functions [95]. This is also essentially what the representer Theorem according to Kimeldorf and Wahba states [133].

The coefficients a_k follow as the solution to the following set of equations:

$$a_k = \gamma L^{'}(y_k - \sum_{l=1}^{N} a_l K(x_k, x_l)) \ , \ k = 1, ..., N. \qquad (3.49)$$

When taking the least squares loss function $L(e) = \frac{1}{2}e^2$ this becomes the linear system

$$[\Omega + I/\gamma]a = y \qquad (3.50)$$

with $y = [y_1; ...; y_N]$ and $\Omega_{kl} = K(x_k, x_l)$ for $k, l = 1, ..., N$ the elements of the kernel matrix. This linear system corresponds to the regularization network solution as derived by Poggio & Girosi [183]. As explained in [73] one can also handle a bias term in this framework. Furthermore one should also note that both parametric and semiparametric versions of the representer theorem exist [133; 206].

In a support vector machines primal-dual optimization formulation context the problem (3.46) corresponds to the following

$$\left[\boxed{\text{P}}: \quad \min_{w,b,e} J_{\text{P}}(w, e) = \frac{1}{2}w^T w + \gamma \sum_{k=1}^{N} L(e_k) \right.$$
$$\left. \text{such that} \quad y_k = w^T \varphi(x_k) + b + e_k, \ k = 1, ..., N \right] \qquad (3.51)$$

where $L(e)$ is a general and differentiable cost function. From the Lagrangian $\mathcal{L}(w, b, e; \alpha) = J_{\text{P}}(w, e) - \sum_{k=1}^{N} \alpha_k \{w^T \varphi(x_k) + b + e_k - y_k\}$ with Lagrange multipliers α_k, one has the following conditions for optimality:

$$\begin{cases} \frac{\partial \mathcal{L}}{\partial w} = 0 & \rightarrow \quad w = \sum_{k=1}^{N} \alpha_k \varphi(x_k) \\[2mm] \frac{\partial \mathcal{L}}{\partial b} = 0 & \rightarrow \quad \sum_{k=1}^{N} \alpha_k = 0 \\[2mm] \frac{\partial \mathcal{L}}{\partial e_k} = 0 & \rightarrow \quad \alpha_k = \gamma L^{'}(e_k), \qquad\qquad k = 1, ..., N \\[2mm] \frac{\partial \mathcal{L}}{\partial \alpha_k} = 0 & \rightarrow \quad w^T \varphi(x_k) + b + e_k - y_k = 0, \quad k = 1, ..., N. \end{cases} \qquad (3.52)$$

Note that $\alpha_k = \gamma L^{'}(e_k)$ is now part of the conditions for optimality, while in the RKHS context this is rather by definition. After elimination of the variables w and e and application of the kernel trick $K(x_k, x_l) = \varphi(x_k)^T \varphi(x_l)$

one gets

$$
\left[\begin{array}{l}
\boxed{D}: \quad \text{solve in } \alpha, b: \\[2mm]
\alpha_k = \gamma L'\left(y_k - \sum_{l=1}^{N} \alpha_l K(x_l, x_k) + b\right), \quad k = 1, ..., N
\end{array}\right]
\tag{3.53}
$$

which is a set of nonlinear equations to be solved in α, b. Alternatively, one may also eliminate α instead of e and solve the nonlinear equations in e. The resulting dual representation of the model is $y(x) = \sum_{k=1}^{N} \alpha_k K(x, x_k) + b$. For the case $L(e) = \frac{1}{2}e^2$ one has the linear system (3.32).

3.6.3 *Gaussian processes*

Link between LS-SVM regression and Gaussian processes

In the RKHS context it is also known that there is a link between the solution to variational problems in RKHS and Gaussian processes, according to Kimeldorf & Wahba [133; 202; 286]. To every RKHS $\mathcal{H}_K$ there corresponds a zero mean Gaussian stochastic process with covariance $K(s,t)$ and there is a one-to-one inner product preserving map between the Hilbert space spanned by the stochastic process and $\mathcal{H}_K$ where the random variable $X(t)$ corresponds to the representer $K_t \in \mathcal{H}_K$.

Gaussian processes have also been studied in the neural networks area within the Bayesian learning context. According to MacKay, one has the following methodology. Consider a given set of N data points $\{x_k, y_k\}_{k=1}^{N}$ and denote $y_{1\_N} = [y_1; ...; y_N] \in \mathbb{R}^N$. Given the input data one can construct a covariance matrix C with $C_{kl} = C(x_k, x_l)$ the kl-entry of the matrix.

Having formed the covariance matrix C the task is then to infer y_{N+1} given the observed vector y_{1_N}. From Bayes rule one has the joint density $P(y_{N+1}, y_{1\_N}) = P(y_{N+1}|y_{1\_N})P(y_{1\_N})$ or

$$
P(y_{N+1}|y_{1\_N}) = \frac{P(y_{N+1}, y_{1\_N})}{P(y_{1\_N})}
\tag{3.54}
$$

with joint density and conditional density considered to be Gaussian. The covariance matrix $C_{N+1} \in \mathbb{R}^{(N+1)\times(N+1)}$ can then be written as a function of $C_N \in \mathbb{R}^{N\times N}$ as follows:

$$
C_{N+1} = \left[\begin{array}{cc} C_N & \theta(x_{N+1}) \\ \theta(x_{N+1})^T & \nu \end{array}\right]
\tag{3.55}
$$

where $\theta(x_{N+1}) = [K(x_{N+1}, x_1); ...; K(x_{N+1}, x_N)]$. One has

$$P(y_{N+1}, y_{1\_N}) \propto \exp\left(-\frac{1}{2} \begin{bmatrix} y_{1\_N} & y_{N+1} \end{bmatrix} C_{N+1}^{-1} \begin{bmatrix} y_{1\_N} \\ y_{N+1} \end{bmatrix}\right) \qquad (3.56)$$

with the following expressions for the inverse matrix

$$C_{N+1}^{-1} = \begin{bmatrix} M & m \\ m^T & \varrho \end{bmatrix} \qquad (3.57)$$

and

$$\begin{aligned} \varrho &= (\nu - \theta^T C_N^{-1} \theta)^{-1} \\ m &= -\varrho C_N^{-1} \theta \\ M &= C_N^{-1} + \frac{1}{\varrho} m m^T. \end{aligned} \qquad (3.58)$$

This gives the posterior distribution

$$P(y_{N+1}|y_{1\_N}) \propto \exp\left(-\frac{1}{2} \frac{(y_{N+1} - \hat{y}_{N+1})^2}{\sigma_{\hat{y}_{N+1}}^2}\right) \qquad (3.59)$$

with

$$\begin{cases} \hat{y}_{N+1} &= \theta^T C_N^{-1} y_{1\_N} \\ \sigma_{\hat{y}_{N+1}}^2 &= \nu - \theta^T C_N^{-1} \theta. \end{cases} \qquad (3.60)$$

The predictive mean at a new point x_{N+1} is given by $\hat{y}_{N+1}$ and $\sigma_{\hat{y}_{N+1}}$ which defines the error bar on the prediction for this new point. Note that one only needs to invert C_N not C_{N+1}.

The link between this Gaussian process formulation and LS-SVM regression can be understood as follows. Take an LS-SVM model with zero bias term $b = 0$. Instead of denoting a new point in a sequential manner, let us denote it as x instead of x_{N+1}. The LS-SVM model can be written as

$$\begin{aligned} \hat{y}(x) &= \sum_{k=1}^{N} \alpha_k K(x, x_k) \\ &= \theta(x)^T \alpha \end{aligned} \qquad (3.61)$$

where $\alpha = [\alpha_1; ...; \alpha_N]$ and $\theta(x) = [K(x, x_1); ...; K(x, x_N)]$. Let us define $C(x_k, x_l) = K(x_k, x_l) + \delta_{kl}/\gamma$. In the zero bias term case the LS-SVM system is

$$[\Omega + I/\gamma]\alpha = y \tag{3.62}$$

with solution $\alpha = [\Omega + I/\gamma]^{-1}y$, where $y = y_{1\_N}$. The LS-SVM model becomes

$$\hat{y}(x) = \theta(x)^T[\Omega + I/\gamma]^{-1}y \tag{3.63}$$

which is the same as (3.60) if one takes $C_N = \Omega + I/\gamma$ where $\Omega_{kl} = K(x_k, x_l)$ for $k, l = 1, ..., N$ in order to have a compatible definition of θ with respect to C_N in (3.55).

Stationary and non-stationary random fields, spectral representation and kriging

The predictions produced by Gaussian processes depend entirely on the covariance matrix C. As for positive definite kernels in SVMs, one has several possible choices for the covariance function and the two are closely related. The Gaussian processes framework is also related to methods of *kriging* (named after D.G. Krige a South African mining engineer), developed within the geostatistics field [2; 50; 137; 153]. For an overview with respect to Gaussian random fields, see e.g. [1; 50]. Random field models are basically specified by expectations and covariances. One makes a distinction between stationary and non-stationary covariance functions. *Stationarity in the wide sense* for a random field means a constant expectation and covariance function $C(t, s) = C(t - s)$ with $\tau = t - s$ and typically $t, s \in \mathbb{R}^n$ [70]. The covariance function $C(t, s)$ is defined as $\text{Cov}(X_t, X_s)$ with $X_t = X(t, \omega)$ a random field on $\mathbb{R}^n$ with $t \in \mathbb{R}^n$, for a given probability space and parameter set T and $X(t, \omega)$ a real valued and measurable function of ω for every $t \in T$. The correlation function is defined as $\rho(t, s) = \text{Corr}(X_t, X_s) = C(t, s)/(\sigma(t)\sigma(s))$ with variance $\sigma^2(t) = C(t, t) = \text{Var}(X_t, X_t)$. Any stationary covariance function has a constant variance such that $C(\tau) = \sigma^2\rho(\tau)$. A stationary random field is called *isotropic* if the covariance function depends on the distance only $C(t, s) = C(\|\tau\|)$ and anisotropic if a weighted norm is used $C(t, s) = C(\|\tau\|_W)$ with matrix W where $\|\tau\|_W = \sqrt{\tau^T W \tau}$. A correlation function is called *separable* if $\rho(\tau) = \rho_1(\tau_1)...\rho_n(\tau_n)$. For stationary correlation functions one has a spectral representation [91].

According to the multidimensional Bochner's Theorem a real function $r(\tau)$ on $\mathbb{R}^n$ is positive definite if and only if it can be represented in the form $r(\tau) = \int_{\mathbb{R}^n} \exp(j\tau k) d^n F(k)$ where $F(\cdot)$ is a non-negative bounded measure. Bochner's theorem says that all positive definite functions have a unique spectral representation. The integral is a Fourier transform of the non-negative F. Because the class of positive definite functions coincides with the class of covariance functions the Wiener-Khintchine theorem can be derived, which states that a real function $\rho(\tau)$ on $\mathbb{R}^n$ is a correlation function if and only if it can be represented in the form

$$\rho(\tau) = \int_{\mathbb{R}^n} \exp(j\tau k) d^n F(k) \tag{3.64}$$

where the function $F(k)$ on $\mathbb{R}^n$ has the properties of an n-dimensional distribution function. According to Matérn one can link correlation functions to characteristic functions. When F is continuous the following spectral density function $f(k) = \partial^n F(k)/\partial k_1...\partial k_n$ exists. One can then write $\rho(\tau) = \int_{\mathbb{R}^n} \exp(j\tau k) f(k) d^n k$. As a result a function $f(k)$ in $\mathbb{R}^n$ is the spectral density function of a stationary correlation function on $\mathbb{R}^n$ if and only if $f(k) \geq 0$ and $\int_{\mathbb{R}^n} f(k) d^n k = 1$, i.e. $f(k)$ has the properties of an n-dimensional probability density. The spectral density function is obtained from the correlation function as

$$f(k) = (2\pi)^{-n} \int_{\mathbb{R}^n} \exp(-j\tau k) \rho(\tau) d^n \tau. \tag{3.65}$$

This gives an explicit method for verifying positive definiteness of a stationary correlation function on $\mathbb{R}^n$ by evaluating the spectral density $f(k)$ and check whether it is non-negative for all $k \in \mathbb{R}^n$.

Stationarity and in particular isotropy impose strong restrictions on the possible definition of a Gaussian random field. However, from stationary random fields it is simple to obtain a wide range of non-stationary random fields. For example, consider an isotropic Gaussian random field X_t with properties

$$\begin{aligned} \mathcal{E}[X_t] &= 0 \\ \text{Cov}(X_t, X_s) &= \rho(\tau) \end{aligned} \tag{3.66}$$

with ρ an isotropic correlation function. One can then consider the Gaussian random field

$$Z_t = \sum_i a_i f_i(t) + X_t \tag{3.67}$$

where a_i are Gaussian random variables independent of X_t and $f_i(t)$ are real valued functions. The expectation and covariance function are

$$
\begin{aligned}
\mathcal{E}[Z_t] &= \sum_i \mathcal{E}[a_i] \, f_i(t) \\
\mathrm{Cov}(Z_t, Z_s) &= \sum_{i,j} \mathrm{Cov}(a_i, a_j) \, f_i(t) f_j(s) + \rho(\tau).
\end{aligned}
\tag{3.68}
$$

Both the expectation and the covariance function have become non-stationary in this case. Here a general Gaussian random function Z_t is modelled in terms of the stationary random field X_t. These insights are used in the kriging method, where one estimates the predictive mean and error bars given a set of observations, resulting into a linear system to be solved.

Automatic relevance determination in Gaussian processes

An interesting choice for C as discussed by MacKay is

$$C(x, z) = \theta_1 \exp\left(-\frac{1}{2} \sum_{i=1}^{n} \frac{(x_i - z_i)^2}{\sigma_i^2} \right) + \theta_2 \tag{3.69}$$

where $x, z \in \mathbb{R}^n$ and x_i, z_i denote the i-th component of these vectors. This can be used for Automatic Relevance Determination (ARD), where σ_i are considered then as length scale parameters to be determined. A very large length scale σ_i means that the output is expected to behave as a constant with respect to the i-th input, and hence would be irrelevant to the model. An extension of this to a non-stationary covariance function is to take spatially varying length scales and work with parameterized functions $\sigma_i(x)$. The following covariance function

$$C(x, z) = \theta_1 \prod_{i=1}^{n} \left(\frac{2\sigma_i(x)\sigma_i(z)}{\sigma_i^2(x) + \sigma_i^2(z)} \right)^{1/2} \exp\left(-\sum_{i=1}^{n} \frac{(x_i - z_i)^2}{\sigma_i^2(x) + \sigma_i^2(z)} \right) \tag{3.70}$$

is positive definite.

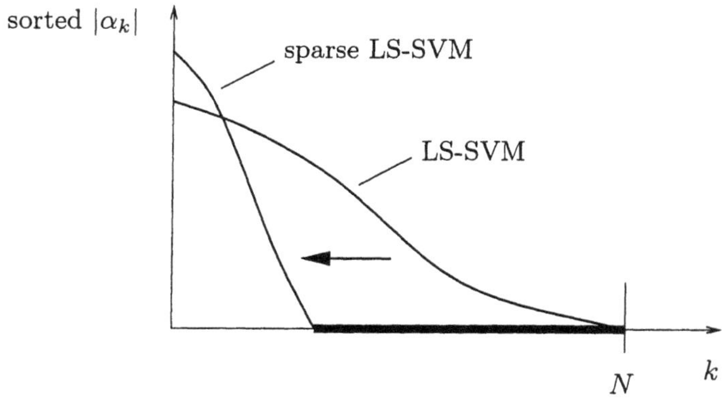

Fig. 3.8   *A simple way for imposing sparseness to LS-SVM models is to apply pruning techniques as known in the neural networks literature.*

3.7   Sparseness by pruning

We have discussed the links between LS-SVM classifiers and a kernel version of Fisher discriminant analysis together with links between LS-SVM regression and regularization networks, estimation in RKHS and Gaussian processes.

One drawback of LS-SVMs in comparison with standard SVMs is the lack of sparseness in the solution vector which is clear from the fact that $\alpha_k = \gamma e_k$. However, we will illustrate here that there are several possible ways to sparsify the LS-SVM system. A simple way is to apply insight of *pruning* from the neural networks literature. For MLPs it is well known that by starting from a huge network and deleting interconnection weights which are less relevant one can improve the generalization performance [23], e.g. by applying optimal brain damage (LeCun in [140]) or optimal brain surgeon (Hassibi & Stork in [106]).

One can proceed as follows. From the linear system one finds the vector of support values. One may apply the simple heuristic that data corresponding to small $|\alpha_k|$ values are less relevant for the construction of the model, in analogy with standard SVMs where zero α_k values do not contribute to the model. One works then in several steps where in each step typically 5% of the data with the smallest support values are removed. One gradually prunes the support value spectrum (Fig. 3.8). In each of these steps one re-estimates the linear system.

Simple LS-SVM pruning algorithm:

(1) Train an LS-SVM based on N points.

(2) Remove a small amount of points (e.g. 5% of the set) with smallest values in the sorted $|\alpha_k|$ spectrum.

(3) Re-train the LS-SVM based on the reduced training set.

(4) Go to (2), unless the user-defined performance index degrades. If the performance becomes worse, one checks whether an additional modification of (γ, σ) (in the RBF kernel case) might improve the performance and modifies the tuning parameters.

This procedure is applicable both to classification and function estimation problems. Note that omitting points implicitly correspond to creating an ϵ-insensitive zone in the underlying cost function which leads to sparseness, which is clear from the condition for optimality $\alpha_k = \gamma L'(e_k)$ in (3.52), because omitting a point is equivalent to imposing a zero value for that point which should correspond to a zero derivative for $L'(e_k)$ around the origin.

The differences with this procedure and obtaining sparseness in standard SVMs are the following: in the LS-SVM pruning case the size of the system shrinks at every stage and one has the possibility to modify the tuning parameters at certain stages while in standard SVMs one has to iterate for the same (γ, σ) until the minimum is reached on the same size of the system. Also note that interior point algorithms can be applied in the case of a general convex cost function as explained in the previous Chapter. In each iteration step one also solves a reduced KKT system which has the same form as one single LS-SVM (see (2.82)).

The simple pruning algorithm shown above is certainly not the most optimal one. More sophisticated pruning techniques can be used which also take into account the kernel matrix. However, it is also easy to apply to larger data sets and shows that in relation to least squares methods there are also ways to impose sparseness and one can overcome the lack of sparseness.

In Figs. 3.9-3.10 some easy toy problems are given of the simple LS-

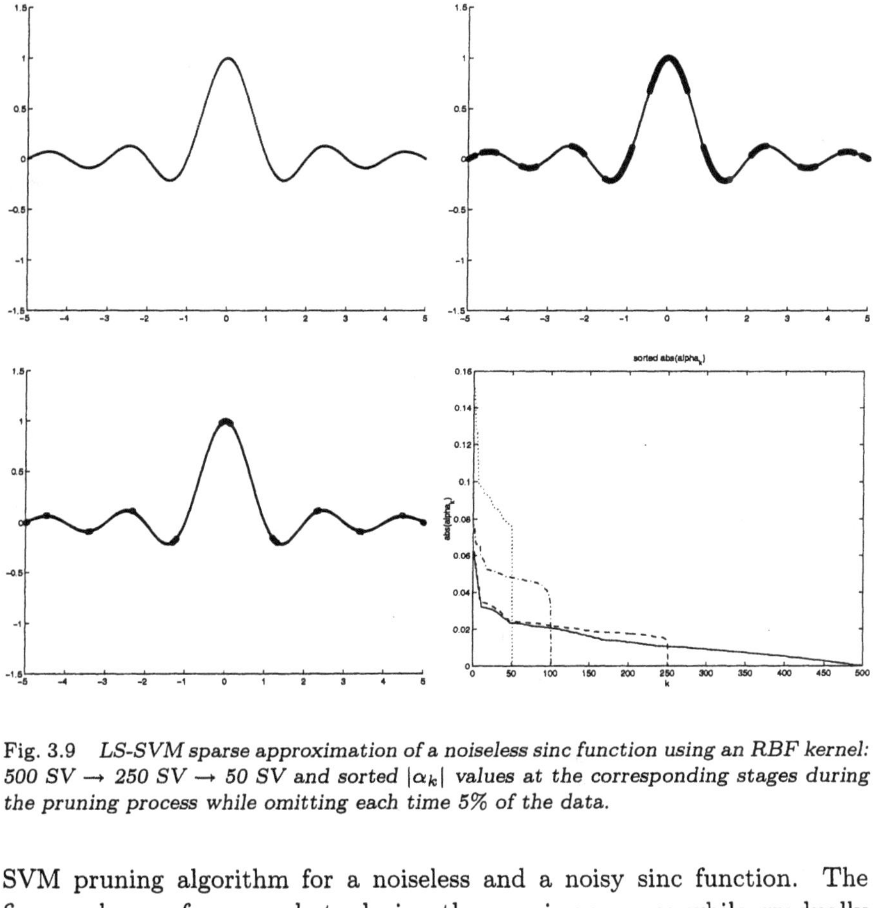

Fig. 3.9 *LS-SVM sparse approximation of a noiseless sinc function using an RBF kernel:*
500 SV → 250 SV → 50 SV and sorted $|\alpha_k|$ values at the corresponding stages during
the pruning process while omitting each time 5% of the data.

SVM pruning algorithm for a noiseless and a noisy sinc function. The
figures show a few snapshots during the pruning process while gradually
removing 5 % of the training data and re-estimate. In these examples the
values of γ and σ for an RBF kernel were kept constant. These values might
be additionally optimized at certain stages in the pruning process, in order
to better ensure a good generalization when re-estimation is done on the
datasets that are becoming smaller. One also observes that higher degrees
of sparseness are usually achieved in the noiseless case.

In Fig. 3.11 an illustration of the simple sparse approximation procedure is given for a binary classification problem where the training data
have been generated by a Gaussian distribution for each class with the
same covariance matrix. According to basic pattern recognition theory the

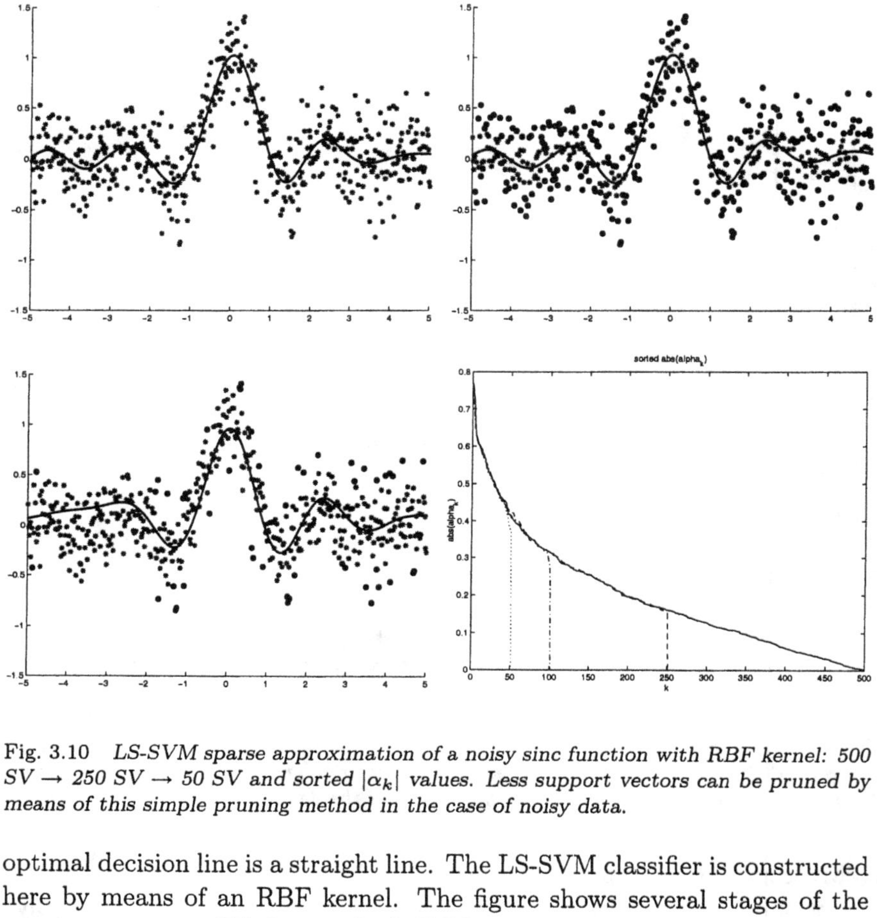

Fig. 3.10   *LS-SVM sparse approximation of a noisy sinc function with RBF kernel: 500 SV* → *250 SV* → *50 SV and sorted* $|\alpha_k|$ *values. Less support vectors can be pruned by means of this simple pruning method in the case of noisy data.*

optimal decision line is a straight line. The LS-SVM classifier is constructed here by means of an RBF kernel. The figure shows several stages of the pruning process. While standard SVMs retain support vectors that are close to decision boundary, in this LS-SVM pruning process one obtains support vectors that are located both close to and far from the decision boundary. This is because the sorting and ranking of the support values is done here on the basis of the absolute values of α. One might also prune the negative values and obtain qualitatively different support vectors with good test set ROC curve performance as observed in [146].

In [264] the sparse approximation method has been successfully applied to 5 binary and 5 multiclass UCI benchmark problems. In Table 3.5 these benchmarking results on UCI datasets are shown.

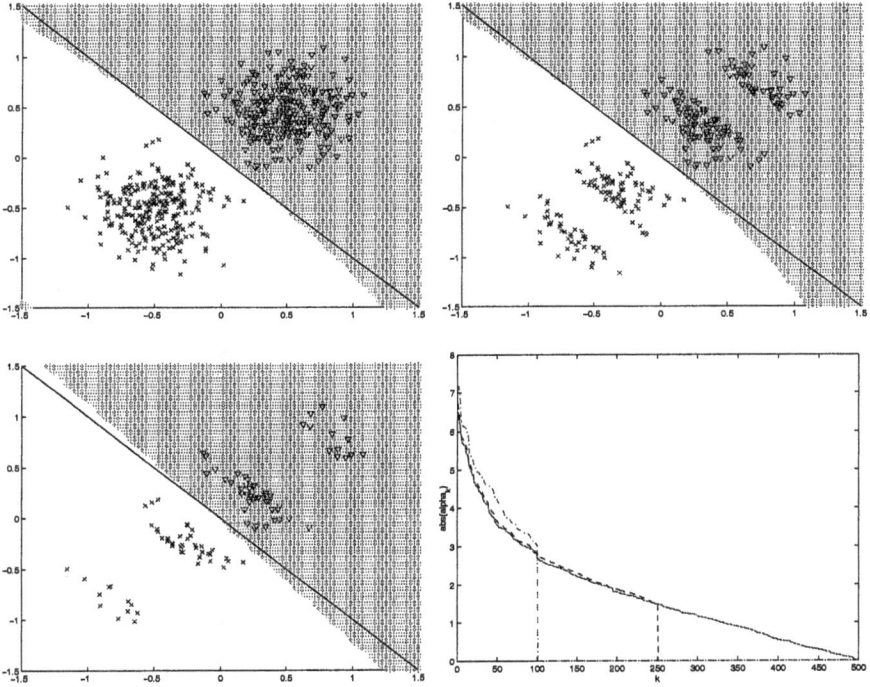

Fig. 3.11   *LS-SVM sparse approximation with RBF kernel illustrated on a classification problem: 500 SV → 250 SV → 50 SV and sorted* $|\alpha_k|$ *values.*

	acr	bld	gcr	hea	ion	bal	cmc	ims	iri	led	AA	AR	P_{ST}
N_{test}	230	115	334	90	117	209	491	770	50	1000			
n	14	6	20	13	33	4	9	18	4	7			
LS-SVM	**87.0**(2.1)	**70.2**(4.1)	**76.3**(1.4)	**84.7**(4.8)	**96.0**(2.1)	94.2(2.2)	**55.7**(2.2)	**96.5**(0.5)	97.6(2.3)	**74.1**(1.3)	**83.2**	**1.3**	**1.000**
LS-SVM$_{SL5\%}$	85.6(2.3)	67.3(4.1)	76.2(1.9)	83.6(4.0)	89.8(1.6)	**95.1**(1.2)	54.4(1.6)	90.7(1.7)	96.8(2.3)	73.6(2.5)	81.4	2.2	0.021
LS-SVM$_{SL1\%}$	85.6(2.3)	67.3(4.1)	75.9(1.5)	82.8(3.6)	89.8(1.6)	**95.5**(1.6)	54.4(1.6)	90.7(1.7)	**97.8**(2.6)	71.0(4.2)	81.1	2.5	0.109
$N_{CV}^{(1:L)}$	460	230	666	180	234	832	1964	9240	200	18000			
$N_{SL5\%}^{(1:L)}$	149	128	344	59	56	104	1418	860	96	4360			
$N_{SL1\%}^{(1:L)}$	149	128	327	57	56	60	1418	860	66	3640			
PPTE$_{SL5\%}$	68%	44%	48%	67%	76%	88%	28%	91%	67%	76%			
PPTE$_{SL1\%}$	68%	44%	51%	68%	76%	93%	28%	91%	52%	80%			

Table 3.5 Sparse approximation of LS-SVMs with RBF kernel by gradually pruning the support value spectrum according to Van Gestel et. al. Pruning is stopped when the cross-validation accuracy decreases significantly at the 5% and 1% significance level (SL 5% and SL 1%), respectively. The following randomized test set performances are reported: no pruning (LS-SVM), pruning till SL 5% (LS-SVM$_{SL5\%}$) and pruning till SL 1% (LS-SVM$_{SL1\%}$). The corresponding number of training examples N_{CV}, $N_{SL5\%}^{(1:L)}$ and $N_{SL1\%}^{(1:L)}$ are shown, together with Percentage of Pruned Training Examples PPTE$_{SL5\%}$ and PPTE$_{SL1\%}$, respectively.

Chapter 4

Bayesian Inference for LS-SVM Models

In the previous Chapter we have discussed basic methods of LS-SVMs for classification and nonlinear function estimation, together with links to other existing methods and frameworks. In the case of RBF kernels, in addition to the calculation of the support values, also extra tuning parameters need to be determined. In this Chapter we present a complete framework for Bayesian inference of LS-SVM classifiers and function estimators with several levels of inference. This will allow us to take probabilistic interpretations of the network outputs and automatically determine the tuning parameters (called hyperparameters in the Bayesian inference context). The framework also enables one to derive error bars on the output and to do input selection using the principle of automatic relevance determination.

4.1 Bayesian inference for LS-SVM classifiers

4.1.1 *Definition of network model, parameters and hyper-parameters*

The original LS-SVM classifier formulation as explained in the previous Chapter is slightly modified into the following formulation:

$$
\left[
\begin{array}{l}
\boxed{P}: \quad \min_{w,b,e_c} J_P(w, e_c) = \quad \mu \frac{1}{2} w^T w + \zeta \frac{1}{2} \sum_{k=1}^{N} e_{c,k}^2 \\[2ex]
\qquad\quad \text{such that} \qquad y_k \left[w^T \varphi(x_k) + b \right] = 1 - e_{c,k}, \ k = 1, ..., N
\end{array}
\right]
$$

$$(4.1)$$

where the error variables are denoted now as $e_{c,k}$ where the index c stands for classification. Two tuning parameters μ, ζ (called hyperparameters) are taken now instead of the single regularization constant γ, which corresponds to $\gamma = \zeta/\mu$. This use of μ, ζ will be more convenient for the statistical interpretation towards the Bayesian inference of these hyperparameters.

The classifier in the primal weight space is given by

$$y(x) = \text{sign}[w^T \varphi(x) + b]. \tag{4.2}$$

When we solve (4.1) we basically aim at maximizing the (soft) margin and minimizing the within class scatter for the two classes with target values $+1$ and -1 for the classes after projection of the input space onto a line, which is clear from the constraints. The latter means that one may consider the problem as a regression problem with targets $+1$ and -1 for the output y. Therefore we can write

$$\sum_{k=1}^{N} e_{c,k}^2 = \sum_{k=1}^{N} (y_k e_{c,k})^2 = \sum_{k=1}^{N} e_k^2 \tag{4.3}$$

where $e_k = y_k e_{c,k}$ and $y_k^2 = 1$. We write (4.1) now as

$$\left[\begin{array}{ll} \boxed{\text{P}}: & \min_{w,b,e} J_P(w,e) = \quad \mu\, E_W + \zeta\, E_D \\[2mm] & \text{such that} \quad\quad e_k = y_k - [w^T \varphi(x_k) + b]\,, \;\; k = 1, ..., N \end{array} \right] \tag{4.4}$$

with

$$\begin{aligned} E_W &= \frac{1}{2} w^T w \\ E_D &= \frac{1}{2} \sum_{k=1}^{N} e_k^2 = \frac{1}{2} \sum_{k=1}^{N} \left(y_k - [w^T \varphi(x_k) + b] \right)^2. \end{aligned} \tag{4.5}$$

This gives the Lagrangian $\mathcal{L}(w, b, e; \alpha) = J_P(w, e) - \sum_{k=1}^{N} \alpha_k (y_k - [w^T \varphi(x_k) +$

$b] - e_k$). From the conditions for optimality we find

$$
\begin{cases}
\frac{\partial \mathcal{L}}{\partial w} = 0 & \rightarrow \quad w = \frac{1}{\mu} \sum_{k=1}^{N} \alpha_k \varphi(x_k) \\
\frac{\partial \mathcal{L}}{\partial b} = 0 & \rightarrow \quad \sum_{k=1}^{N} \alpha_k y_k = 0 \\
\frac{\partial \mathcal{L}}{\partial e_k} = 0 & \rightarrow \quad \alpha_k = \zeta e_k, & k = 1, ..., N \\
\frac{\partial \mathcal{L}}{\partial \alpha_k} = 0 & \rightarrow \quad w^T \varphi(x_k) + b - y_k + e_k = 0, & k = 1, ..., N
\end{cases}
\tag{4.6}
$$

which results in the following dual problem

$$
\left[\boxed{D} : \text{ solve in } \alpha, b : \right.
\left.
\begin{bmatrix} 0 & 1_v^T \\ \hline 1_v^T & \frac{1}{\mu}\Omega + \frac{1}{\zeta}I \end{bmatrix}
\begin{bmatrix} b \\ \alpha \end{bmatrix} =
\begin{bmatrix} 0 \\ y \end{bmatrix}
\right]
\tag{4.7}
$$

with $y = [y_1; ...; y_N]$, $1_v = [1; ...; 1]$ and $\Omega_{kl} = K(x_k, x_l) = \varphi(x_k)^T \varphi(x_l)$ for $k, l = 1, ..., N$. In the dual space the classifier becomes

$$
y(x) = \text{sign}[\frac{1}{\mu} \sum_{k=1}^{N} \alpha_k K(x, x_k) + b]
\tag{4.8}
$$

with support values α_k and bias term b the solution to (4.7).

4.1.2 Bayes rule and levels of inference

As a typical example in this Chapter we will take the RBF kernel. In that case the total unknown parameter vector will be defined as $\Theta = [w; b; \mu; \zeta; \sigma]$ where σ denotes the width of the kernel. A model will be denoted as $\mathcal{H}_i$ (the i-th model of a model set) with $i = 1, ..., n_{\mathcal{H}}$ with $n_{\mathcal{H}}$ the number of models in the set. In the case of an RBF kernel the different models $\mathcal{H}_i$ mean that one tries several values of σ for the RBF kernel. We will also denote this as $\mathcal{H}_\sigma$ to stress that the model is parameterized by σ. The given training data set is denoted as $\mathcal{D} = \{x_k, y_k\}_{k=1}^{N}$. Note that w may become infinite dimensional. Throughout this Chapter we will formally calculate with infinite dimensional vectors as in finite dimensional cases. This is motivated by the fact that all calculations are done in the dual space after applying the kernel trick.

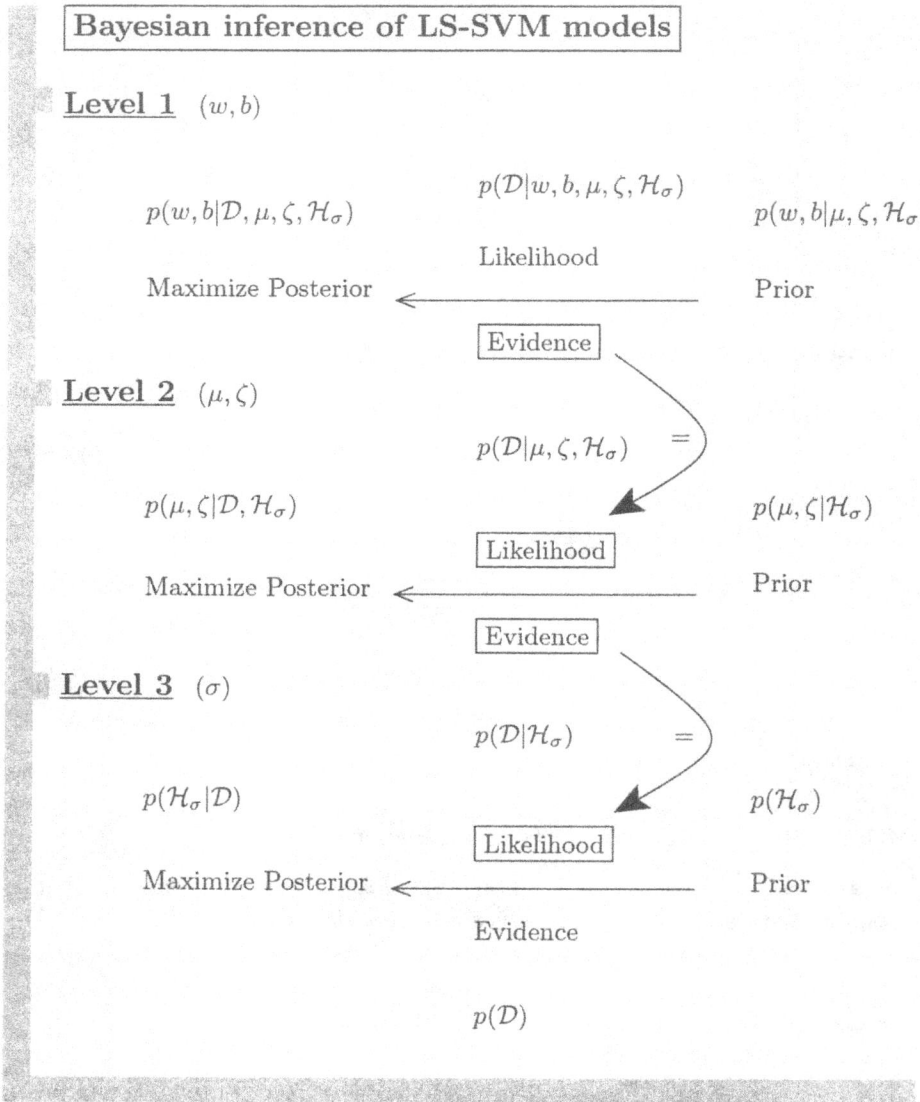

     Bayesian inference of LS-SVM models

Level 1 (w, b)

$p(w, b | \mathcal{D}, \mu, \zeta, \mathcal{H}_\sigma)$ $p(\mathcal{D} | w, b, \mu, \zeta, \mathcal{H}_\sigma)$ $p(w, b | \mu, \zeta, \mathcal{H}_\sigma)$

     Likelihood

Maximize Posterior ←—————— Prior

     Evidence

Level 2 (μ, ζ)

     $p(\mathcal{D} | \mu, \zeta, \mathcal{H}_\sigma)$ $=$

$p(\mu, \zeta | \mathcal{D}, \mathcal{H}_\sigma)$ $p(\mu, \zeta | \mathcal{H}_\sigma)$

     Likelihood

Maximize Posterior ←—————— Prior

     Evidence

Level 3 (σ)

     $p(\mathcal{D} | \mathcal{H}_\sigma)$ $=$

$p(\mathcal{H}_\sigma | \mathcal{D})$ $p(\mathcal{H}_\sigma)$

     Likelihood

Maximize Posterior ←—————— Prior

     Evidence

     $p(\mathcal{D})$

Fig. 4.1   *Illustration of Bayesian inference for LS-SVM models with several levels of inference.*

As already briefly explained in the first Chapter, the Bayesian inference process is done at several levels. The primal weight space parameters w, b are inferred at Level 1. The hyperparameters μ, ζ are inferred at Level 2 and the RBF kernel parameters are inferred at Level 3. Note that in the case of a linear kernel one has no Level 3 because there are no additional tuning parameters for the kernel.

In Fig. 4.1 a conceptual interpretation is given of the Bayesian inference procedure:

- *Level 1* [inference of parameters w, b]

$$p(w, b | \mathcal{D}, \mu, \zeta, \mathcal{H}_\sigma) = \frac{p(\mathcal{D} | w, b, \mu, \zeta, \mathcal{H}_\sigma)}{p(\mathcal{D} | \mu, \zeta, \mathcal{H}_\sigma)} p(w, b | \mu, \zeta, \mathcal{H}_\sigma) \qquad (4.9)$$

- *Level 2* [inference of hyperparameters μ, ζ]

$$p(\mu, \zeta | \mathcal{D}, \mathcal{H}_\sigma) = \frac{p(\mathcal{D} | \mu, \zeta, \mathcal{H}_\sigma)}{p(\mathcal{D} | \mathcal{H}_\sigma)} p(\mu, \zeta | \mathcal{H}_\sigma) \qquad (4.10)$$

- *Level 3* [inference of kernel parameter σ and model comparison]

$$p(\mathcal{H}_\sigma | \mathcal{D}) = \frac{p(\mathcal{D} | \mathcal{H}_\sigma)}{p(\mathcal{D})} p(\mathcal{H}_\sigma). \qquad (4.11)$$

At each of these levels one has

$$\text{Posterior} = \frac{\text{Likelihood}}{\text{Evidence}} \times \text{Prior} \qquad (4.12)$$

and the likelihood at a certain level equals the evidence at the previous level. In this way, by gradually integrating out the parameters at different levels, the subsequent levels are linked to each other.

An important aspect of Bayesian methods is that the assumptions are made very explicit in the prior. If one makes a certain assumption for the prior at a given level it makes in fact clear in which sense the model is weak. In general, a considerable amount of work is done for Bayesian methods on the choice of the prior [19].

4.1.3 *Probabilistic interpretation of LS-SVM classifiers (Level 1)*

Computation of maximum posterior

Let us now discuss the derivations for each of the different levels as explained by Van Gestel *et al.* in [266]. The Bayesian learning framework is applied in a similar fashion as the methods developed by MacKay for classical MLPs [149; 150; 151]. However, one of the advantages of the present LS-SVM models is that they are linear in w, b, and on the other hand only nonlinear in typically one tuning parameter, such as σ of the RBF kernel, while for MLPs it is nonlinear in a (huge) interconnection matrix of the hidden layer.

At Level 1 we have

$$p(w, b|\mathcal{D}, \mu, \zeta, \mathcal{H}_\sigma) = \frac{p(\mathcal{D}|w, b, \mu, \zeta, \mathcal{H}_\sigma)}{p(\mathcal{D}|\mu, \zeta, \mathcal{H}_\sigma)} p(w, b|\mu, \zeta, \mathcal{H}_\sigma) \qquad (4.13)$$

where the evidence $p(\mathcal{D}|\mu, \zeta, \mathcal{H}_\sigma)$ is a normalizing constant, which can be determined by taking the integral over all possible values of w, b of the posterior and setting this integral to one.

At this level the prior corresponds to the regularization term and the sum of the squared error variables to the likelihood. For the prior we assume that it is separable between w and b (assuming w, b to be statistically independent), i.e. that we can write $p(w, b|\mu, \zeta, \mathcal{H}_\sigma) = p(w|\mu, \zeta, \mathcal{H}_\sigma)p(b|\mu, \zeta, \mathcal{H}_\sigma)$ and that we may simplify this to $p(w, b|\mu, \zeta, \mathcal{H}_\sigma) = p(w|\mu, \mathcal{H}_\sigma)p(b|\sigma_b, \mathcal{H}_\sigma)$ where we let $\sigma_b \to \infty$ to approximate a uniform distribution. When $\sigma_b \to \infty$ one gets

$$
\begin{aligned}
p(w, b|\mu, \mathcal{H}_\sigma) &= \left(\tfrac{\mu}{2\pi}\right)^{n_h/2} \exp\left(-\mu \tfrac{1}{2} w^T w\right) \tfrac{1}{\sqrt{2\pi}\sigma_b} \exp\left(-\tfrac{1}{2}\tfrac{b^2}{\sigma_b^2}\right) \\
&\propto \left(\tfrac{\mu}{2\pi}\right)^{n_h/2} \exp\left(-\mu \tfrac{1}{2} w^T w\right)
\end{aligned}
\qquad (4.14)
$$

where n_h denotes the dimension of the feature space for the mapping $\varphi(\cdot)$: $\mathbb{R}^n \to \mathbb{R}^{n_h}$. We formally calculate with w as in a finite dimensional setting.

Assuming that the data are independent one can write for the likelihood

$$
\begin{aligned}
p(\mathcal{D}|w,b,\zeta,\mathcal{H}_\sigma) &= \prod_{k=1}^{N} p(x_k,y_k|w,b,\zeta,\mathcal{H}_\sigma) \\
&\propto \prod_{k=1}^{N} p(e_k|w,b,\zeta,\mathcal{H}_\sigma) \\
&= \prod_{k=1}^{N} \sqrt{\frac{\zeta}{2\pi}} \exp\left(-\frac{1}{2}\zeta e_k^2\right)
\end{aligned}
\tag{4.15}
$$

where the second step is by assumption. We can then take the interpretation of LS-SVM classifiers as explained in the link with kernel Fisher discriminant analysis. In this case the variance around the targets $+1$ and -1 is assumed to be $1/\zeta$.

Application of Bayes' rule for the posterior gives

$$
p(w,b|\mathcal{D},\mu,\zeta,\mathcal{H}_\sigma) \propto \exp\left(-\mu\frac{1}{2}w^T w - \zeta\frac{1}{2}\sum_{k=1}^{N} e_k^2\right).
\tag{4.16}
$$

One then aims at finding w,b that maximize the posterior which means that one minimizes $J_P(w,e_c)$. One rather maximizes the logarithm of the posterior with maximum posterior solution denoted as $w_{\mathrm{MP}},b_{\mathrm{MP}}$. It is important to note that when taking the logarithm of this posterior one obtains a quadratic form in the vector $[w;b]$, which is due to the fact that the LS-SVM model is linear in the unknowns w,b. After straightforward calculation one obtains

$$
p(w-w_{\mathrm{MP}},b-b_{\mathrm{MP}}|\mathcal{D},\mu,\zeta,\mathcal{H}_\sigma) = \frac{1}{\sqrt{(2\pi)^{(n_h+1)}\det Q}} \exp\left(-\frac{1}{2}g^T Q^{-1} g\right)
\tag{4.17}
$$

with $g = [w-w_{\mathrm{MP}};b-b_{\mathrm{MP}}]$ and $Q = \mathrm{Cov}([w;b],[w;b])$ and Hessian $H = Q^{-1}$. This covariance matrix is related to the Hessian of the (quadratic) cost function as follows

$$
Q = H^{-1} = \begin{bmatrix} (\mu I + \zeta G)^{-1} & -(\mu I + \zeta G)^{-1}H_{12}H_{22}^{-1} \\ -H_{22}^{-1}H_{12}^T(\mu I + \zeta G)^{-1} & H_{22}^{-1} + H_{22}^{-1}H_{12}^T(\mu I + \zeta G)^{-1}H_{12}H_{22}^{-1} \end{bmatrix}
\tag{4.18}
$$

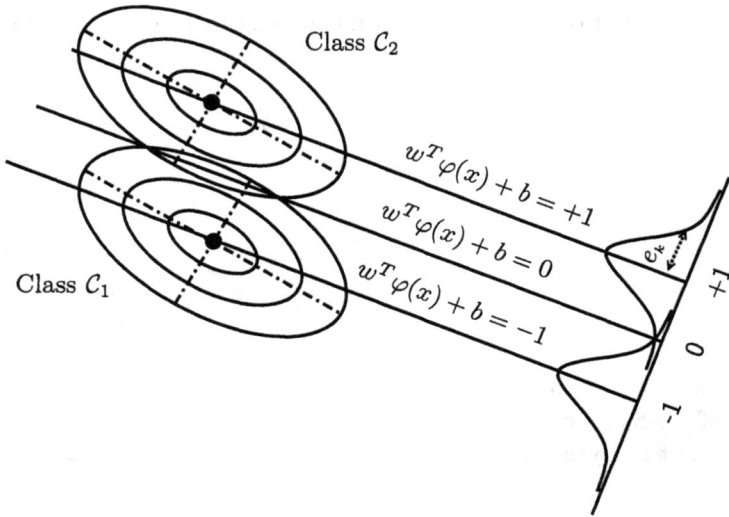

Fig. 4.2   *As explained in the kernel Fisher discriminant interpretation of LS-SVM clas-*
sifiers, the constraints can be interpreted as projecting the input space onto a one-
dimensional axis where one aims at minimizing the within scatter for Class C_1 and Class
C_2.

where

$$H_{11} = \mu I + \zeta \Upsilon \Upsilon^T$$
$$H_{12} = \zeta \Upsilon 1_v$$
$$H_{22} = \zeta N$$

and $\Upsilon = [\varphi(x_1), ..., \varphi(x_N)]$, $G = \Upsilon M_c \Upsilon^T$ and $M_c = I - (1/N)1_v 1_v^T$ a
symmetric centering matrix which is idempotent. Note that in the case of
classical MLPs, the cost function is non-convex and this Hessian is related
then to a quadratic approximation of the cost function at a local minimum
[149; 150; 151].

Class probabilities for the LS-SVM classifier by applying Bayesian
decision theory in the target space

In the previous Chapter we have explained the link between LS-SVM classifiers and kernel Fisher discriminant analysis. We have discussed how the set
of constraints can be related to projecting multivariate data to univariate

data in the one-dimensional target space where one considers targets $+1$ and -1 for the two classes. In the introductory Chapter it was discussed in the context of Bayesian decision theory that the optimal decision boundary between two classes was a linear separating hyperplane when the class conditional densities were assumed to be Gaussian and the class covariance matrices for the two classes are the same. A different covariance matrix led to quadratic decision surfaces. Let us now apply Bayesian decision theory to the one-dimensional target space with projected variables $w^T \varphi(x) + b$.

Due to the use of least squares and the link with Fisher discriminant analysis the following assumption is meaningful:

$$p(x|y = +1, w, b, \varsigma, \mathcal{H}_\sigma) = \sqrt{\frac{1}{2\pi\varsigma^{-1}}} \exp\left(-\frac{1}{2}\frac{e^2}{\varsigma^{-1}}\right) \tag{4.19}$$

and

$$p(x|y = -1, w, b, \varsigma, \mathcal{H}_\sigma) = \sqrt{\frac{1}{2\pi\varsigma^{-1}}} \exp\left(-\frac{1}{2}\frac{e^2}{\varsigma^{-1}}\right). \tag{4.20}$$

One has respectively

$$\begin{aligned} e_k &= w^T\left(\varphi(x_k) - \hat{\mu}^{(1)}\right), \quad k = 1, ..., N_1 \\ e_l &= w^T\left(\varphi(x_l) - \hat{\mu}^{(2)}\right), \quad l = 1, ..., N_2 \end{aligned} \tag{4.21}$$

where $\hat{\mu}^{(1)} = (1/N_1)\sum_{k=1}^{N_1}\varphi(x_k)$, $\hat{\mu}^{(2)} = (1/N_2)\sum_{l=1}^{N_2}\varphi(x_l)$ and N_1, N_2 denote the number of data points of class 1 and 2, with running indices k and l, respectively.

Given a new point x the predictive means $m_{(1)}, m_{(2)}$ and variances $\sigma_{(1)}^2, \sigma_{(2)}^2$ for the error variable e around the ± 1 targets are

$$\begin{aligned} m_{(1)}(x) &= w_{\text{MP}}^T(\varphi(x) - \mu^{(1)}) \simeq \frac{1}{\mu}\sum_{k=1}^{N}\alpha_k K(x, x_k) - \hat{\mu}_{d1} \\ m_{(2)}(x) &= w_{\text{MP}}^T(\varphi(x) - \mu^{(2)}) \simeq \frac{1}{\mu}\sum_{k=1}^{N}\alpha_k K(x, x_k) - \hat{\mu}_{d2} \end{aligned} \tag{4.22}$$

with $\hat{\mu}_{d1} = (1/N_1)\sum_{k=1}^{N_1}\sum_{j=1}^{N}\alpha_j K(x_j, x_k)$ and $\hat{\mu}_{d2} = (1/N_2)\sum_{l=1}^{N_2}\sum_{j=1}^{N}\alpha_j K(x_j, x_l)$ and

$$\begin{aligned} \sigma_{(1)}^2(x) &= [\varphi(x) - \mu^{(1)}]H_{11}^{-1}[\varphi(x) - \mu^{(1)}] \\ \sigma_{(2)}^2(x) &= [\varphi(x) - \mu^{(2)}]H_{11}^{-1}[\varphi(x) - \mu^{(2)}]. \end{aligned} \tag{4.23}$$

One obtains the following expression for the variances after some straight-forward matrix algebra and applying the kernel trick:

$$
\sigma^2_{(1)}(x) = \frac{1}{\mu}K(x,x) - \frac{2}{\mu N_1}\sum_{k=1}^{N_1} K(x,x_k) + \frac{1}{\mu N_1^2}1_v^T\Omega_{11}1_v
$$
$$
- \frac{\zeta}{\mu}(\theta(x)^T - \frac{1}{N_1}1_v^T\Omega_{1*})M_c(\mu I + \zeta\Omega_c)^{-1}M_c(\theta(x) - \frac{1}{N_1}\Omega_{*1}1_v)
$$

(4.24)

and

$$
\sigma^2_{(2)}(x) = \frac{1}{\mu}K(x,x) - \frac{2}{\mu N_2}\sum_{l=1}^{N_2} K(x,x_l) + \frac{1}{\mu N_2^2}1_v^T\Omega_{22}1_v
$$
$$
- \frac{\zeta}{\mu}(\theta(x)^T - \frac{1}{N_2}1_v^T\Omega_{2*})M_c(\mu I + \zeta\Omega_c)^{-1}M_c(\theta(x) - \frac{1}{N_2}\Omega_{*2}1_v)
$$

(4.25)

with $\Omega_{11} \in \mathbb{R}^{N_1 \times N_1}, \Omega_{22} \in \mathbb{R}^{N_2 \times N_2}, \Omega_{1*} \in \mathbb{R}^{N_1 \times N}, \Omega_{*1} \in \mathbb{R}^{N \times N_1}, \Omega_{2*} \in \mathbb{R}^{N_2 \times N}, \Omega_{*2} \in \mathbb{R}^{N \times N_2}, \Omega \in \mathbb{R}^{N \times N}$, where it is clear from the notation which training data vectors are selected to form the corresponding sub-kernel matrix. M_c denotes the centering matrix, $\Omega_c = M_c\Omega M_c$ the centered Gram matrix and $\theta(x) = [K(x,x_1); ...; K(x,x_N)]$. The expressions (4.24) and (4.25) can also be expressed in terms of the eigenvalues of the centered kernel matrix [266].

By marginalization of (4.19) and (4.20) over w, b (in fact it is more accurate to marginalize also over the hyperparameters and the kernel parameters, but this has less influence on the result as explained in [151]), one obtains

$$
p(x|y = +1, \mathcal{D}, \mu, \zeta, \mathcal{H}_\sigma) = \frac{1}{\sqrt{2\pi}}\frac{1}{\sqrt{\zeta^{-1} + \sigma^2_{(1)}(x)}}\exp\left(-\frac{1}{2}\frac{m^2_{(1)}(x)}{\zeta^{-1} + \sigma^2_{(1)}(x)}\right)
$$

$$
p(x|y = -1, \mathcal{D}, \mu, \zeta, \mathcal{H}_\sigma) = \frac{1}{\sqrt{2\pi}}\frac{1}{\sqrt{\zeta^{-1} + \sigma^2_{(2)}(x)}}\exp\left(-\frac{1}{2}\frac{m^2_{(2)}(x)}{\zeta^{-1} + \sigma^2_{(2)}(x)}\right).
$$

(4.26)

Application of Bayesian decision theory gives

$$
P(y|x, \mathcal{D}, \mu, \zeta, \mathcal{H}_\sigma) = \frac{p(x|y, \mathcal{D}, \mu, \zeta, \mathcal{H}_\sigma)}{p(x|\mathcal{D}, \mu, \zeta, \mathcal{H}_\sigma)} P(y|\mathcal{D}, \mu, \zeta, \mathcal{H}_\sigma)
$$

(4.27)

where $P(y)$ denotes the prior class probability. One has $p(x|\mathcal{D}, \mu, \zeta, \mathcal{H}_\sigma) = P(y = +1)p(x|y = +1, \mathcal{D}, \mu, \zeta, \mathcal{H}_\sigma) + P(y = -1)p(x|y = -1, \mathcal{D}, \mu, \zeta, \mathcal{H}_\sigma)$. One selects then the class with maximal posterior class probability. When assuming $\sigma_{(1)}^2 = \sigma_{(2)}^2$ and defining $\sigma_e^2 = \sqrt{\sigma_{(1)}^2 \sigma_{(2)}^2}$ one obtains the following classifier

$$y(x) = \text{sign}[\frac{1}{\mu} \sum_{k=1}^{N} \alpha_k K(x, x_k) - \frac{\mu_{d1} + \mu_{d2}}{2} + \frac{\zeta^{-1} + \sigma_e^2(x)}{\mu_{d1} - \mu_{d2}} \log \frac{P(y = +1)}{P(y = -1)}].$$
$$(4.28)$$

Moderated outputs by using the softmax function

A simple way to obtain a probabilistic interpretation of the LS-SVM classifier is to apply the softmax function to the latent variable $z = w^T \varphi(x) + b$. This approach is standard in neural networks and statistics [23; 107; 151; 287]. It results in

$$P(y = +1|x, w, b, \mathcal{D}, \mu, \zeta, \mathcal{H}_\sigma) = \frac{1}{1 + \exp\left(-(w^T \varphi(x) + b)\right)}$$

$$P(y = -1|x, w, b, \mathcal{D}, \mu, \zeta, \mathcal{H}_\sigma) = 1 - P(y = +1|x, w, b, \mathcal{D}, \mu, \zeta, \mathcal{H}_\sigma)$$

$$= \frac{1}{1 + \exp\left(w^T \varphi(x) + b\right)}.$$
$$(4.29)$$

The functional form of this expression is consistent with a Gaussian assumption of the error variables around the targets $+1$ and -1.

Unbalanced data sets

It may often happen in real-life situations that the given training data set is unbalanced in the sense that $N_1 \gg N_2$ or $N_2 \gg N_1$. The decision rule (4.28) obtained after applying Bayesian decision theory provides a method for taking into account this unbalancing. It is then meaningful to choose the prior class probabilities as

$$P(y = +1) = \frac{N_1}{N_1 + N_2} \quad \text{and} \quad P(y = -1) = \frac{N_2}{N_1 + N_2}. \quad (4.30)$$

A minimal requirement of any classifier is in fact that it should be able to perform better than the majority rule which decides $y = +1$ if $N_1 > N_2$ and else $y = -1$. This classifier has a training set performance of $N_1/(N_1 + N_2)$.

The LS-SVM classifier (4.28) from Bayesian decision theory contains a

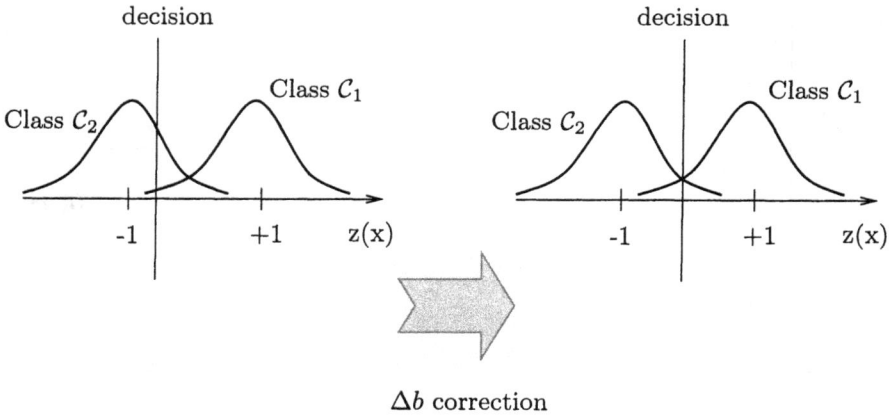

Δb correction

Fig. 4.3 *Especially in the case of small and/or unbalanced data sets the estimation of the bias term from the linear system for the LS-SVM classifier will be unreliable. A bias term correction is needed then. In order to understand the problem it is often meaningful to visualize the empirical distribution of the latent variable z as a univariate distribution in the target space.*

bias term correction. It is of the form

$$y(x) = \text{sign}[\frac{1}{\mu} \sum_{k=1}^{N} \alpha_k K(x, x_k) + \Delta b(x)] \qquad (4.31)$$

with bias term correction $\Delta b(x)$. The idea of a constant bias term correction $\Delta b(x) = \Delta b$ has been discussed in [74] to improve the validation set performance. In [169] the probabilities $p(e|y, w_{\text{MP}}, b_{\text{MP}}, \mu, \zeta, \mathcal{H}_\sigma)$ were estimated using leave-one-out cross-validation given the obtained SVM classifier and the corresponding decision was made in a similar fashion as (4.28). A simple density estimation algorithm was used and no Gaussian assumptions were made, while on the other hand no marginalization over the model parameters was performed. Also for small data sets a bias term correction is usually recommended such as occurring e.g. in classification of microarray data and biomedical data sets including the classification of brain tumours from MRS signals [148].

4.1.4 *Inference of the hyperparameters (Level 2)*

Calculation of posterior

Let us now study the problem of inference for the hyperparameters μ and ζ. The posterior in the hyperparameters is written as follows

$$p(\mu, \zeta | \mathcal{D}, \mathcal{H}_\sigma) = \frac{p(\mathcal{D} | \mu, \zeta, \mathcal{H}_\sigma)}{p(\mathcal{D} | \mathcal{H}_\sigma)} p(\mu, \zeta | \mathcal{H}_\sigma). \tag{4.32}$$

We assume a uniform distribution for $\log \mu$ and $\log \zeta$ and assume $p(\mu, \zeta | \mathcal{H}_\sigma)$ $= p(\mu | \mathcal{H}_\sigma) p(\zeta | \mathcal{H}_\sigma)$. Hence we can write

$$p(\mu, \zeta | \mathcal{D}, \mathcal{H}_\sigma) \propto p(\mathcal{D} | \mu, \zeta, \mathcal{H}_\sigma), \tag{4.33}$$

meaning that when we aim at maximizing the posterior we need to maximize $p(\mathcal{D} | \mu, \zeta, \mathcal{H}_\sigma)$, which is the evidence of the previous Level 1. Using the expressions derived at Level 1, we can derive the following for the posterior

$$p(\mu, \zeta | \mathcal{D}, \mathcal{H}_\sigma) \propto \exp\left(-J_{\mathrm{P}}(w_{\mathrm{MP}}, b_{\mathrm{MP}})\right) \times \sqrt{\mu^{n_h} \zeta^N} \sqrt{\det H^{-1}} \tag{4.34}$$

where $J_{\mathrm{P}}(w, b) = J_{\mathrm{P}}(w_{\mathrm{MP}}, b_{\mathrm{MP}}) + \frac{1}{2}([w; b] - [w_{\mathrm{MP}}; b_{\mathrm{MP}}])^T H([w; b] - [w_{\mathrm{MP}}; b_{\mathrm{MP}}])$ and $J_{\mathrm{P}}(w_{\mathrm{MP}}, b_{\mathrm{MP}}) = \mu E_W(w_{\mathrm{MP}}) + \zeta E_D(w_{\mathrm{MP}}, b_{\mathrm{MP}})$. Note that $\det H^{-1}$ characterizes the volume of an ellipsoid centered around $[w_{\mathrm{MP}}; b_{\mathrm{MP}}]$ with matrix H for the posterior at Level 1.

By taking the negative logarithm of this expression, one can see that maximizing the posterior corresponds to solving the following optimization problem in the hyperparameters:

$$\min_{\mu, \zeta} J(\mu, \zeta) = \mu E_W(w_{\mathrm{MP}}) + \zeta E_D(w_{\mathrm{MP}}, b_{\mathrm{MP}})$$
$$+ \frac{1}{2} \sum_{i=1}^{N_{\mathrm{eff}}} \log(\mu + \zeta \lambda_{G,i}) - \frac{N_{\mathrm{eff}}}{2} \log \mu - \frac{N-1}{2} \log \zeta, \tag{4.35}$$

where $\det H^{-1} = 1/\det H$ with $\det H = N \mu^{n_h - N_{\mathrm{eff}}} \zeta \prod_{i=1}^{N_{\mathrm{eff}}} (\mu + \zeta \lambda_{G,i})$ and $\lambda_{G,i}$ are the eigenvalues of matrix G and N_{eff} is the number of non-zero eigenvalues of the matrix $M_c \Omega M_c$. Note that the eigenvalues of the centered Gram matrix also play a role in kernel PCA [203].

After some calculations one can show that this problem can be written as

$$
\min_{\mu,\zeta} J(\mu,\zeta) = \frac{1}{2} y^T M_c \left(\frac{1}{\mu} \Omega_c + \frac{1}{\zeta} I \right)^{-1} M_c y \\
+ \frac{1}{2} \log \det(\frac{1}{\mu} \Omega_c + \frac{1}{\zeta} I) - \frac{1}{2} \log \frac{1}{\zeta}.
$$
(4.36)

Up to the centering and contribution of the bias term b of the model, this expression corresponds to Level 2 cost function as considered in Gaussian processes [292].

Effective number of parameters and characterization of solution

The effective number of parameters as considered in neural networks [23; 151] is equal to $d_{\text{eff}} = \sum_i \lambda_{i,u}/\lambda_{i,r}$ where $\lambda_{i,u}, \lambda_{i,r}$ denote the eigenvalues of the Hessian for the unregularized cost function $J_{P,u} = \zeta E_D$ and the regularized cost function $J_{P,r} = \mu E_W + \zeta E_D$, respectively. For the LS-SVM classifier model, the effective number of parameters can be expressed as

$$
d_{\text{eff}} = 1 + \sum_{i=1}^{N_{\text{eff}}} \frac{\zeta \lambda_{G,i}}{\mu + \zeta \lambda_{G,i}} = 1 + \sum_{i=1}^{N_{\text{eff}}} \frac{\gamma \lambda_{G,i}}{1 + \gamma \lambda_{G,i}}
$$
(4.37)

with $\gamma = \zeta/\mu$. The term $+1$ appears due to the fact that no regularization is done on the bias term b. Note that one always has $N_{\text{eff}} \leq N-1$ and hence $d_{\text{eff}} \leq N$, even in the case of high dimensional feature spaces. The optimal values of μ, ζ can be found by setting $\partial J(\mu,\zeta)/\partial \mu = 0$, $\partial J(\mu,\zeta)/\partial \zeta = 0$ evaluated at the maximum posterior solution. This gives the following set of equations in μ, ζ:

$$
\begin{cases}
2\mu \, E_W(w_{\text{MP}};\mu,\zeta) = d_{\text{eff}}(\mu,\zeta) - 1 \\
2\zeta \, E_D(w_{\text{MP}},b_{\text{MP}};\mu,\zeta) = N - d_{\text{eff}}(\mu,\zeta).
\end{cases}
$$
(4.38)

In the training of MLP neural networks this implicit set of equations is iteratively solved [151]. In the context of LS-SVM models one rather prefers to directly minimize $J(\mu,\zeta)$ in the unknowns μ, ζ instead of solving these equations.

Instead of solving an optimization problem in μ and ζ and may also solve a scalar optimization problem in $\gamma = \zeta/\mu$:

$$\min_{\gamma} J(\gamma) = \sum_{i=1}^{N-1} \log \left(\lambda_{G,i} + \frac{1}{\gamma} \right) + (N-1) \log \left(E_W(w_{\mathrm{MP}}) + \gamma E_D(w_{\mathrm{MP}}, b_{\mathrm{MP}}) \right).$$

(4.39)

At the optimum $(\partial J(\gamma)/\partial\gamma = 0)$ one obtains the following implicit equation in γ:

$$\gamma = \frac{N - d_{\mathrm{eff}}(\gamma)}{d_{\mathrm{eff}}(\gamma) - 1} \frac{E_W(w_{\mathrm{MP}}; \gamma)}{E_D(w_{\mathrm{MP}}, b_{\mathrm{MP}}; \gamma)}$$

(4.40)

with

$$
\begin{aligned}
E_D(w_{\mathrm{MP}}, b_{\mathrm{MP}}) &= \tfrac{1}{2\gamma^2} y^T M_c V_G (D_G + I/\gamma)^{-2} V_G^T M_c y \\
E_W(w_{\mathrm{MP}}) &= \tfrac{1}{2} y^T M_c V_G D_G (D_G + I/\gamma)^{-2} V_G^T M_c y \\
E_W(w_{\mathrm{MP}}) + \gamma E_D(w_{\mathrm{MP}}, b_{\mathrm{MP}}) &= \tfrac{1}{2} y^T M_c V_G (D_G + I/\gamma)^{-1} V_G^T M_c y,
\end{aligned}
$$

(4.41)

and eigenvalue decomposition $\Omega_c v_{G,i} = \lambda_{G,i} v_{G,i}$ with eigenvalues $\lambda_{G,i}$ and eigenvectors $v_{G,i}$ with $i = 1, ..., N_{\mathrm{eff}} \leq N - 1$. One constructs $V_G = [v_{G,1}, ..., v_{G,N_{\mathrm{eff}}}] \in \mathbb{R}^{N \times N_{\mathrm{eff}}}$, $D_G = \mathrm{diag}(\lambda_{G,1}, ..., \lambda_{G,N_{\mathrm{eff}}})$ and $U_G = [\lambda_{G,1}^{-1/2} v_{G,1}, ..., \lambda_{G,N_{\mathrm{eff}}}^{-1/2} v_{G,N_{\mathrm{eff}}}]$. Note that $\mathrm{rank}(M_c) = N - 1$.

4.1.5 Inference of kernel parameters and model comparison (Level 3)

Ranking based on evidence and Bayes factor

The parameters w, b and the hyperparameters μ, ζ have been determined at the previous levels. In the case of an RBF kernel one still needs to infer then the width σ of the kernel at Level 3.

The posterior at Level 3 equals

$$p(\mathcal{H}_\sigma | \mathcal{D}) = \frac{p(\mathcal{D} | \mathcal{H}_\sigma)}{p(\mathcal{D})} p(\mathcal{H}_\sigma).$$

(4.42)

One assumes a uniform prior for $p(\mathcal{H}_\sigma)$, meaning that all RBF kernel models with different σ value are equally likely a priori. As a result one has then

$$p(\mathcal{H}_\sigma | \mathcal{D}) \propto p(\mathcal{D} | \mathcal{H}_\sigma)$$

(4.43)

meaning that the Level 3 posterior is proportional to the Level 2 evidence. When assuming equal prior probabilities for two models with values σ_1 and σ_2, the models are compared and ranked according to the maximal evidence. This can be understood as follows. By application of Bayes rule to the two models one obtains

$$\frac{p(\mathcal{H}_{\sigma_1}|D)}{p(\mathcal{H}_{\sigma_2}|D)} = \frac{p(D|\mathcal{H}_{\sigma_1})}{p(D|\mathcal{H}_{\sigma_2})} \cdot \frac{p(\mathcal{H}_{\sigma_1})}{p(\mathcal{H}_{\sigma_2})}. \tag{4.44}$$

When $p(\mathcal{H}_{\sigma_1}) = p(\mathcal{H}_{\sigma_2})$ then

$$\frac{p(\mathcal{H}_{\sigma_1}|D)}{p(\mathcal{H}_{\sigma_2}|D)} = \frac{p(D|\mathcal{H}_{\sigma_1})}{p(D|\mathcal{H}_{\sigma_2})} = \text{Bayes factor.} \tag{4.45}$$

This Bayes factor characterizes the improvement of model $\mathcal{H}_{\sigma_1}$ with respect to $\mathcal{H}_{\sigma_2}$.

Occam factor: from maximum likelihood to maximal evidence

In order to find the Level 3 posterior, let us first explain the notion of Occam factor. In general for a parameter vector θ of model $\mathcal{H}_\sigma$ to be inferred at a certain level, one can write

$$p(\theta|D, \mathcal{H}_\sigma) = \frac{p(D|\theta, \mathcal{H}_\sigma)}{p(D|\mathcal{H}_\sigma)} p(\theta|\mathcal{H}). \tag{4.46}$$

From the normalization $\int p(\theta|D, \mathcal{H}_\sigma) d\theta = 1$ one finds that the evidence equals

$$p(D|\mathcal{H}_\sigma) = \int p(D|\theta, \mathcal{H}_\sigma) p(\theta|\mathcal{H}_\sigma) d\theta \tag{4.47}$$

meaning that one marginalizes over the likelihood and the prior.

The Laplace method [193; 143; 251] consists of making the following approximation

$$
\begin{aligned}
p(D|\mathcal{H}_\sigma) &= \int \exp\left(-h(\theta)\right) d\theta \\
&\simeq \exp(-h(\theta_{\text{MP}})) \int \exp\left(-\frac{1}{2}(\theta - \theta_{\text{MP}})^T H (\theta - \theta_{\text{MP}})\right) d\theta \\
&= \exp(-h(\theta_{\text{MP}})) \sqrt{(2\pi)^{n_p}} \sqrt{\det H^{-1}}
\end{aligned}
$$

$$\tag{4.48}$$

with Hessian H evaluated at the maximal posterior and $\theta \in \mathbb{R}^{n_p}$ and $\exp(-h(\theta)) = p(\mathcal{D}|\theta, \mathcal{H}_\sigma)p(\theta|\mathcal{H}_\sigma)$. In the context of MLP neural networks this has been interpreted as follows by MacKay [151] by assuming that the posterior is sharply peaked at the maximum posterior solution θ_{MP}:

$$p(\mathcal{D}|\mathcal{H}_\sigma) \simeq p(\mathcal{D}|\theta_{\mathrm{MP}}, \mathcal{H}_\sigma) \times p(\theta_{\mathrm{MP}}|\mathcal{H}_\sigma) \sqrt{(2\pi)^{n_p}} \sqrt{\det H^{-1}}$$

$$\mathrm{Evidence} \simeq \mathrm{Likelihood}|_{\theta=\theta_{\mathrm{MP}}} \times \mathrm{Occam\ factor}$$

(4.49)

with H the Hessian of a quadratic function (or approximation) around θ_{MP}. In the case of a one-dimensional variable θ one can write for the Occam factor $p(\theta_{\mathrm{MP}}|\mathcal{H}_\sigma)\sigma_{\theta|\mathcal{D}} = \sigma_{\theta|\mathcal{D}}/\sigma_\theta$ which illustrates that the Occam factor is equal to the ratio of the posterior accessible volume of $\mathcal{H}_\sigma$'s parameter space to the prior accessible volume, where σ_θ and $\sigma_{\theta|\mathcal{D}}$ denote the width of the Gaussian related to the prior and posterior, respectively.

Computation of Level 3 posterior

We are in a position now to do the Level 3 inference. Assuming a separable Gaussian prior $p(\mu_{\mathrm{MP}}, \zeta_{\mathrm{MP}}|\mathcal{H}_\sigma)$ and standard deviations σ_μ, σ_ζ of the Gaussian distributions for the prior and $\sigma_{\mu|\mathcal{D}}, \sigma_{\zeta|\mathcal{D}}$ for the posterior. A considerable advantage in the case of LS-SVM models is that one can find exact expressions for this integral (4.48) on Level 2 and only on Level 3 one has to make approximations, while in the case of MLP neural networks one has to make approximations on both levels.

The reasoning on Level 3 is then as follows

$$\begin{aligned}\mathrm{Posterior}_{\mathcal{L}_3} &\propto \mathrm{Likelihood}_{\mathcal{L}_3} \\ &= \mathrm{Evidence}_{\mathcal{L}_2}\end{aligned}$$

(4.50)

where $\mathcal{L}_2, \mathcal{L}_3$ denote Level 2 and 3, respectively. In addition one has

$$\mathrm{Evidence}_{\mathcal{L}_2} \propto \mathrm{Likelihood}_{\mathcal{L}_2}|_{(\mu,\zeta)=(\mu_{\mathrm{MP}},\zeta_{\mathrm{MP}})} \times \mathrm{Occam\ factor}_{\mathcal{L}_2}. \quad (4.51)$$

This becomes

$$p(\mathcal{D}|\mathcal{H}_\sigma) \propto p(\mathcal{D}|\mu_{\mathrm{MP}}, \zeta_{\mathrm{MP}}, \mathcal{H}_\sigma) \frac{\sigma_{\mu|\mathcal{D}}\sigma_{\zeta|\mathcal{D}}}{\sigma_\mu \sigma_\zeta} \quad (4.52)$$

where the following approximations are meaningful according to [151]: $\sigma_{\mu|\mathcal{D}}^2 \simeq 2/(d_{\mathrm{eff}} - 1)$ and $\sigma_{\zeta|\mathcal{D}}^2 \simeq 2/(N - d_{\mathrm{eff}})$. Finally, one can rank the models then according to

$$p(\mathcal{D}|\mathcal{H}_\sigma) \propto \sqrt{\frac{\mu_{\mathrm{MP}}^{N_{\mathrm{eff}}} \zeta_{\mathrm{MP}}^{N-1}}{(d_{\mathrm{eff}} - 1)(N - d_{\mathrm{eff}}) \prod_{i=1}^{N_{\mathrm{eff}}} (\mu_{\mathrm{MP}} + \zeta_{\mathrm{MP}} \lambda_{G,i})}}. \tag{4.53}$$

Link with model selection criteria

A similar reasoning can be done at the previous Level of inference. One has

$$\begin{aligned} \mathrm{Posterior}_{\mathcal{L}_2} &\propto \mathrm{Likelihood}_{\mathcal{L}_2} \\ &= \mathrm{Evidence}_{\mathcal{L}_1} \end{aligned} \tag{4.54}$$

where $\mathcal{L}_1, \mathcal{L}_2$ denote Level 1 and 2, respectively. For the evidence one can write

$$\begin{aligned} \mathrm{Evidence}_{\mathcal{L}_1} &= \mathrm{Likelihood}_{\mathcal{L}_1}|_{(w,b)=(w_{\mathrm{MP}},b_{\mathrm{MP}})} \times \mathrm{Occam\ factor}_{\mathcal{L}_1} \\ &\propto \mathrm{Likelihood}_{\mathcal{L}_1}|_{(w,b)=(w_{\mathrm{MP}},b_{\mathrm{MP}})} \times \frac{\sqrt{\det H^{-1}}}{\sqrt{1/\mu^{n_h}}}. \end{aligned} \tag{4.55}$$

Instead of only maximizing the likelihood (as in maximum likelihood methods), in Bayesian learning one aims at maximizing the evidence. By taking the negative logarithm of this equation, one minimizes a sum of two terms where the first term characterizes the model fit and the second term the model complexity. Further approximations of the complexity term lead to the well-known AIC (An Information Criterion) by Akaike and BIC (Bayesian Information Criterion) by Schwarz, where the latter is related to Akaike's information criterion B [193].

4.1.6 *Design of the LS-SVM classifier in the Bayesian evidence framework*

Based upon the theoretical insights outlined in the previous Sections, the design of the LS-SVM classifier is done as follows:

Algorithm of Bayesian inference for LS-SVM classifiers:

(1) Normalize and standardize the inputs to zero mean and unit variance.

(2) Select a model $\mathcal{H}_i$ by choosing a kernel K_i which may have a kernel parameter, e.g. σ_i in the RBF kernel case, with $i = 1, ..., n_{\mathcal{H}}$ number of models to be compared.

(3) Compute the effective number of parameters d_{eff}, from the eigenvalue decomposition of the centered Gram matrix.

(4) Find the optimal hyperparameters $\mu_{\text{MP}}, \zeta_{\text{MP}}$ by solving the scalar optimization problem in $\gamma = \zeta/\mu$, related to maximizing the Level 2 posterior.

(5) Use the expression for Level 3 model comparison based on the evidence (4.53).

(6) Refine the kernel tuning parameters, e.g. σ_i in the RBF kernel case and go back to step 2 as long as one can improve according to the criterion (4.53).

Note that in the case of a linear kernel, one has only Level 1 & 2 and no Level 3 inference. Once the classifier has been trained in this way, decision making is done as follows.

Bayesian decision making for LS-SVM classifiers:

(1) Normalize inputs in the same way as for the training procedure.

(2) Compute the predictive mean and variance of the input pattern to be classified according to (4.22) and (4.24), (4.25). This mean and variance are for the projected variable on the one-dimensional target space, keeping in mind that the LS-SVM classifier aims at minimizing the within scatter around targets $+1$ and -1.

(3) Calculate the posterior class probability $P(y|x, \mathcal{D})$ according to (4.26) for given prior class probabilities $P(y = +1), P(y = -1)$. Often these will be unknown. A meaningful choice in that case is $P(y = +1) = N_1/N$ and $P(y = -1) = N_2/N$.

(4) Assign the given input pattern to the class with maximal posterior class probability (4.26). The decision threshold will usually

be taken then as 1/2. However, it will often be insightful to vary the threshold over the whole range of possible threshold values. Varying the threshold leads to computing an ROC curve for the classifier.

4.1.7   *Example and benchmarking results*

Illustrative example

Let us apply the Bayesian LS-SVM classifier now to a well-known example, the Ripley synthetic dataset [193] according to [266]. The dataset consists of a training and test set of $N = 250$ and $N_{\text{test}} = 1000$ data points, respectively. There are two inputs ($n = 2$) and each class is an equal mixture of two normal distributions with the same covariance matrices. Both classes have the same prior probability, i.e. $P(y = +1) = P(y = -1) = 1/2$. The training data are visualized in Fig. 4.4. The inputs are normalized to zero mean and unit variance.

In this example we apply an RBF kernel. Assuming a flat prior on the value of $\log \sigma$, the optimal σ_{MP} was selected according to the algorithm outlined in the previous Section, resulting into $\sigma_{\text{MP}} = 1.3110$. This yields a test set performance of 90.6% for the LS-SVM classifier. Gaussian processes for classification performed slightly worse (but not significantly) on this example with test set performance 89.9%. In Fig. 4.4 the posterior class probability $P(y|x, \mathcal{D})$ is shown with respect to a given range for the two input variables, together with the corresponding contour plots of equal posterior class probability.

Benchmarking results

Benchmarking results of LS-SVM classifiers within the Bayesian evidence framework are shown in Table 4.1, according to the work of Van Gestel *et al.* reported in [266]. The test set classification performance of the LS-SVM classifier with RBF-kernel with Bayesian inference of the hyperparameters was assessed on 10 publically available binary classification datasets. The results of LS-SVM and SVM classification were compared with GP regression (basically LS-SVM without bias term) where the hyperparameter and kernel parameter were determined by 10-fold cross-validation (CV10). The

	n	N	N_{test}	N_{tot}	LS-SVM (BayM)	LS-SVM (Bay)	LS-SVM (CV10)	SVM (CV10)	GP (Bay)	GP$_b$ (Bay)	GP (CV10)
bld	6	230	115	345	69.4(2.9)	69.4(3.1)	69.4(3.4)	69.2(3.5)	69.2(2.7)	68.9(3.3)	69.7(4.0)
cra	6	133	67	200	96.7(1.5)	96.7(1.5)	96.9(1.6)	95.1(3.2)	96.4(2.5)	94.8(3.2)	96.9(2.4)
gcr	20	666	334	1000	73.1(3.8)	73.5(3.9)	75.6(1.8)	74.9(1.7)	76.2(1.4)	75.9(1.7)	75.4(2.0)
hea	13	180	90	270	83.6(5.1)	83.2(5.2)	84.3(5.3)	83.4(4.4)	83.1(5.5)	83.7(4.9)	84.1(5.2)
ion	33	234	117	351	95.6(0.9)	96.2(1.0)	95.6(2.0)	95.4(1.7)	91.0(2.3)	94.4(1.9)	92.4(2.4)
pid	8	512	256	768	77.3(3.1)	77.5(2.8)	77.3(3.0)	76.9(2.9)	77.6(2.9)	77.5(2.7)	77.2(3.0)
rsy	2	250	1000	1250	90.2(0.7)	90.2(0.6)	89.6(1.1)	89.7(0.8)	90.2(0.7)	90.1(0.8)	89.9(0.8)
snr	60	138	70	208	76.7(5.6)	78.0(5.2)	77.9(4.2)	76.3(5.3)	78.6(4.9)	75.7(6.1)	76.6(7.2)
tit	3	1467	734	2201	78.8(1.1)	78.7(1.1)	78.7(1.1)	78.7(1.1)	78.5(1.0)	77.2(1.9)	78.7(1.2)
wbc	9	455	228	683	95.9(0.6)	95.7(0.5)	96.2(0.7)	96.2(0.8)	95.8(0.7)	93.7(2.0)	96.5(0.7)
AP					83.7	83.9	84.1	83.6	83.7	83.2	83.7
AR					2.3	2.5	2.5	3.8	3.2	4.2	2.6
P_{ST}					1.000	0.754	1.000	0.344	0.754	0.344	0.508

Table 4.1 Benchmarking results according to Van Gestel et al. of Bayesian LS-SVM classifiers showing good performance in comparison with Gaussian processes for classification and standard SVM classifiers. For the case (BayM) the moderated output was applied to an LS-SVM classifier with Bayesian inference of the hyper- and kernel parameters. (Bay) is similar to (BayM) but the formulation with sign function is used for the decision. An RBF kernel was used for all models. The model GP_b has a bias term within the kernel function. AA (Average Accuracy), AR (Average rank) and P_{ST} (Probability of equal medians using the Sign Test) are taken over all domains. Best performances are underlined and denoted in bold face, performances not significantly different at the 5% level are denoted in bold face, performances significantly different at the 1% level are emphasized.

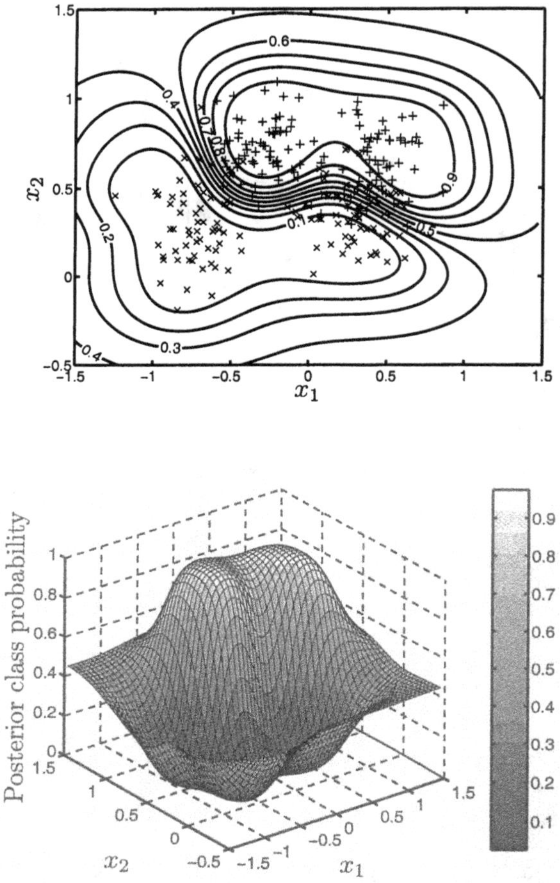

Fig. 4.4   *Illustrative example on the Ripley data set for an LS-SVM classifier trained within the Bayesian evidence framework: (Top) training data set with data points of class 1 depicted as + and points of class 2 as x. The figure also shows contour plots of equal posterior class probability; (Bottom) posterior class probability $P(y|x, \mathcal{D})$ shown with respect to a given range for the two input variables.*

Bupa Liver Disorders (**bld**), the Statlog German Credit (**gcr**), the Statlog Heart Disease (**hea**), the John Hopkins University Ionosphere (**ion**), the Pima Indians Diabetes (**pid**), the Sonar (**snr**) and the Wisconsin Breast Cancer (**wbc**) datasets were retrieved from the UCI benchmark repository

[24]. The Ripley synthetic dataset (rsy) and Leptograpsus crabs (cra) are described in [193]. The Titanic data (tit) was obtained from Delve. Each dataset was split into a training set (2/3) and test set (1/3), except for the rsy dataset, where $N = 250$ and $N_{\text{test}} = 1000$ was taken. Each data set was randomized ten times in order to reduce possible sensitivities in the test set performances to the choice of training and test set. For each randomization, the algorithm of Bayesian inference for LS-SVM classifiers from the previous Section was applied to estimate optimal hyperparameters, together with the application of Bayesian decision theory. As shown in the Table this results in good test set performance on all tested data sets. The differences with methods of Gaussian processes for classification and standard SVM methods are statistically not significant.

4.2 Bayesian inference for LS-SVM regression

So far we have outlined a Bayesian framework with three levels of inference for LS-SVM classifiers. An essential difference with the Gaussian processes approach to this problem is that in the LS-SVM classifier case one rather takes a network perspective with explicit and very clear primal-dual interpretations, linking the work in this way to SVM methodology [139]. The primal unknown parameters, hyperparameters and kernel parameters are inferred at different and increasing levels of inference which is usually not the case in Gaussian processes where the kernel parameters are optimized at Level 2. The bias term is also included there within the kernel and rather considered as an additional tuning parameter instead of a parameter to be inferred at Level 1. Another difference is that in the Bayesian LS-SVM classifier we take in fact a regression approach with targets $+1$ and -1, while in Gaussian processes one relies on sampling techniques due to the fact that no regression interpretation is made for the classification problem.

As a result, this means that the Bayesian inference LS-SVM approach to nonlinear function estimation is very similar to the classification approach. Furthermore also for regularization networks a Bayesian interpretation has been made [73]. However, inference with 3 levels of inference in a similar style as Bayesian learning for classical MLP networks has not been investigated for this approach. This basically forms the goal of the Bayesian evidence framework for LS-SVM regressors and has been studied by Van Gestel *et al.* in [265]. This method has also been successfully applied to problems of financial time series predictions where both a time series model and a volatility model have been estimated by LS-SVMs [265].

4.2.1 *Probabilistic interpretation of LS-SVM regressors (Level 1): predictive mean and error bars*

Calculation of maximum posterior

The considered model in primal weight space is

$$y(x) = w^T \varphi(x) + b \qquad (4.56)$$

with $w \in \mathbb{R}^{n_h}$, $b \in \mathbb{R}$ and $\varphi(\cdot)$ the mapping to the high dimensional feature space. The given training data set is denoted as $\mathcal{D} = \{x_k, y_k\}_{k=1}^N$.

We consider here the more general problem of non-constant variance of the noise (hetero-skedastic noise), meaning that we take hyperparameters $\zeta_{1\_N} = [\zeta_1; ...; \zeta_N]$ instead of one single value ζ. In the primal weight space one solves the optimization problem

$$\left[\boxed{\text{P}} : \quad \min_{w,b,e} J_P(w,e) = \mu E_W + \sum_{k=1}^N \zeta_k E_{D,k} \right.$$
$$\left. \text{such that} \quad e_k = y_k - [w^T \varphi(x_k) + b], \ k = 1, ..., N \right] \qquad (4.57)$$

with

$$\begin{aligned} E_W &= \frac{1}{2} w^T w \\ E_{D,k} &= \frac{1}{2} e_k^2 = \frac{1}{2} \left(y_k - [w^T \varphi(x_k) + b] \right)^2. \end{aligned} \qquad (4.58)$$

This gives the Lagrangian $\mathcal{L}(w, b, e; \alpha) = J_P(w, e) - \sum_{k=1}^N \alpha_k(y_k - [w^T$

$\varphi(x_k) + b] - e_k)$. From the conditions for optimality we find

$$
\begin{cases}
\frac{\partial \mathcal{L}}{\partial w} = 0 & \rightarrow \quad w = \frac{1}{\mu} \sum_{k=1}^{N} \alpha_k \varphi(x_k) \\[2ex]
\frac{\partial \mathcal{L}}{\partial b} = 0 & \rightarrow \quad \sum_{k=1}^{N} \alpha_k y_k = 0 \\[2ex]
\frac{\partial \mathcal{L}}{\partial e_k} = 0 & \rightarrow \quad \alpha_k = \zeta_k e_k, \qquad\qquad\qquad k = 1, ..., N \\[2ex]
\frac{\partial \mathcal{L}}{\partial \alpha_k} = 0 & \rightarrow \quad w^T \varphi(x_k) + b - y_k + e_k = 0, \qquad k = 1, ..., N
\end{cases}
\tag{4.59}
$$

which gives the following dual problem

$$
\left[\boxed{D} : \quad \text{solve in } \alpha, b : \right.
$$
$$
\left.
\begin{bmatrix} 0 & 1_v^T \\ \hline 1_v^T & \frac{1}{\mu}\Omega + D_\zeta^{-1} \end{bmatrix}
\begin{bmatrix} b \\ \alpha \end{bmatrix}
=
\begin{bmatrix} 0 \\ y \end{bmatrix}
\right]
\tag{4.60}
$$

with $D_\zeta = \text{diag}([\zeta_1; ...; \zeta_N])$, $y = [y_1; ...; y_N]$, $1_v = [1; ...; 1]$ and $\Omega_{kl} = K(x_k, x_l) = \varphi(x_k)^T \varphi(x_l)$ for $k, l = 1, ..., N$. In the dual space the model becomes

$$
y(x) = \frac{1}{\mu} \sum_{k=1}^{N} \alpha_k K(x, x_k) + b
\tag{4.61}
$$

with support values α_k and bias term b the solution to (4.60).

The Level 1 posterior equals

$$
p(w, b | \mathcal{D}, \mu, \zeta_{1\_N}, \mathcal{H}_\sigma) = \frac{p(\mathcal{D} | w, b, \mu, \zeta_{1\_N}, \mathcal{H}_\sigma)}{p(\mathcal{D} | \mu, \zeta_{1\_N}, \mathcal{H}_\sigma)} p(w, b | \mu, \zeta_{1\_N}, \mathcal{H}_\sigma).
\tag{4.62}
$$

We take the same assumptions for the prior as in the classifier case. On the other hand the likelihood becomes now

$$
p(\mathcal{D} | w, b, \mu, \zeta_{1\_N}, \mathcal{H}_\sigma) = \prod_{k=1}^{N} \sqrt{\frac{1}{2\pi \zeta_k^{-1}}} \exp\left(-\frac{1}{2} \frac{e_k^2}{\zeta_k^{-1}} \right).
\tag{4.63}
$$

Hence $1/\zeta_k$ is to be considered as the variance of the noise e_k. After straightforward calculation, the posterior becomes

$$
p(w - w_{\text{MP}}, b - b_{\text{MP}} | \mathcal{D}, \mu, \zeta_{1\_N}, \mathcal{H}_\sigma) = \frac{1}{\sqrt{(2\pi)^{(n_h+1)} \det Q}} \exp\left(-\frac{1}{2} g^T Q^{-1} g \right)
\tag{4.64}
$$

with $g = [w - w_{MP}; b - b_{MP}]$ and $Q = \text{Cov}([w; b], [w; b])$ with Hessian $H = Q^{-1}$. This covariance matrix is related to the Hessian of the (quadratic) cost function as follows

$$Q = H^{-1} = \begin{bmatrix} (\mu I + \zeta G)^{-1} & -(\mu I + \zeta G)^{-1} H_{12} H_{22}^{-1} \\ -H_{22}^{-1} H_{12}^T (\mu I + \zeta G)^{-1} & H_{22}^{-1} + H_{22}^{-1} H_{12}^T (\mu I + \zeta G)^{-1} H_{12} H_{22}^{-1} \end{bmatrix}$$

$$(4.65)$$

where

$$\begin{aligned} H_{11} &= \mu I + \Upsilon D_\zeta \Upsilon^T \\ H_{12} &= \Upsilon D_\zeta 1_v \\ H_{22} &= \sum_{k=1}^N \zeta_k =: s_\zeta \end{aligned}$$

and $\Upsilon = [\varphi(x_1), ..., \varphi(x_N)]$, $G = \Upsilon M_c \Upsilon^T$ and centering matrix $M_c = I - (1/N)1_v 1_v^T$.

Moderated output of LS-SVM regressor

The predictive mean of a point x and the error bar on this estimate is given by

$$\hat{y}(x) = w_{MP}^T \varphi(x) + b_{MP} \tag{4.66}$$

where w_{MP}, b_{MP} satisfy the conditions for optimality related to (4.57) and α, b are the solution to the linear KKT system (4.60). The variance on the prediction is

$$\sigma_{\hat{y}}^2(x) = 1/\zeta(x) + \sigma_z^2(x) \tag{4.67}$$

with

$$\begin{aligned} \sigma_z^2(x) &= \mathcal{E}[(z - z_{MP})^2] = \mathcal{E}[((w^T \varphi(x) + b) - (w_{MP}^T \varphi(x) + b_{MP}))^2] \\ &= \psi(x)^T H^{-1} \psi(x) \end{aligned}$$

$$(4.68)$$

with $\psi(x) = [\varphi(x); 1]$. The value $1/\zeta(x)$ is the variance of the noise corresponding to input x. According to [265] one obtains

$$\sigma_z^2(x) = \theta(x)^T U_G Q_D U_G^T \theta(x) + \frac{1}{\mu} K(x, x) - \frac{2}{s_\zeta} \theta(x)^T U_G Q_D U_G^T \Omega D_\zeta 1_v$$

$$+ \frac{2}{\mu s_\zeta} \theta(x)^T D_\zeta 1_v + \frac{1}{s_\zeta} + \frac{1}{s_\zeta^2} 1_v^T D_\zeta \Omega U_G Q_D U_G^T \Omega D_\zeta 1_v + \frac{1}{\mu s_\zeta^2} 1_v^T D_\zeta \Omega D_\zeta 1_v$$

$$(4.69)$$

with $Q_D = (\mu I + D_G)^{-1} - \mu^{-1} I$, $\theta(x) = [K(x, x_1); ...; K(x, x_N)]$ as in the classifier case and U_G, D_G are related to the eigenvalue decomposition

$$(D_\zeta - \frac{1}{s_\zeta} D_\zeta 1_v 1_v^T D_\zeta) \Omega \ \nu_{G,i} = \lambda_{G,i} \nu_{G,i}, \quad i = 1, ..., N_{\text{eff}} \leq N - 1 \quad (4.70)$$

and $D_G = \text{diag}([\lambda_{G,1}; ...; \lambda_{G,N_{\text{eff}}}])$, $U_G = [(\nu_{G,1} \ \Omega \ \nu_{G,1})^{1/2} \nu_{G,1} \ ; ...; (\nu_{G,N_{\text{eff}}} \ \Omega \ \nu_{G,N_{\text{eff}}})^{1/2} \nu_{G,N_{\text{eff}}}]$.

In the application of financial time series prediction this value was estimated by means of an additional LS-SVM volatility model [265]. In the case of a constant noise variance assumption (homo-skedastic noise) one assumes the constant value $1/\zeta$.

4.2.2 Inference of hyperparameters (Level 2)

One can proceed at Level 2 in a similar way as for the classifier case. The posterior in the hyperparameters is

$$p(\mu, \zeta_{1\_N}|\mathcal{D}, \mathcal{H}_\sigma) = \frac{p(\mathcal{D}|\mu, \zeta_{1\_N}, \mathcal{H}_\sigma)}{p(\mathcal{D}|\mathcal{H}_\sigma)} p(\mu, \zeta_{1\_N}|\mathcal{H}_\sigma). \quad (4.71)$$

We assume a uniform prior distribution for $\log \mu$ and $\log \zeta_{1\_N}$ and assume $p(\mu, \zeta_{1\_N}| \mathcal{H}_\sigma) = p(\mu|\mathcal{H}_\sigma)p(\zeta_{1\_N}|\mathcal{H}_\sigma)$. Hence

$$p(\mu, \zeta_{1\_N}|\mathcal{D}, \mathcal{H}_\sigma) \propto p(\mathcal{D}|\mu, \zeta_{1\_N}, \mathcal{H}_\sigma), \quad (4.72)$$

and

$$p(\mu, \zeta_{1\_N}|\mathcal{D}, \mathcal{H}_\sigma) \propto \exp\left(-J_P(w_{\text{MP}}, b_{\text{MP}})\right) \times \sqrt{\mu^{n_h} \prod_{k=1}^{N} \zeta_k \sqrt{\det H^{-1}}}$$

$$(4.73)$$

where $J_P(w, b) = J_P(w_{\text{MP}}, b_{\text{MP}}) + \frac{1}{2}([w; b] - [w_{\text{MP}}; b_{\text{MP}}])^T H([w; b] - [w_{\text{MP}}; b_{\text{MP}}])$ and $J_P(w_{\text{MP}}, b_{\text{MP}}) = \mu E_W(w_{\text{MP}}) + \zeta E_D(w_{\text{MP}}, b_{\text{MP}})$.

By taking the negative logarithm of this expression, one gets the following optimization problem in the hyperparameters:

$$\min_{\mu, \zeta_{1\_N}} J(\mu, \zeta_{1\_N}) = \mu E_W(w_{\text{MP}}) + \sum_{k=1}^{N} \zeta_k E_{D,k}(w_{\text{MP}}, b_{\text{MP}})$$

$$+ \frac{1}{2} \sum_{i=1}^{N_{\text{eff}}} \log(\mu + \lambda_{G,i}) - \frac{N_{\text{eff}}}{2} \log \mu - \frac{1}{2} \sum_{k=1}^{N} \log \zeta_k + \frac{1}{2} \log(\sum_{k=1}^{N} \zeta_k).$$

$$(4.74)$$

When having non-constant values ζ_k this becomes a heavy nonlinear optimization problem in the case of many given data points. The case of $\zeta_k = \zeta$ for all k leads to a similar procedure as in the classification problem case [265].

4.2.3 *Inference of kernel parameters and model comparison (Level 3)*

The Level 3 inference is similar to the classification case. Considering the case of a constant ζ hyperparameter value one can write

$$P(D|\mathcal{H}_\sigma) \propto P(D|\mu_{\text{MP}}, \zeta_{\text{MP}}, \mathcal{H}_\sigma) \frac{\sigma_{\mu|D}\sigma_{\zeta|D}}{\sigma_\mu \sigma_\zeta} \qquad (4.75)$$

with the same assumptions as made in the classification case. This leads to the following evidence according to which the models can be ranked

$$P(D|\mathcal{H}_\sigma) \propto \sqrt{\frac{\mu_{\text{MP}}^{N_{\text{eff}}} \zeta_{\text{MP}}^{N-1}}{(d_{\text{eff}} - 1)(N - d_{\text{eff}}) \prod_{i=1}^{N_{\text{eff}}} \mu_{\text{MP}} + \zeta_{\text{MP}} \lambda'_{G,i}}} \qquad (4.76)$$

where $M_c \Omega \nu_{G,i} = \lambda'_{G,i} \nu_{G,i}$ for $i = 1, ..., N_{\text{eff}} \leq N - 1$ and $d_{\text{eff}} = 1 + \sum_{i=1}^{N_{\text{eff}}} \zeta \lambda'_{G,i}/(\mu + \zeta \lambda'_{G,i})$.

An illustration of (4.76) on the example of a noisy sinc function is given in Fig. 4.5. It illustrates the tuning of the width of an RBF kernel by minimizing $-\log(P(D|\mathcal{H}_\sigma))$ and shows the probabilistic interpretation of the output from Level 1.

4.3 Input selection by automatic relevance determination

Input selection is an important problem in many real-life applications, both for classification and regression problems. Given measured data we would like to know automatically which inputs are more or less relevant with respect to the estimated model. As discussed in the previous Chapter for Gaussian processes, one possible way to do this is Automatic Relevance Determination (ARD). Essentially this can be understood as considering a weighted norm, which was also discussed by Poggio & Girosi in [183]. In the RBF kernel case one takes then

$$K(x, z) = \exp\left(-(x - z)^T S^{-1} (x - z)\right) \qquad (4.77)$$

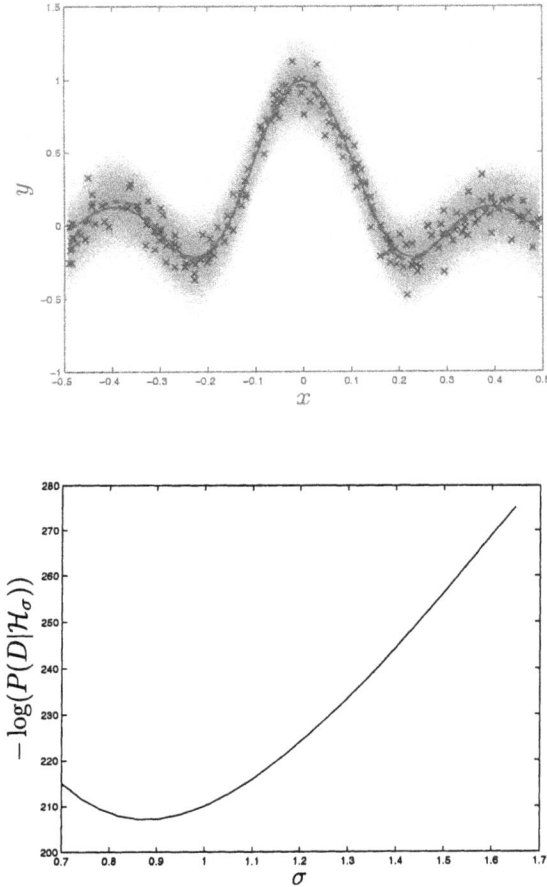

Fig. 4.5 *Bayesian inference of LS-SVM with RBF kernel on a noisy sinc function: (Top) Probabilistic interpretation of the output at Level 1: estimated output (dashed line) and error bars; (Bottom) Tuning of kernel parameter at Level 3:* $-\log(P(D|\mathcal{H}_\sigma))$ *with respect to the width* σ *of the RBF kernel.*

where $S = \mathrm{diag}([s_1^2; ...; s_n^2])$. Instead of taking a diagonal matrix S one may also take a general symmetric and positive definite matrix. However, the advantage of taking a diagonal matrix is that when inferring these kernel parameters, one can directly associate the i-th diagonal element of S to the i-th input. Taking a more general choice for S may lead to higher dimensionality reduction, but the reduced variables cannot be directly associated to the individual inputs. A similar problem occurs with PCA analysis methods. The input selection is done then by inferring the elements of S at Level 3. From the optimal values one will be able to see which inputs are more relevant than others: large values $1/s_i^2$ will indicate a highly relevant input i, while a small value indicates that this input is less relevant in comparison with the other inputs.

In general there exist several selection procedures for the inputs. The following algorithm gradually removes less relevant inputs by starting from the original input vector. One may also work in the opposite direction by gradually adding inputs, starting from a single input. Such methods are often called backward and forward input selection, respectively. The method to be preferred will typically depend on the specific application under study.

LS-SVM input selection by ARD:

- Normalize the inputs to zero mean and unit variance

- Start with equal elements $s_1 = s_2 = ... = s_n$ or $S = s^2 I$. This kernel parameter is optimized by Level 3 inference based on the evidence formula (4.76). The result of this optimization is taken as starting point for initializing a nonlinear optimization problem for the diagonal matrix S by optimizing the same evidence formula at Level 3.

- Inspect the results from Level 3 inference and remove the least relevant input. Reduce the dimensionality of the problem and set $n := n - 1$.

- Go to step 1 and remove inputs as long as one improves according to the Level 3 evidence formula (4.76).

This algorithm can be both applied to regression and classification problems. Several other methods for input selection have been investigated in the literature, e.g. for gene selection from microarray data [102] where sensitivity analysis and correlation coefficients have been applied. In fact when selecting a subset of inputs from the given set of n inputs there exist many possible combinations. This number of possible combinations grows drastically as the number of inputs increases which makes this problem hard.

One can also formulate the whole problem then as a global optimization problem and apply a feature wrapper method [297], which on the other hand is time-consuming.

Chapter 5

Robustness

In the previous Chapters basic methods and Bayesian learning for LS-SVM models were discussed. The use of least squares and equality constraints for the models results into simpler formulations but on the other hand has potential drawbacks such as the lack of sparseness but also lack of robustness. It was already illustrated that one can overcome the lack of sparseness for example by applying pruning methods. In this Chapter we will discuss ways to enhance the robustness of LS-SVM models for nonlinear function estimation by incorporating methods from robust statistics. Weighted LS-SVM versions are explained in order to cope with outliers in the data and non-Gaussian noise distributions. The use of robust statistics towards hyperparameter selection by cross-validation is a second illustration of employing such techniques to further improve and robustify the estimates. For weighted LS-SVM robust scale estimators are used while for robust cross-validation robust location estimators such as trimmed mean are applied.

5.1 Noise model assumptions and robust statistics

It is well-known in statistics that the least squares estimator has the good property of being a minimum variance estimator. However, given a model description of the form

$$y_k = f(x_k) + e_k, \quad k = 1, ..., N \qquad (5.1)$$

with training data $\{x_k, y_k\}_{k=1}^{N}$, true model f and error variables e_k, least squares estimators are the best linear unbiased estimator (b.l.u.e.) under the Gauss-Markov conditions (zero mean noise, homoscedastic (constant and finite variance) noise and uncorrelated noise). On the other hand, in many real-life situations, the data may be corrupted by additional outliers or by a non-Gaussian noise distribution. In such cases other cost functions instead of least squares are preferable. For linear parametric models estimated by least squares it is known that L_2 estimators have a low breakdown point [69]. This means that a small amount of contamination (e.g. outliers) may destroy the good quality of the least squares estimate.

In robustness studies, one therefore widely uses a Tukey contamination scheme [259] with super model

$$F_\epsilon(e) = (1 - \epsilon)F_0(e) + \epsilon H(e), \ 0 \leq \epsilon \leq 1, \tag{5.2}$$

where the noise model is a mixture of the nominal distribution F_0 and a distribution H describing the kind of contamination with respect to the nominal distribution. In other words, the ideal noise model F_0 is perturbed by H. Here $(1 - \epsilon)$ percent of the data are assumed to be generated by F_0 and ϵ percent by H.

In the context of LS-SVMs for function estimation the nominal density will be Gaussian and the following mixture model is often taken

$$p_\epsilon(e) = (1 - \epsilon)\mathcal{N}(0, \sigma_e^2) + \epsilon\mathcal{N}(0, \kappa^2\sigma_e^2) \tag{5.3}$$

where $p_\epsilon(e)$ denotes the mixture model with ϵ percent amount of contamination, $\mathcal{N}(0, \sigma_e^2)$ the normal distribution with zero mean and standard deviation σ_e and $\kappa > 1$. Another typical example is

$$p_\epsilon(e) = (1 - \epsilon)\mathcal{N}(0, \sigma_e^2) + \epsilon\mathcal{L}_a(0, \lambda) \tag{5.4}$$

where $\mathcal{L}_a$ denotes the Laplace (or double exponential) density. The heavier the tails of the contaminating density the more difficult it becomes for the estimator (assume it has been optimally designed with respect to the nominal density) to cope with this contamination.

5.2 Weighted LS-SVMs

In standard SVM methods one may in principle choose any convex cost function (Fig. 5.1). One starts by choosing the cost function and next takes an efficient algorithm for convex optimization (such as interior point methods) in order to compute the SVM solution. However, in a maximum

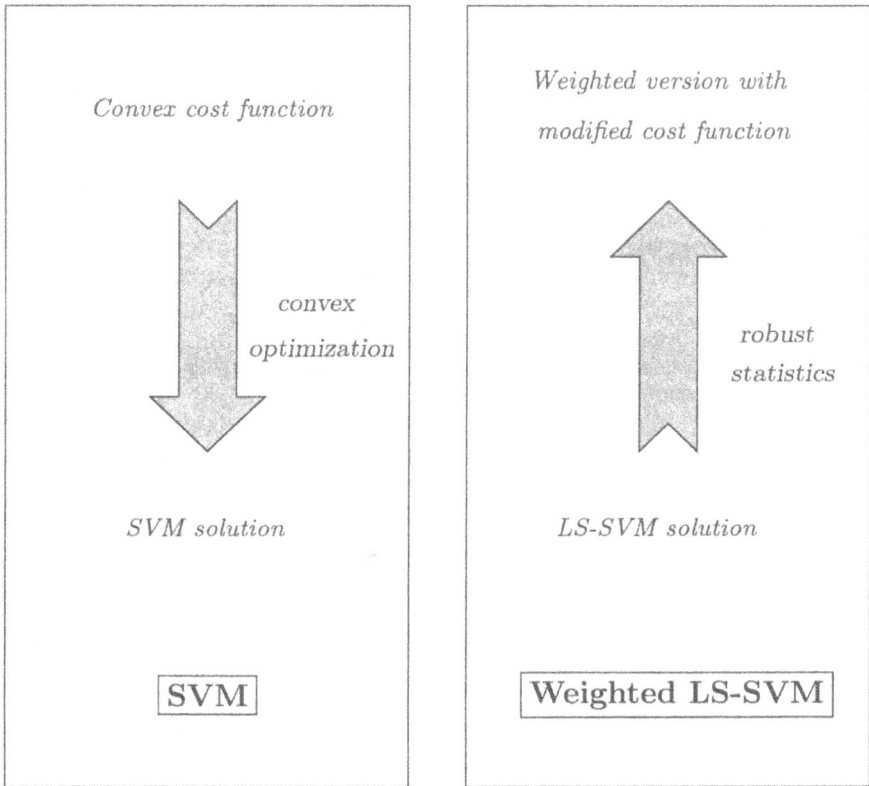

Fig. 5.1   *In standard SVMs one takes a top-down approach by choosing a convex cost function and finds the SVM solution by convex optimization. On the other hand one can also follow a bottom-up approach by starting with LS-SVM and define a weighting in terms of the error distribution based on robust statistics which implicitly corresponds to a modified cost function. In this way one aims at implicitly finding a most suitable cost function in view of robust statistics.*

likelihood sense, a certain cost function (or loss function) is optimal for a given noise model [219], such that the cost function equals

$$J(e) = -\log p(e). \tag{5.5}$$

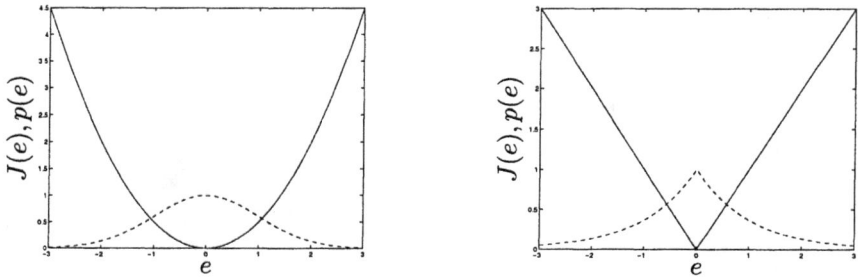

Fig. 5.2   (Left) In a maximum likelihood setting the least squares cost function (full line)
is optimal in case of a Gaussian noise distribution (dashed line); (Right) for a Laplacian
noise distribution the L_1 estimator is optimal.

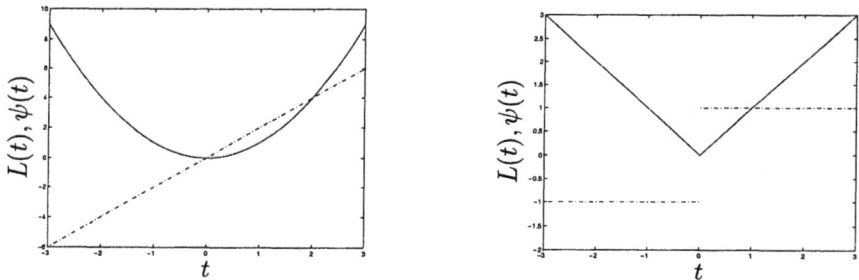

Fig. 5.3   (Left) Score function $\psi(t)$ (dashdot line) of the L_2 norm $L(t)$ (solid line);
(Right) score function of the L_1 norm. From the score function it is clear that for the
L_2 norm the influence of outliers is not reduced.

For example the Gaussian noise model (Fig. 5.2)

$$p(e) = \exp(-\frac{1}{2}e^2) \qquad (5.6)$$

corresponds to least squares with cost function

$$J(e) = \frac{1}{2}e^2 \qquad (5.7)$$

and the Laplacian noise model

$$p(e) = \exp(-|e|) \qquad (5.8)$$

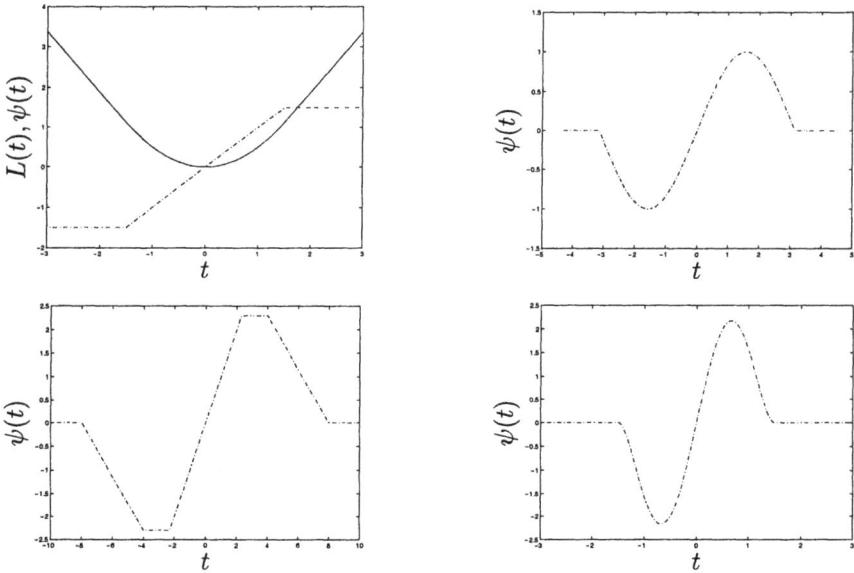

Fig. 5.4   *(Top-Left) Score function $\psi(t)$ (dashdot line) for the Huber loss function $L(t)$
(solid line); (Bottom-Left) score function for Hampel's modification to the Huber loss
function.   This M-estimator has the property of completely rejecting outliers; (Top-
Right) Andrew's score function; (Bottom-Right) Tukey's biweight loss function and the
corresponding score function. Note that the corresponding loss functions are non-convex
for Hampel's, Andrew's and Tukey's.*

to the L_1-estimator with

$$J(e) = |e|. \tag{5.9}$$

On the other hand the noise distribution is often not known beforehand and
hence one does not know the most suitable cost function beforehand neither.
In the case of a mixture model for the noise distribution the optimal cost
function in this maximum likelihood setting will deviate from least squares.
In the case of outliers on the data or non-Gaussian noise distributions with
heavy tails it becomes important then to reduce the influence of the outliers
or points of the tails on the estimate. For this purpose several cost functions
have been designed. Well-known examples are the Huber loss function,
Hampel's modification to it and Tukey's biweight loss function. Note that
Hampel's and Tukey's loss function are non-convex cost functions. Hence,

one cannot plug in this cost function in the standard SVM theory which is based on convex optimization.

When applying interior point algorithms with a given convex cost function, at each iteration step one solves a reduced Karush-Kuhn-Tucker (KKT) system which has the same form as one single LS-SVM. Hence, solving a standard SVM with any convex cost function can be interpreted as solving a sequence of LS-SVM systems with an iterative weighting in each step. Therefore, instead of taking a top-down approach one may equally well try to follow a bottom-up approach where one starts from least squares, compute an LS-SVM model in a first step and then apply one or more weighted LS-SVM steps where the weighting is done in such a way that the estimate improves in view of robust statistics.

In order to obtain a robust estimate based upon an existing LS-SVM solution, in a subsequent step, one can weight the error variables $e_k = \alpha_k/\gamma$ by weighting factors v_k. This leads to the optimization problem:

$$
\left[\boxed{\text{P}}: \quad \min_{w^\star, b^\star, e^\star} J_\text{P}(w^\star, e^\star) = \frac{1}{2} w^{\star T} w^\star + \gamma \frac{1}{2} \sum_{k=1}^{N} v_k e_k^{\star 2} \right.
$$
$$
\left. \text{such that} \qquad y_k = w^{\star T} \varphi(x_k) + b^\star + e_k^\star, \ k = 1, ..., N. \right]
$$
$$
\tag{5.10}
$$

The Lagrangian becomes

$$
\mathcal{L}(w^\star, b^\star, e^\star; \alpha^\star) = J_\text{P}(w^\star, e^\star) - \sum_{k=1}^{N} \alpha_k^\star \{ w^{\star T} \varphi(x_k) + b^\star + e_k^\star - y_k \}. \tag{5.11}
$$

The unknown variables for this weighted LS-SVM problem are denoted by the $\star$ symbol. From the conditions for optimality and elimination of $w^\star, e^\star$ one obtains the KKT system:

$$
\left[\boxed{\text{D}}: \quad \text{solve in } \alpha^\star, b^\star : \right.
$$
$$
\left. \begin{bmatrix} 0 & 1_v^T \\ \hline 1_v & \Omega + V_\gamma \end{bmatrix} \begin{bmatrix} b^\star \\ \hline \alpha^\star \end{bmatrix} = \begin{bmatrix} 0 \\ \hline y \end{bmatrix} \right] \tag{5.12}
$$

where the diagonal matrix V_γ is given by

$$
V_\gamma = \text{diag}([\frac{1}{\gamma v_1}; ...; \frac{1}{\gamma v_N}]). \tag{5.13}
$$

The choice of the weights v_k is determined based upon the error variables $e_k = \alpha_k/\gamma$ resulting from the unweighted LS-SVM case. Robust estimates are obtained then (see [56; 196]) e.g. by taking

$$
v_k = \begin{cases}
1 & \text{if} \quad |e_k/\hat{s}| \le c_1 \\
\dfrac{c_2 - |e_k/\hat{s}|}{c_2 - c_1} & \text{if} \quad c_1 \le |e_k/\hat{s}| \le c_2 \\
10^{-4} & \text{otherwise}
\end{cases}
\tag{5.14}
$$

where $\hat{s}$ is a robust scale estimator, more precisely a robust estimate of the standard deviation of the LS-SVM error variables e_k:

$$
\hat{s} = \frac{\text{IQR}}{2 \times 0.6745}.
\tag{5.15}
$$

The interquartile range IQR is the difference between the 75th percentile and 25th percentile. In the estimate $\hat{s}$ one takes into account how much the estimated error distribution deviates from a Gaussian distribution. Another robust estimate of the standard deviation is $\hat{s} = 1.483 \, \text{MAD}(x_k)$, where MAD stands for the median absolute deviation [105]. The cost function in the unweighted LS-SVM formulation is optimal under the assumption of a normal Gaussian distribution for e_k. The procedure corrects for this assumption in order to obtain a robust estimate when this distribution is not normal. The procedure can be repeated iteratively, but in practice one single additional weighted LS-SVM step will often be sufficient. One assumes that e_k has a symmetric distribution which is usually the case when (γ, σ) are well-determined by an appropriate model selection method. The constants c_1, c_2 are typically chosen as $c_1 = 2.5$ and $c_2 = 3$ [196]. This is a reasonable choice taking into account the fact that for a Gaussian distribution there will be very few residuals larger than $2.5\hat{s}$. Another possibility is to determine c_1, c_2 from a density estimation of the e_k distribution. Using these weightings one can correct for y-outliers or for a non-Gaussian instead of Gaussian error distribution.

The following simple algorithm for weighted LS-SVM was proposed in [243]:

Weighted LS-SVM algorithm:

(1) Given training data $\{x_k, y_k\}_{k=1}^{N}$, estimate an unweighted LS-SVM. Compute $e_k = \alpha_k/\gamma$ from the solution vector.

(2) Compute a robust estimate of the standard deviation $\hat{s}$ based on the empirical e_k distribution.

(3) Determine the weights v_k based upon e_k and $\hat{s}$.

(4) Solve the Weighted LS-SVM system, giving the model $y(x) = \sum_{k=1}^{N} \alpha_k^{\star} K(x, x_k) + b^{\star}$.

An important notion in robust estimation is the breakdown point of an estimator [9; 69]. Roughly speaking it is the smallest fraction of contamination of a given data set that can result in an estimate which is arbitrarily far away from the estimated parameter vector obtained from the uncontaminated data set. It is well known that the least squares estimate in linear parametric regression without regularization has a low breakdown point. At this point the unweighted LS-SVM has better properties due to the fact that least squares is only applied to finding α (which corresponds to estimating the output layer and not the tuning parameters (γ, σ) in the case of an RBF kernel). Although unweighted LS-SVM is globally already quite robust for an RBF kernel (but locally not), the breakdown point is further improved by applying the weighted LS-SVM step afterwards.

In Figs. 5.5-5.6 the weighted LS-SVM improvements are illustrated for a synthetic data set of a noisy sinc function with outliers on the data and in Figs. 5.7-5.8 for a non-Gaussian noise distribution, where in both cases a comparison is made between the unweighted and the weighted LS-SVM. In Figs. 5.9-5.10 the weighted LS-SVM is applied to real-life data sets of the motorcycle data and the Boston housing data, respectively, with improved results.

5.3   Robust cross-validation

In [59; 60] robust statistics has recently been applied in relation to cross-validation methods. In weighted LS-SVM the error variable e is the studied random variable. Towards cross-validation methods one may consider the value of the cost function as a random variable, so in the case of L_2 cross-validation consider e^2 as a random variable to which techniques of robust location estimation are applied then.

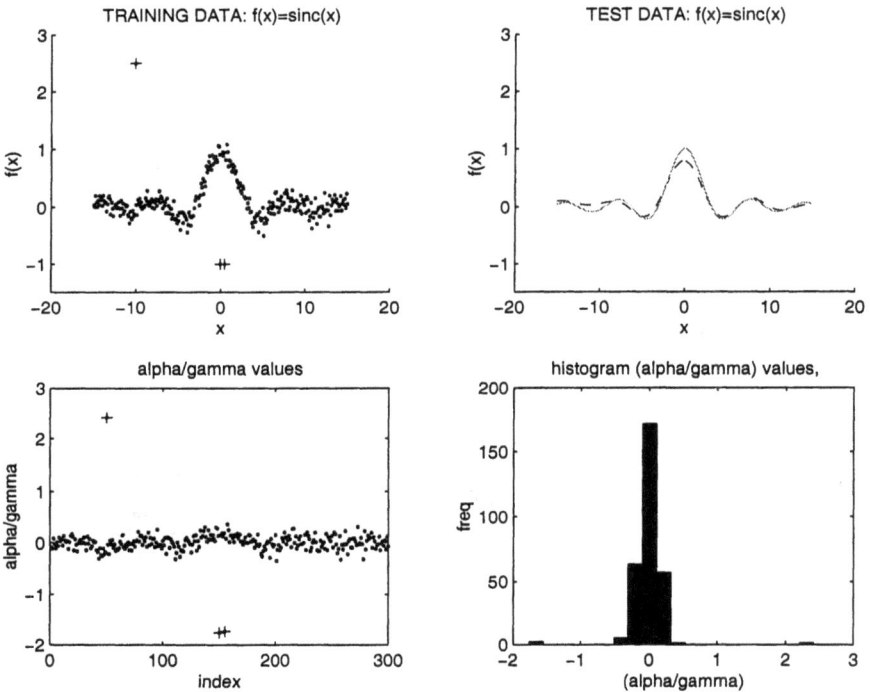

Fig. 5.5 *Estimation of a sinc function by LS-SVM with RBF kernel, given 300 training data points, corrupted by zero mean Gaussian noise and 3 outliers (denoted by '+'). (Top-Left) Training data set; (Top-Right) resulting LS-SVM model evaluated on an independent test set: (solid line) true function, (dashed line) LS-SVM estimate; (Bottom-Left) $e_k = \alpha_k/\gamma$ values; (Bottom-Right) histogram for distribution of e_k values, which is non-Gaussian due to the 3 outliers.*

The expected value of a random variable is its average value and can be viewed as an indication of the central value of the density function. Therefore it is often referred to as a location parameter. Given a sample of data, usually the mean is used then to estimate this value. However, also other location estimators exist such as the median, the β-trimmed mean and the β-Winsorized mean. These are more robust for location parameter estimation than using the mean. In general two important classes of location estimators are M-estimators and L-estimators.

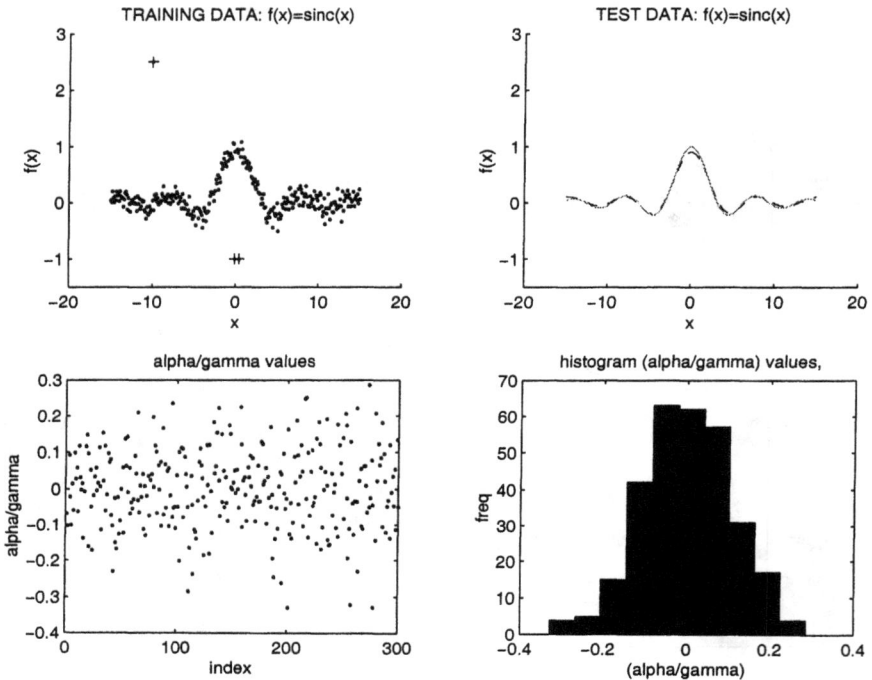

Fig. 5.6   *(Continued) Weighted LS-SVM applied to the results of the previous figure. The e_k distribution becomes Gaussian and the generalization performance on the test data improves.*

5.3.1   *M-Estimators*

Let $x_1, ..., x_N$ be a random sample from a distribution F with density $p(x - \xi)$, where ξ is the location parameter. Assume that F is a symmetric unimodal distribution, then ξ is the center of symmetry to be estimated. The M-estimator $\hat{\xi}_N$ of the location parameter is defined then as a solution to the minimization problem

$$\hat{\xi}_N = \arg \min_{\xi} \sum_{k=1}^{N} L(x_k - \xi), \tag{5.16}$$

where $L(t)$ is an even non-negative loss function (often called the contrast function [180]). $L(x_k - \xi)$ is the measure of discrepancy between the ob-

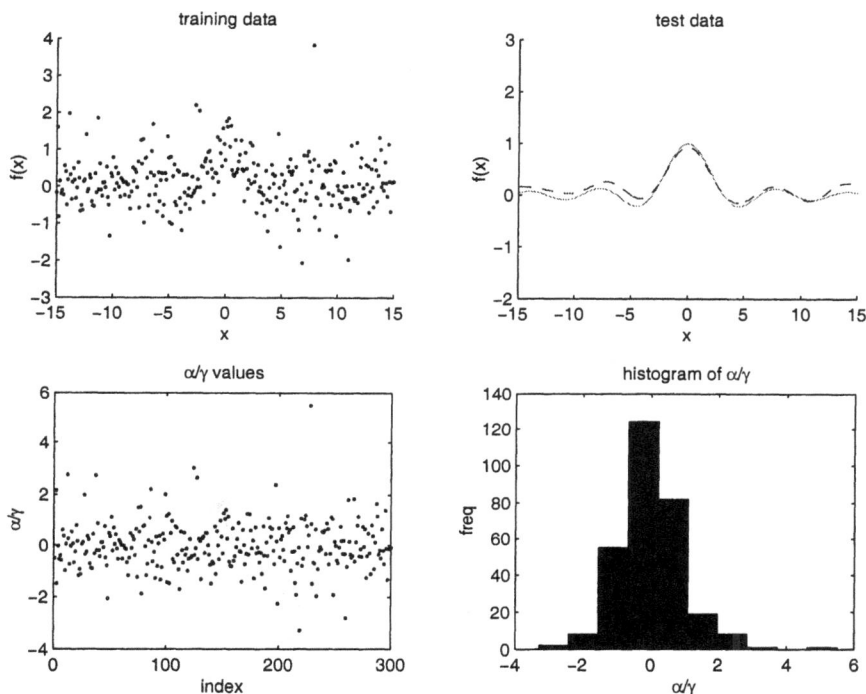

Fig. 5.7 *Estimation of a sinc function by LS-SVM with RBF kernel, given 300 training data points, corrupted by a central t-distribution with heavy tails. (Top-Left) Training data set; (Top-Right) resulting LS-SVM model evaluated on an independent test set: (solid line) true function, (dashed line) LS-SVM estimate; (Bottom-Left) $e_k = \alpha_k/\gamma$ values; (Bottom-Right) histogram for distribution of e_k values, which is non-Gaussian.*

servation x_k and the estimated center. M-estimators are formulated in terms of the so-called score function which is the derivative of the contrast function denoted here as $\psi(t) = dL(t)/dt$. The M-estimator is defined as a solution to the following implicit equation

$$\sum_{k=1}^{N} \psi\left(x_k - \hat{\xi}_N\right) = 0 \qquad (5.17)$$

which is usually calculated by iteratively re-weighted least squares.

Well-known examples of location parameter M-estimators are:

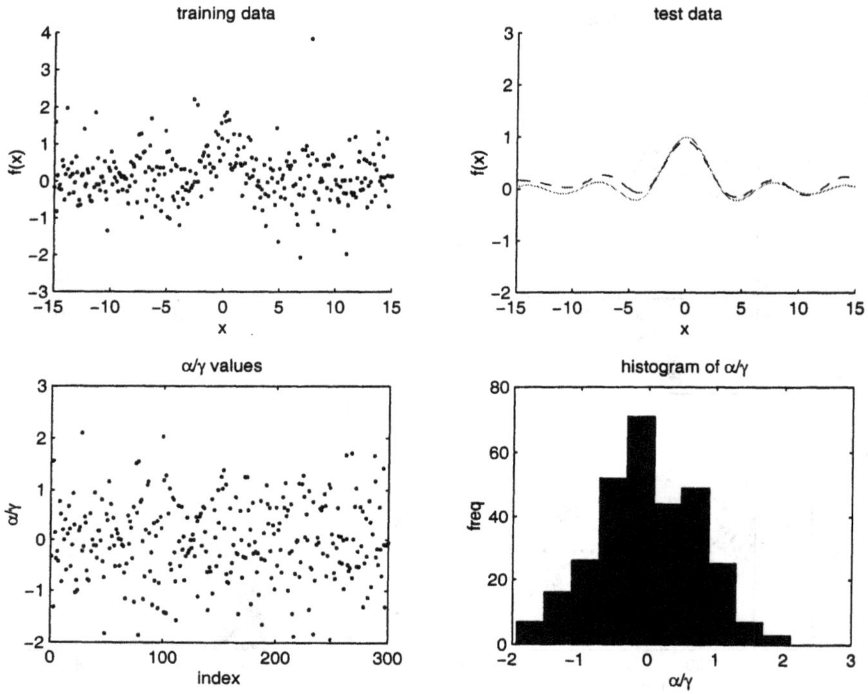

Fig. 5.8 *(Continued) Weighted LS-SVM applied to the results of the previous figure. The e_k distribution becomes Gaussian and the generalization performance on the test data improves.*

(1) For $L(t) = t^2$, one obtains the least squares solution by minimization of $\sum_{k=1}^{N} \|x_k - \xi\|_2^2$. The corresponding score function is $\psi(t) = 2t$. For this score function ψ, the M-estimate is the sample mean.

(2) For $L(t) = |t|$, one obtains the least absolute values by minimization of $\sum_{k=1}^{N} |x_k - \xi|$. The corresponding score function is

$$\psi(t) = \begin{cases} -1, & t < 0 \\ 0, & t = 0 \\ 1, & t > 0. \end{cases} \tag{5.18}$$

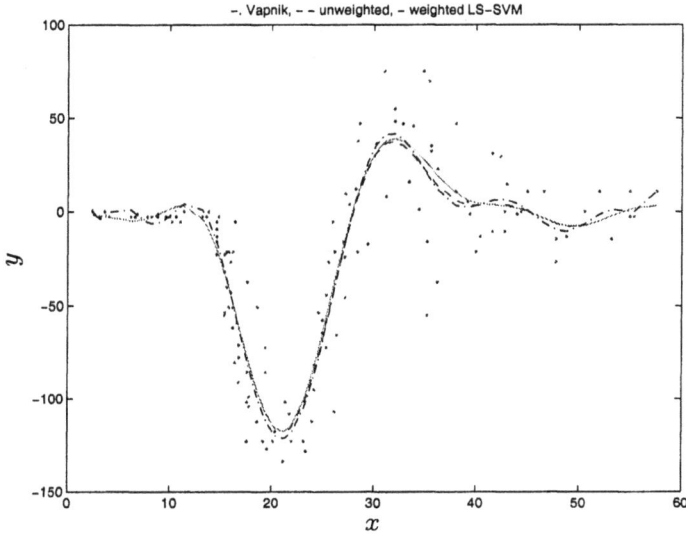

Fig. 5.9 *Motorcycle dataset: comparison between standard SVM, unweighted LS-SVM and weighted LS-SVM. The standard SVM suffers more from boundary effects.*

The corresponding M-estimate is the sample median.

(3) Huber [119] considered minimization of $\sum_{k=1}^{N} L(x_k - \xi)$ with

$$L(t) = \left\{ \begin{array}{ll} \frac{1}{2}t^2, & |t| \leq c \\ c|t| - \frac{1}{2}c^2, & |t| > c. \end{array} \right. \tag{5.19}$$

The corresponding score function is

$$\psi(t) = \left\{ \begin{array}{ll} -c, & t < -c \\ t, & |t| \leq c \\ c, & t > c. \end{array} \right. \tag{5.20}$$

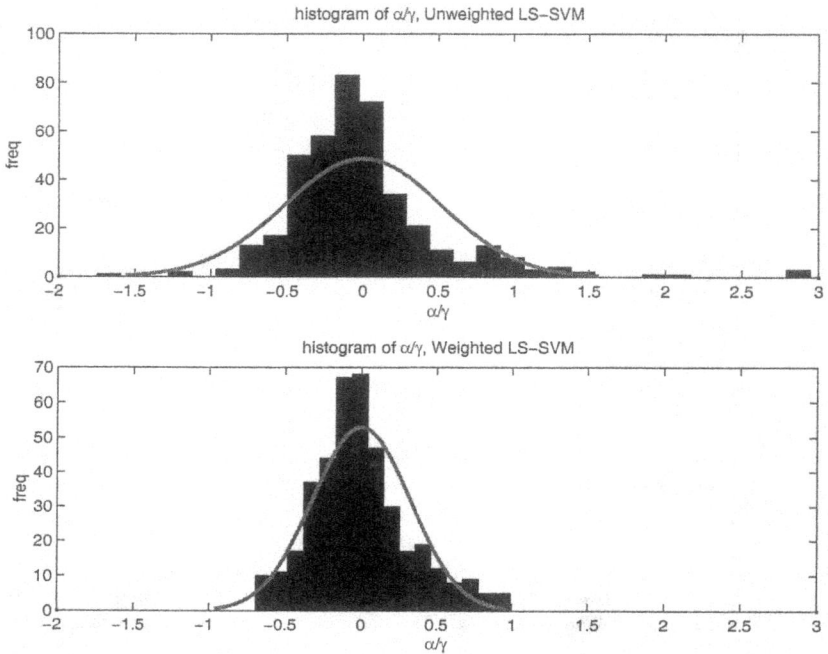

Fig. 5.10   *Illustration of weighted LS-SVM to the Boston housing data: (Top) empirical error distribution from unweighted LS-SVM; (Bottom) empirical error distribution after weighted LS-SVM.*

(4) Hampel [105] suggested a modification to the Huber estimator:

$$\psi(t) = \begin{cases} t, & 0 \le |t| \le a \\ a \, \text{sign}\,(t)\,, & a \le |t| \le b \\ a \left(\frac{c-|t|}{c-b} \right) \text{sign}\,(t)\,, & b \le |t| \le c \\ 0, & |t| > c, \end{cases} \tag{5.21}$$

making $\psi(t)$ zero for $|t|$ sufficiently large. This M-estimator has the property of completely rejecting outliers.

(5) Andrews proposed

$$\psi(t) = \sin(t)\delta_{[-\pi,\pi]}(t), \tag{5.22}$$

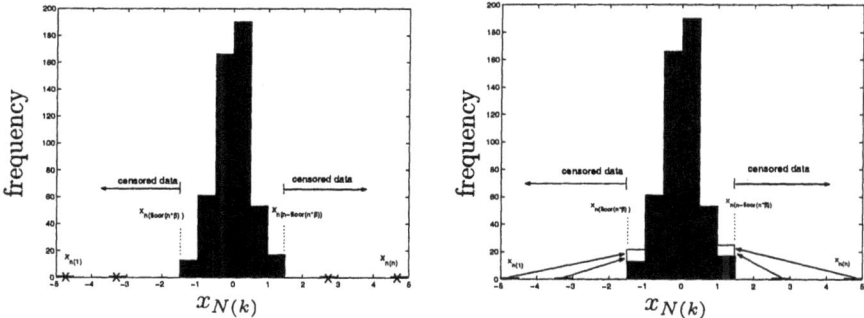

Fig. 5.11   *Schematic representation on a symmetric distribution of (Left) the β-trimmed mean and (Right) the β-Winsorized mean. In the trimmed mean the outliers are removed, while in the Winsorized mean the influence is reduced.*

where

$$\delta_{[-\pi,\pi]} = \begin{cases} 1 & \text{if } t \in [-\pi, \pi] \\ 0 & \text{otherwise.} \end{cases} \tag{5.23}$$

(6) Tukey proposed a smoother score function, the biweight

$$\psi(t) = t\left(a^2 - t^2\right)^2 \delta_{[-a,a]}(t). \tag{5.24}$$

These contrast functions and the corresponding score functions are illustrated in Figs. 5.3-5.4.

5.3.2   *L-Estimators*

L-estimators were originally proposed by Daniel [55] around 1920 and forgotten for many years, with a revival in robustness studies. For a random sample $x_1, ..., x_N$ of a distribution F, the ordered sample values denoted by $x_{N(1)} \leq ... \leq x_{N(N)}$ are called the order statistics.

In general, a linear combination of order statistics (or a linear combination of transformed order statistics), gives an *L*-statistic of the form

$$\hat{\xi}_N = \sum_{k=1}^{N} c_{N(k)} a\left(x_{N(k)}\right) \tag{5.25}$$

for a choice of constants $c_{N(1)}, ..., c_{N(N)}$ with $\sum_{k=1}^{N} c_{N(k)} = 1$ and $a(\cdot)$ a fixed function. The simplest example of an *L*-statistic is the sample mean.

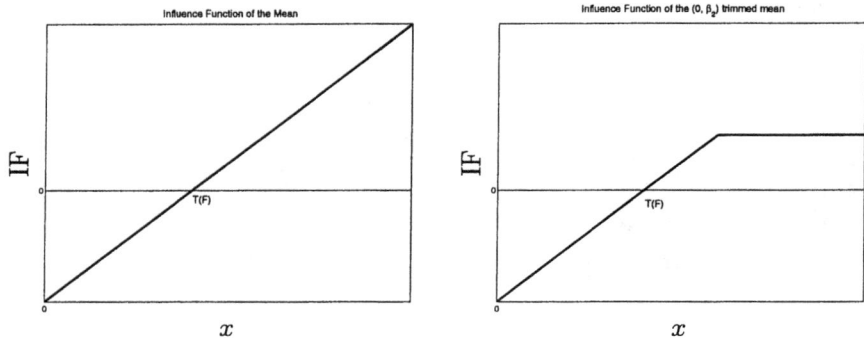

Fig. 5.12   *Influence Function IF of the mean and the* $(0, \beta_2)$ *trimmed mean. (Left) the IF of the mean is unbounded. This means that an added observation at a large distance from* $T(F_N)$ *gives a large value in absolute sense for the IF; (Right) The IF of the* $(0, \beta_2)$ *trimmed mean is both continuous and bounded, which is a desirable property to achieve robustness.*

An interesting compromise between the mean and the median (corresponding to making a trade-off between robustness and asymptotic efficiency) is obtained by the β-trimmed mean, defined as

$$\hat{\mu}_{(\beta)} = \frac{1}{N - 2g} \sum_{k=g+1}^{N-g} x_{N(k)}, \tag{5.26}$$

where the trimming proportion β is selected so that $g = \lfloor N\beta \rfloor$ and $a(x_{N(k)}) = x_{N(k)}$. The β-trimmed mean is a linear combination of the order statistics given zero weight to a number of g extreme observations at each end. It gives equal weight $1/(N-2g)$ to the number of $(N-2g)$ central observations.

When the distribution is no longer symmetric, it may be preferable to trim asymmetrically if the tail is expected to be heavier in one direction than the other. If the trimming proportions are β_1 at the left and β_2 at the right, the (β_1, β_2)-trimmed mean is defined as

$$\hat{\mu}_{(\beta_1, \beta_2)} = \frac{1}{N - (g_1 + g_2)} \sum_{k=g_1+1}^{N-g_2} x_{N(k)}, \tag{5.27}$$

where β_1 and β_2 are selected so that $g_1 = \lfloor N\beta_1 \rfloor$ and $g_2 = \lfloor N\beta_2 \rfloor$. The (β_1, β_2)-trimmed mean is a linear combination of the order statistics giving zero weight to g_1 and g_2 extreme observations at each end and equal weight $1/(N - g_1 - g_2)$ to the $(N - g_1 - g_2)$ central observations.

Another L-estimator is the β-Winsorized mean. In the symmetric case it is defined as

$$\hat{\mu}_{W(\beta)} = \frac{1}{N}(gx_{N(g+1)} + \sum_{k=g+1}^{N-g} x_{N(k)} + gx_{N(N-g)}), \qquad (5.28)$$

where $0 < \beta < 0.5$. While the β-trimmed mean censors the smallest and largest $g = \lfloor N\beta \rfloor$ observations, the β-Winsorized means replaces each of them by the values of the smallest and the largest uncensored values. This is illustrated in Fig. 5.11.

5.3.3 *Efficiency-robustness trade-off*

For the Huber score function $c = 1.5$ is typically a good choice and $a = 1.7, b = 3.4, c = 8.5$ for Hampel's score function [9]. For asymmetric distributions on the other hand, the computation of the Huber type M-estimators requires rather complex iterative algorithms and its convergence cannot be guaranteed in some important cases [122; 156]. Consider for example the case where the noise e has a normal distribution. The distribution of e^2 in relation to a cross-validation cost function will be no longer symmetric then. Therefore, it is more convenient to apply the (β_1, β_2)-trimmed mean which is applicable to asymmetric distributions.

In order to get a deeper understanding into robustness properties of location parameters, one defines the notion of influence function (IF) which aims at giving a formal characterization of the bias caused by an outlier. The breakdown point of the location estimator is a measure of robustness in how much contaminated data a location estimator can tolerate before it becomes useless, in relation to the supermodel (5.2).

Let F be a fixed distribution and $T(F)$ a statistical functional defined on a set $\mathcal{F}$ of distributions satisfying some regularity conditions [105; 119]. Statistics which are representable as functionals $T(F_N)$ of the sample distribution F_N are called statistical functions. For the variance σ^2 the relevant functional is $T(F) = \int (x - \int xdF(x))^2 dF(x)$. The estimator $T_N = T(F_N)$ of $T(F)$ denotes the functional of the sample distribution F_N.

The influence function $\text{IF}(x; T, F)$ is defined then as

$$\text{IF}\,(x; T, F) = \lim_{\epsilon \downarrow 0} \frac{T\left[(1 - \epsilon)\,F + \epsilon \delta_x\right] - T\,(F)}{\epsilon} \qquad (5.29)$$

and the breakdown point ϵ^* of the estimator $T_N = T(F_N)$ for the functional $T(F)$ at F is defined by

$$\epsilon^*(T, F) = \sup \left\{ \epsilon : \sup_{F:F=(1-\epsilon)F_0+\varepsilon H} |T(F) - T(F_0)| < \infty \right\}. \qquad (5.30)$$

Here δ_x denotes the pointmass 1 at x. The IF reflects the bias caused by adding a few outliers at the point x, normalized by the amount ϵ of contamination. This kind of differentiation of statistical functionals is a differentiation in the sense of von Mises [78]. From the influence function several robustness measures can be defined such as the gross error sensitivity, the local shift sensitivity and the rejection point [105].

Examples of influence functions are:

(1) IF *of sample mean:*
    For the mean one has $T\,(F) = \int x dF\,(x)$ and

$$\text{IF}\,(x; T, F) \quad = \quad x - T(F_N). \qquad (5.31)$$

As shown in Fig. 5.12 this IF is unbounded in $\mathbb{R}$.

(2) IF *of the sample* (β_1, β_2)-*trimmed mean:*
    In this case the IF can be expressed in terms of the quantile function of a cumulative distribution function F which is the generalized inverse $F^- : (0, 1) \to \mathbb{R}$ (if F is continuous and strictly increasing F^- corresponds to F^{-1}) given by

$$F^-(q) = \inf\,\{x : F(x) \geq q\}. \qquad (5.32)$$

In the absence of information concerning the underlying distribution function F of the sample, the empirical distribution function F_N and the empirical quantile function F_N^- are reasonable estimates for F and F^- respectively. The empirical quantile function is related to the order statistics $x_{N(1)} \leq \dots \leq x_{N(N)}$ of the sample through

$$F^-(q) = x_{N(k)}, \quad \text{for } q \in \left(\frac{k-1}{N}, \frac{k}{N}\right). \qquad (5.33)$$

Using the influence function of the qth quantile functional $F^-(q)$, one obtains as influence function $\text{IF}(x; \mu_{(0,\beta_2)}, F)$ for the $(0,\beta_2)$-trimmed mean

$$\text{IF}(x; \mu_{(0,\beta_2)}, F) = \begin{cases} \frac{x - \beta_2 F^-(1-\beta_2)}{1-\beta_2} - \mu_{(0,\beta_2)}, & 0 \leq x \leq F^-(1-\beta_2) \\ F^-(1-\beta_2) - \mu_{(0,\beta_2)}, & F^-(1-\beta_2) < x. \end{cases}$$

(5.34)

The IF of the $(0,\beta_2)$-trimmed mean is shown in Fig. 5.12. Note that this influence function becomes bounded now, which is an essential property to achieve robustness.

Given two estimates $T_1(F_N)$ and $T_2(F_N)$ of a location parameter $T(F)$, the best choice of the two estimates is then the one for which the sampling distribution is highly concentrated around the true parameter value. The efficiency of $T_2(F_N)$ relative to $T_1(F_N)$ is defined to be

$$\text{Eff}(T_2(F_N), T_1(F_N)) = \frac{\text{Var}(T_1(F_N)) + (\mathcal{E}(T_1(F_N)) - T(F))^2}{\text{Var}(T_2(F_N)) + (\mathcal{E}(T_2(F_N)) - T(F))^2}$$

(5.35)

or in the case of two unbiased estimators

$$\text{Eff}(T_2(F_N), T_1(F_N)) = \frac{\text{Var}(T_1(F_N))}{\text{Var}(T_2(F_N))}.$$

(5.36)

Figure 5.13 illustrates the good asymptotic efficiency-robustness trade-off of the trimmed mean. While estimation by the mean is efficient (when the true noise is Gaussian) but non-robust in general and the median is robust in general but non-efficient when the true noise is non-Laplacian, the trimmed mean aims at combining the best properties of both and achieves a good efficiency-robustness trade-off.

5.3.4 *A robust and efficient cross-validation score function*

A commonly used method for hyperparameter selection is cross-validation such as the cross-validation criterion introduced by Stone [228] and the generalized cross-validation criterion of Craven & Wahba [49]. Results on L_2, L_1 cross-validation statistical properties have become available [300]. However, the condition $\mathcal{E}\left[e_k^2\right] < \infty$ (respectively, $\mathcal{E}\left[|e_k|\right] < \infty$) is necessary for establishing weak and strong consistency for L_2 (respectively, L_1) cross-validated estimators. On the other hand, when there are outliers in the y

observations (or if the distribution of the random errors has a heavy tail so that $\mathcal{E}\left[|e_k|\right] = \infty$), then it becomes difficult to obtain good asymptotic results for the L_2 (L_1) cross-validation criterion. This motivates the use of a robust cross-validation score function with incorporation of ideas from robust statistics based on the trimmed mean instead of cross-validation based on the mean.

Let $\{z_k = (x_k, y_k)\}_{k=1}^{N}$ denote an independently identically distributed (i.i.d.) random sample from a population with distribution function $F(z)$ and $F_N(z)$ the empirical estimate of $F(z)$. We estimate then quantities of the form

$$T_N = \int L(z, F_N(z)) \, dF(z), \tag{5.37}$$

with $L(\cdot)$ the loss function and $\mathcal{E}[T_N]$ can be estimated by cross-validation. One splits the data randomly into V disjoint sets of nearly equal size with the size of the v-th group equal to m_v and assuming that $\lfloor N/V \rfloor \leq m_v \leq \lfloor N/V \rfloor + 1$ for all v. A general form of the V-fold cross-validated estimate of T_N is given by

$$\text{CV}_{V-\text{fold}}\left(\theta\right) = \sum_{v=1}^{V} \frac{m_v}{N} \int L\left(z, F_{(N-m_v)}(z)\right) dF_{m_v}(z), \tag{5.38}$$

where $F_{(N-m_v)}(z)$ is the empirical estimate of $F(z)$ based on $(N - m_v)$ observations outside group v and $F_{m_v}(z)$ is the empirical estimate of $F(z)$ based on m_v observations within group v.

When using the trimmed mean as a location parameter one obtains then

$$\text{CV}_{V-\text{fold}}^{\text{Robust}}\left(\theta\right) = \sum_{v=1}^{V} \frac{m_v}{N} \int_0^{F^-(1-\beta_2)} L\left(z, F_{(N-m_v)}(z)\right) dF_{m_v}(z). \tag{5.39}$$

This procedure can be repeated a number of times by permuting and splitting the data repeatedly, e.g. a number of r times, into V groups. Each time, the robust V-fold cross-validation score function is calculated. The result is then averaged over the r estimates:

$$\text{CV}_{V-\text{fold}}^{\text{Repeated Robust}} = \frac{1}{r} \sum_{j=1}^{r} \text{CV}_{V-\text{fold},j}^{\text{Robust}}. \tag{5.40}$$

This procedure reduces the variance of the score function. The method is illustrated on a number of examples on a synthetic noisy sinc function where the nominal noise distribution is contaminated by outliers and distributions with heavy tails Figs. 5.14-5.15. In these cases the use of a robust cross-validation score function outperforms the classical L_2 and L_1 cross-validation procedures in terms of the efficiency-robustness trade-off. For the cases where L_2 and L_1 estimators are optimal the method based on trimmed mean remains efficient and robust. In Fig. 5.16 boxplots are shown that visualize the performance of the use of the robust cross-validation score function.

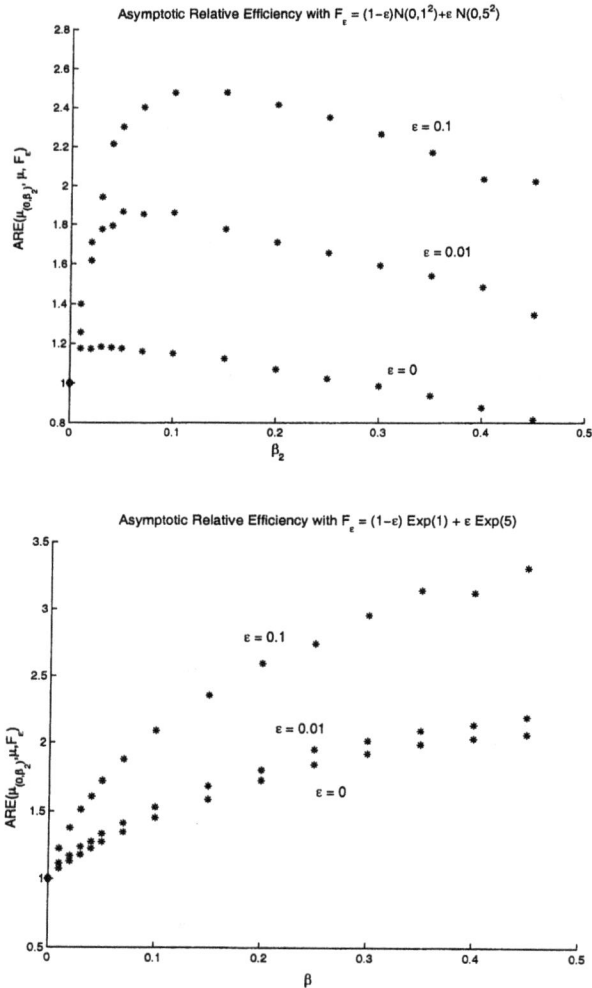

Fig. 5.13   The behaviour of the asymptotic relative efficiency of the trimmed mean
with respect to the mean as a function of the contamination ϵ, studied for a sample
of $N = 500$ by a Monte Carlo technique; (Top) efficiency for the case that F is a
symmetric distribution and several values of β_2. This plot shows that a moderate amount
of trimming can provide much better protection against contamination with heavy tails
in comparison with the mean as a location estimator; (b) efficiency for the case of an
asymmetric distribution F.

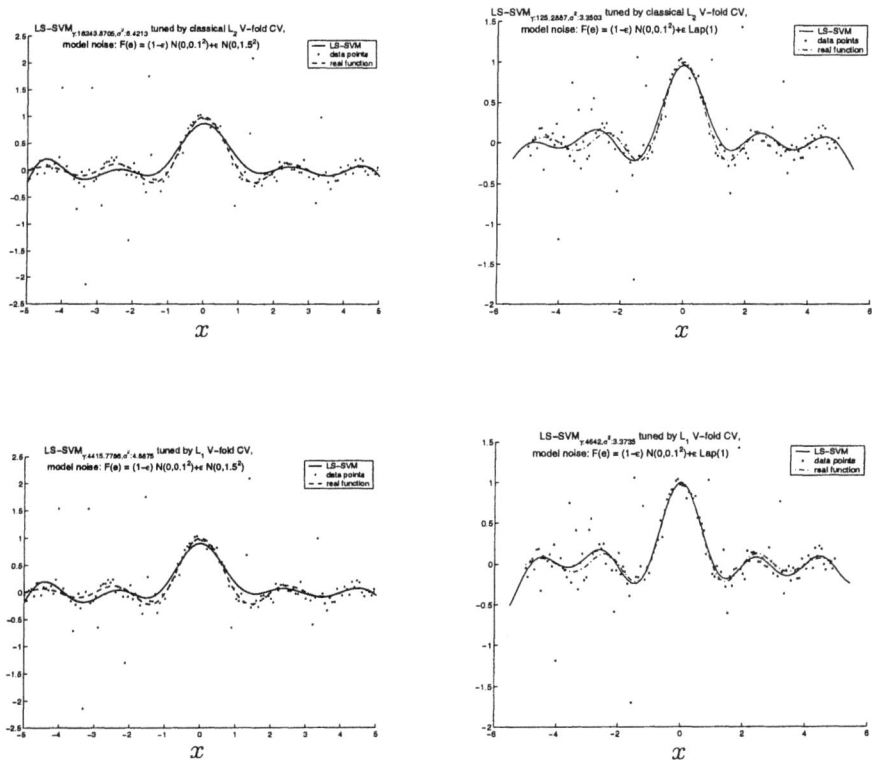

Fig. 5.14 *(Left) Comparison of L_2 V-fold CV (Left-Top), L_1 V-fold CV (Left-Bottom) and repeated robust CV (Next Figure-Left) for a noise model $p_\epsilon(x) = (1 - \epsilon)\mathcal{N}(0, \sigma^2) + \epsilon \mathcal{N}(0, \kappa^2 \sigma^2)$ with $\epsilon = 0.15$, $\sigma^2 = 0.1^2$ and $\kappa^2 = 15$, where LS-SVMs with RBF kernel are tuned in each of the cases; (Right) Comparison of L_2 V-fold CV (Right-Top), L_1 V-fold CV (Reight-Bottom) and repeated robust CV (Next Figure-Right) for a noise model $p_\epsilon(x) = (1 - \epsilon)\mathcal{N}(0, \sigma^2) + \epsilon \mathcal{L}_a(0, \lambda)$ with $\epsilon = 0.15$, $\sigma^2 = 0.1^2$ and $\lambda = 1$ where LS-SVMs with RBF kernel are tuned in each of the cases.*

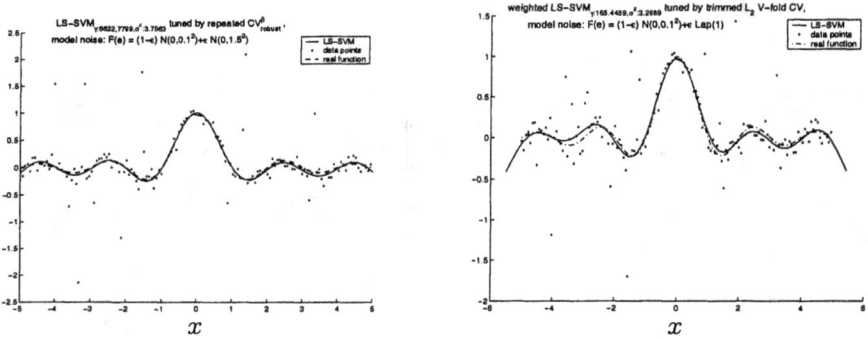

Fig. 5.15   *(Continued) Repeated robust CV which performs better than the methods shown in the previous figure.*

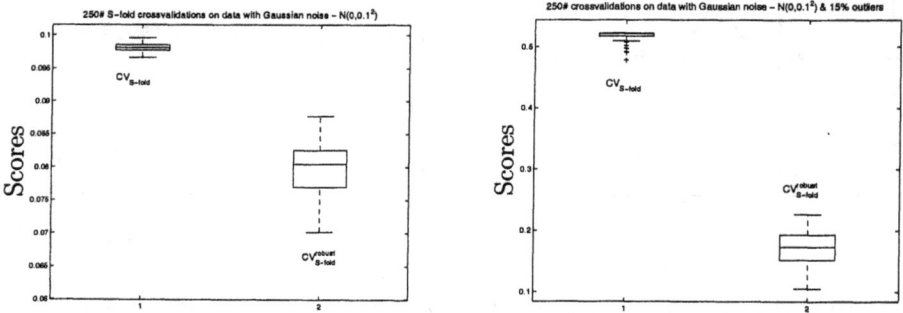

Fig. 5.16   *Boxplots that illustrate the improvements of using a robust cross-validation score function: (Left) The data are generated by a sinc function with Gaussian noise; at the left on this figure the results by L_2 cross-validation are shown and at the right the robust procedure; (Right) In this case Gaussian noise with 15% outliers has been added. Significant improvements by using the robust cross-validation score function are clearly visible now from the boxplots.*

Chapter 6

Large Scale Problems

In this Chapter we discuss a number of approaches in order to solve LS-SVM problems for function estimation and classification in the case of large data sets. We explain the Nyström method as proposed in the context of Gaussian processes and incomplete Cholesky factorization for low rank approximation. Then a new technique of fixed size LS-SVM is presented. In this fixed size LS-SVM method one solves the primal problem instead of the dual, after estimating the map to the feature space φ based upon the eigenfunctions obtained from kernel PCA, which is explained in more detail in the next Chapter. This method gives explicit links between function estimation and density estimation, exploits the primal-dual formulations, and addresses the problem of how to actively select suitable support vectors instead of taking random points as in the Nyström method. Next we explain methods that aim at constructing a suitable basis in the feature space. Furthermore, approaches for combining submodels are discussed such as committee networks and nonlinear and multilayer extensions of this approach.

6.1 Low rank approximation methods

6.1.1 *Nyström method*

Suppose one takes a linear kernel. We already mentioned that one can in fact equally well solve then the primal problem as the dual problem. In fact solving the primal problem is more advantageous for larger data sets while solving the dual problem is more suitable for large dimensional input

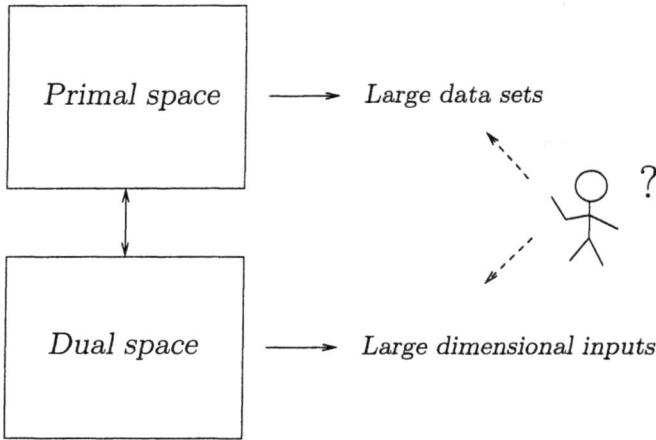

Fig. 6.1 *For linear support vector machines the dual problem is suitable for solving problems with large dimensional input spaces while the primal problem is convenient towards large data sets. However, for nonlinear SVMs one has no expression for $\varphi(x)$, as a result one can only solve the dual problem in terms of the related kernel function. In the method of Fixed Size LS-SVM the Nyström method is used to estimate eigenfunctions. After obtaining estimates for $\varphi(x)$ and linking primal-dual formulations, the computation of w, b is done in the primal space.*

spaces because the unknowns are $w \in \mathbb{R}^n$ and $\alpha \in \mathbb{R}^N$, respectively, where n denotes the dimension of the input space and N the number of given training data points. For example in the linear function estimation case one has

$$\min_{w,b} \frac{1}{2} w^T w + \gamma \frac{1}{2} \sum_{k=1}^{N} \left(y_k - (w^T x_k + b)\right)^2 \qquad (6.1)$$

by elimination of the error variables e_k, which one can immediately solve. In this case the mapping $\varphi(\cdot)$ becomes $\varphi(x_k) = x_k$ and there is no need to solve the dual problem in the support values α, certainly not for large data sets.

For the nonlinear case on the other hand the situation is much more complicated. For many choices of the kernel, $\varphi(\cdot)$ may become infinite dimensional and hence also the w vector. However, one may still try in this case to find meaningful estimates for $\varphi(x_k)$.

A procedure to find such estimates is implicitly done by the Nyström method, which is well known in the area of integral equations [14; 63] and has been successfully applied in the context of Gaussian processes by Williams & Seeger in [294]. The method is related to finding a low rank approximation to the given kernel matrix by randomly choosing M rows/columns of the kernel matrix. Let us denote the *big* kernel matrix by $\Omega_{(N,N)} \in \mathbb{R}^{N \times N}$ and the *small* kernel matrix based on the random subsample $\Omega_{(M,M)} \in \mathbb{R}^{M \times M}$ with $M < N$ (in practice often $M \ll N$). Consider the eigenvalue decomposition of the *small* kernel matrix $\Omega_{(M,M)}$

$$\Omega_{(M,M)} \overline{U} = \overline{U} \overline{\Lambda} \tag{6.2}$$

where $\overline{\Lambda} = \mathrm{diag}([\overline{\lambda}_1; ...; \overline{\lambda}_M])$ contains the eigenvalues and $\overline{U} = [\overline{u}_1 ... \overline{u}_M] \in \mathbb{R}^{M \times M}$ the corresponding eigenvectors. This is related to eigenfunctions $\phi_i(x)$ and eigenvalues λ_i of the integral equation

$$\int K(x, x') \phi_i(x) p(x) dx = \lambda_i \phi_i(x') \tag{6.3}$$

as follows

$$
\begin{aligned}
\hat{\lambda}_i &= \tfrac{1}{M} \overline{\lambda}_i \\
\hat{\phi}_i(x_k) &= \sqrt{M}\, \overline{u}_{ki} \\
\hat{\phi}_i(x') &= \tfrac{\sqrt{M}}{\overline{\lambda}_i} \sum_{k=1}^{M} \overline{u}_{ki} K(x_k, x')
\end{aligned}
\tag{6.4}
$$

where $\hat{\lambda}_i$ and $\hat{\phi}_i$ are estimates to λ_i and ϕ_i, respectively, for the integral equation, and $\overline{u}_{ki}$ denotes the ki-th entry of the matrix $\overline{U}$. This can be understood from sampling the integral by M points $x_1, x_2, ..., x_M$. For the *big* kernel matrix one has the eigenvalue decomposition

$$\Omega_{(N,N)} \tilde{U} = \tilde{U} \tilde{\Lambda}. \tag{6.5}$$

Furthermore, as explained in [294] one has

$$
\begin{aligned}
\tilde{\lambda}_i &= \tfrac{N}{M} \overline{\lambda}_i \\
\tilde{u}_i &= \sqrt{\tfrac{N}{M}} \tfrac{1}{\overline{\lambda}_i} \Omega_{(N,M)} \overline{u}_i.
\end{aligned}
\tag{6.6}
$$

One can then show that

$$\Omega_{(N,N)} \simeq \Omega_{(N,M)} \Omega_{(M,M)}^{-1} \Omega_{(M,N)} \tag{6.7}$$

where $\Omega_{(N,M)}$ is the $N \times M$ block matrix taken from $\Omega_{(N,N)}$.

These insights are used then for solving in an approximate sense the linear system

$$(\Omega_{(N,N)} + I/\gamma)\alpha = y \qquad (6.8)$$

without bias term in the model, as considered in Gaussian process regression problems. By applying the Sherman-Morrison-Woodbury formula [98] one obtains [294]:

$$\alpha = \gamma \left(y - \tilde{U}(\frac{1}{\gamma}I + \tilde{\Lambda}\tilde{U}^T\tilde{U})^{-1}\tilde{\Lambda}\tilde{U}^T y \right) \qquad (6.9)$$

where $\tilde{U}, \tilde{\Lambda}$ are calculated from (6.6) based upon $\overline{U}, \overline{\Lambda}$ from the small matrix. In LS-SVM classification and regression one usually considers a bias term which leads to centering of the kernel matrix. For application of the Nyström method the eigenvalue decomposition of the centered kernel matrix is then taken. Finally, further characterizations of the error for approximations to a kernel matrix have been investigated in [206; 226].

The Nyström method approach has been applied to the Bayesian LS-SVM framework at the second level of inference while solving the level 1 problems without Nyström method approximation by the conjugate gradient method [273]. In Table 6.1 this is illustrated on three data sets **cra** (leptograpsus crab), **rsy** (Ripley synthetic data), **hea** (heart disease) according to [273]. In [273] it has also been illustrated that larger data sets such as the UCI adult data set, a successful approximation by the Nyström method can be made at the second level of inference by taking a subsample of 100 data points in order to determine (γ, σ) instead of the whole training data set size 33000.

6.1.2 *Incomplete Cholesky factorization*

The Sherman-Morrison-Woodbury formula

$$(D + VV^T)^{-1} = D^{-1} - D^{-1}V(I + V^TD^{-1}V)^{-1}V^TD^{-1} \qquad (6.10)$$

which is used within the Nyström method has also been widely used in the context of interior point algorithms for linear programming. However, as pointed out by Fine & Scheinberg in [79] it may lead to numerical difficulties. In [79] it has been illustrated with a simple example how the computed

cra	Test set Perf.	$\log \sigma$	$\log \gamma$
100%	95.5	1.25	5.79
75%	95.6 (0.47)	1.15 (0.10)	5.65 (0.28)
50%	95.6 (0.63)	1.26 (0.09)	6.05 (0.22)
25%	95.5 (0.00)	1.60 (0.03)	4.99 (0.50)
10%	95.5 (0.00)	1.57 (0.11)	4.77 (0.32)
rsy	Test set Perf.	$\log \sigma$	$\log \gamma$
100%	90.6	0.26	0.47
75%	90.6 (0.04)	0.26 (0.00)	0.50 (0.03)
50%	90.4 (0.04)	0.69 (0.00)	0.49 (0.03)
25%	90.5 (0.06)	0.19 (0.05)	0.55 (0.03)
10%	90.6 (0.08)	0.20 (0.07)	0.89 (0.17)
hea	Test set Perf.	$\log \sigma$	$\log \gamma$
100%	85.6	0.61	2.28
75%	85.6 (0.00)	2.23 (0.03)	0.59 (0.02)
50%	85.6 (0.00)	2.26 (0.07)	0.60 (0.02)
25%	85.6 (0.00)	2.25 (0.07)	0.62 (0.01)
10%	85.6 (0.00)	2.22 (0.06)	0.64 (0.02)

Table 6.1 *Application of the Nyström approximation in the Bayesian LS-SVM framework at the second level of inference. Shown are performances and inferred hyperparameters for different proportions 10% ... 100% of the training data N. The standard deviation on 10 randomizations of the Nyström sample is given between parentheses.*

solutions may not even be close to the correct solution when applying the Woodbury formula (when the limited machine precision is taken into account). Small numerical perturbations due to limited machine precision may cause numerical instabilities resulting into computed solutions that are far from the true solution.

A numerically stable method for kernel based methods has been studied in [79], which is based on a low rank update of the Cholesky factorization [98] of the kernel matrix for solving kernel based learning problems of the form $\Omega \alpha = y$ with kernel matrix $\Omega = D + VV^T$. It is often the case that the exact low-rank representation of the kernel matrix Ω is not given or even does not exist when the rank of the kernel matrix is large. The best that one may hope then is to find a good approximation together with keeping the rank low. Therefore, one can compute an *incomplete* Cholesky factorization $\Omega = GG^T$ with G a lower triangular matrix where some columns of G are

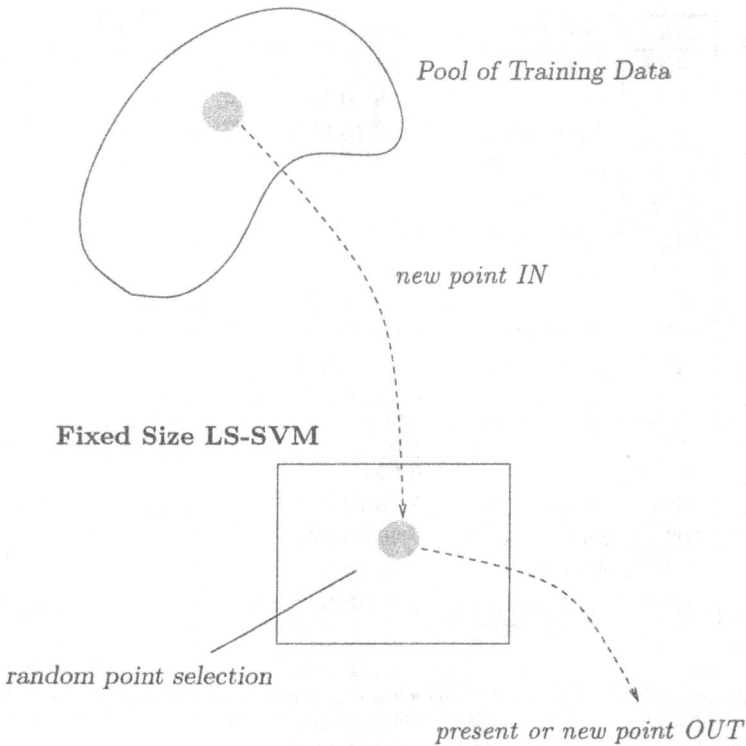

Fig. 6.2   *Fixed size LS-SVM: the number of support vectors is pre-fixed beforehand and the support vectors are actively selected from the pool of training data. After estimating eigenfunctions the model is computed in the primal space with calculation of w, b. In the working set of support vectors a point is randomly selected and replaced by a randomly selected point from the training data if this new point improves an entropy criterion which is related to density estimation and kernel PCA.*

zero. This procedure can also be applied to almost singular matrices by skipping pivots that are below a certain threshold [98]. In [79] this method has been successfully applied to SVMs.

6.2   Fixed Size LS-SVMs

In the Nyström method an approximate solution to the linear system is computed based upon a random subsample of the given training data set.

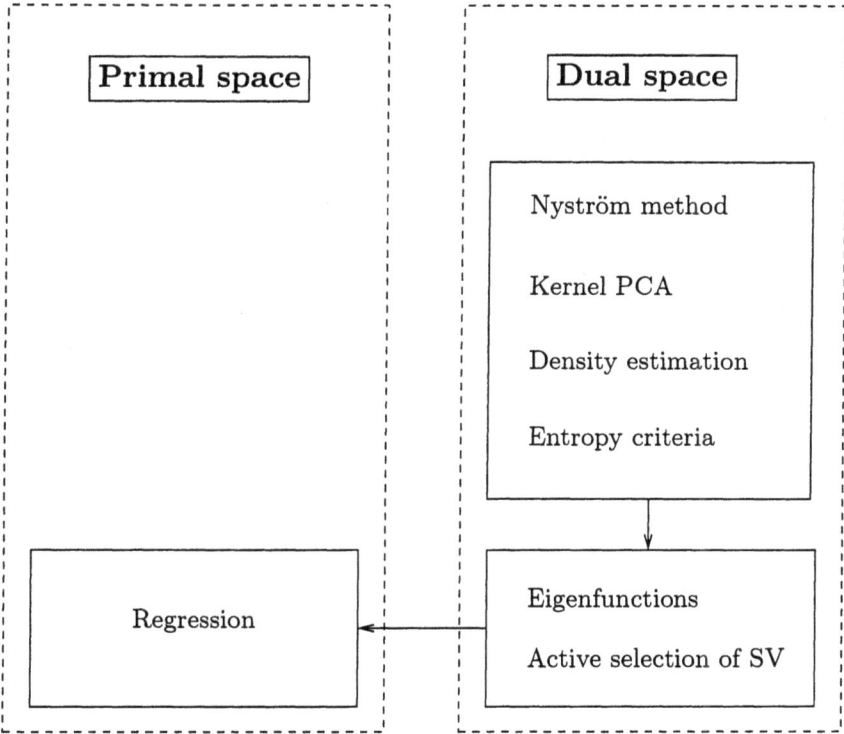

Fig. 6.3   *In the method of fixed size LS-SVM the Nyström method, kernel PCA and density estimation are linked to estimation of eigenfunctions and active selection of support vectors. Regression in the primal space is done in a next step.*

However, an important open question at this point is whether the random points can be selected in another way than just trying out several random subsamples and comparing these solutions on a separate validation set. We explain now how active selection of such support vectors can be done in relation to the Nyström method for a *fixed size* LS-SVM.

6.2.1   *Estimation in primal weight space*

As we discussed in (6.11), for large data sets it would be advantageous if one could solve the problem in the primal space (Fig. 6.1). However, in the nonlinear case we would then need an explicit expression for φ which is in principle only implicitly determined by the kernel trick. On the other hand, if one succeeds in finding a meaningful estimate of φ one could solve the following ridge regression problem in the primal weight space with unknowns w, b

$$\min_{w \in \mathbb{R}^{n_h}, b \in \mathbb{R}} \frac{1}{2} w^T w + \gamma \frac{1}{2} \sum_{k=1}^{N} \left(y_k - (w^T \varphi(x_k) + b) \right)^2. \tag{6.11}$$

Such an estimate can be made after obtaining the eigenfunctions. One chooses as fixed size M ($M \leq N$ and typically $M \ll N$) for a working set of support vectors where the value of M is related to the Nyström subsample. Using the expression (6.4) one obtains

$$\varphi_i(x') = \sqrt{\tilde{\lambda}_i}\, \hat{\phi}_i(x') = \frac{\sqrt{M}}{\sqrt{\tilde{\lambda}_i}} \sum_{k=1}^{M} u_{ki} K(x_k, x'), \tag{6.12}$$

assuming $M \leq N_{\text{eff}}$. Hence one constructs the $M \times M$ kernel matrix, takes its eigenvalue decomposition, and computes the eigenfunctions based upon the eigenvalue decomposition of the kernel matrix which gives the expression for $\varphi(x')$ evaluated at any point x'. This can be applied both to function estimation and classification problems.

The model takes the form

$$
\begin{aligned}
y(x) &= w^T \varphi(x) + b \\
&= \sum_{i=1}^{M} w_i \frac{\sqrt{M}}{\sqrt{\tilde{\lambda}_i}} \sum_{k=1}^{M} u_{ki} K(x_k, x) + b.
\end{aligned}
\tag{6.13}
$$

The support values corresponding to the number of M support vectors are then

$$\alpha_k = \sum_{i=1}^{M} w_i \frac{\sqrt{M}}{\sqrt{\tilde{\lambda}_i}} u_{ki} \tag{6.14}$$

if we represent the model as

$$y(x) = \sum_{k=1}^{M} \alpha_k K(x_k, x) + b. \tag{6.15}$$

This approach gives explicit links between the primal and the dual representation. However, the approximation is based on a random selection of the support vectors, but does not provide an algorithm for making a good selection of the support vectors.

6.2.2 Active selection of support vectors

In order to make a more suitable selection of the support vectors instead of a random selection, one can relate the Nyström method to kernel principal component analysis, density estimation and entropy criteria, as discussed by Girolami in [94]. These links will be explained in more detail in the Chapter on unsupervised learning and support vector machine formulations to kernel PCA.

In [94] an analysis is done of the quadratic Renyi entropy

$$H_R = -\log \int p(x)^2 dx \qquad (6.16)$$

in relation to kernel PCA and density estimation with

$$\int \hat{p}(x)^2 dx = \frac{1}{N^2} 1_v^T \Omega 1_v \qquad (6.17)$$

where $1_v = [1; 1; ...; 1]$ and a normalized kernel is assumed with respect to density estimation. One chooses a fixed size M then and actively selects points from the pool of training data as candidate support vectors (Fig. 6.2). In the working set of support vectors a point is randomly selected and replaced by a randomly selected point from the training data set if the new point improves the entropy criterion. This leads to the following fixed size LS-SVM algorithm.

Fixed size LS-SVM algorithm:

(1) Given normalized and standardized training data $\{x_k, y_k\}_{k=1}^N$ with inputs $x_k \in \mathbb{R}^n$, outputs $y_k \in \mathbb{R}$ and N training data.

(2) Choose a working set with size M and impose in this way a number of M support vectors (typically $M \ll N$).

(3) Randomly select a support vector x^* from the working set of M support vectors.

(4) Randomly select a point x^{t*} from the N training data and replace x^* by x^{t*} in the working set. If the entropy increases by taking the point x^{t*} instead of x^* then this point x^{t*} is accepted for the working set of M support vectors, otherwise the point x^{t*} is rejected (and returned to the training data pool) and the support vector x^* stays in the working set.

(5) Calculate the entropy value for the present working set.

(6) Stop if the change in entropy value (6.17) is small or the number of iterations is exceeded, otherwise go to (3).

(7) Estimate w, b in the primal space after estimating the eigenfunctions from the Nyström approximation.

Illustrative examples are shown in Figs. 6.4-6.8 on a noisy sinc function and a double spiral classification problem, both with RBF kernels. A bad initial choice of the support vectors is intentionally taken, after which a self-organizing process is taking place. Corresponding plots for the evolution of the entropy criterion are shown. An important aspect of the algorithm is that the determination of the support vectors is done *independently* of the regression problem of w, b after estimating φ. In other words, not in every iteration step one needs to estimate w, b. This can be done after a suitable set of support vectors has been found. This also means that in this algorithm the tuning parameters (γ, σ) can be determined separately from each other. First, a good choice of σ is determined in relation to the chosen number of support vectors M. Then one may apply e.g. Bayesian learning in order to automatically determine γ. Note that also in classical radial basis function network learning, methods have been developed where one separates the training of the centers from the training of the output weights. In RBF networks the centers are often determined by means of clustering [39; 40] or using a vector quantization method eventually in combination with self-organizing maps [134] to visualize the input space of the network. In a fixed size LS-SVM this is done through the explicit *primal-dual interpretations* (Fig. 6.3). Because the model is estimated in the primal space the estimation problem becomes parametric. In this sense, one may also consider applying other methods such as Bayesian learning to the determination of the width of the kernel instead of applying the entropy criterion which only takes into account the input space of the network.

The fixed size LS-SVM method is also suitable for *adaptive* signal processing applications where on-line updating of w, b and recursive methods for the eigenvalue decomposition can be applied. Both for recursive least squares and singular value decomposition updating, various efficient algorithms have been developed in the past [109; 138; 165; 166; 167; 199; 298], which can be used at this point. Also for transductive inference the use of fixed size LS-SVM is very natural due to the fact that the search of support vectors is done in an unsupervised way. In *transductive inference* [90; 278; 279; 281] one is rather interested in finding a model which is performing well on a specific future data point (or set of data points) than a general model (as obtained by inductive inference). In other words in transduction one is interested in inference from particular to particular. The future points on which the model should perform well are used in an unlabeled way (for classifiers) or without output target values (for regression), i.e. in an unsupervised way, to improve the quality of the model. If for fixed size LS-SVM one knows future data (in addition to the given training data) on which the model should perform well, then one may take these points into account (without class labels or target values) for possible selection of support vectors.

In Figs. 6.4-6.5 $N = 200$ training data were generated for a sinc function with zero mean white Gaussian noise with standard deviation 0.1. A fixed size $M = 10$ was taken and an RBF kernel with $\sigma^2 = 3$. In Fig. 6.6 $N = 20000$ were generated for a noisy sinc, with the same model. In Fig. 6.7 $N = 400$ training data were generated and a fixed size LS-SVM model with RBF kernel was taken for $M = 20$.

An application of Fixed Size LS-SVM with RBF kernel to the UCI adult data set with a training set of 22000 data and a validation set of 11000 data gives the following validation set performance indication: $M = 20$: 79.25 ($\sigma^2 = 100$), 79.00 ($\sigma^2 = 1000$); $M = 50$: 81.25 ($\sigma^2 = 100$), 80.00 ($\sigma^2 = 1000$); $M = 1000$: 84.23 ($\sigma^2 = 100$), 84.16 ($\sigma^2 = 1000$). The value of γ is determined by Bayesian inference.

6.3 Basis construction in the feature space

Another approach to large scale problems is to find a suitable set of basis vectors in the feature space in order to build and span a large subspace

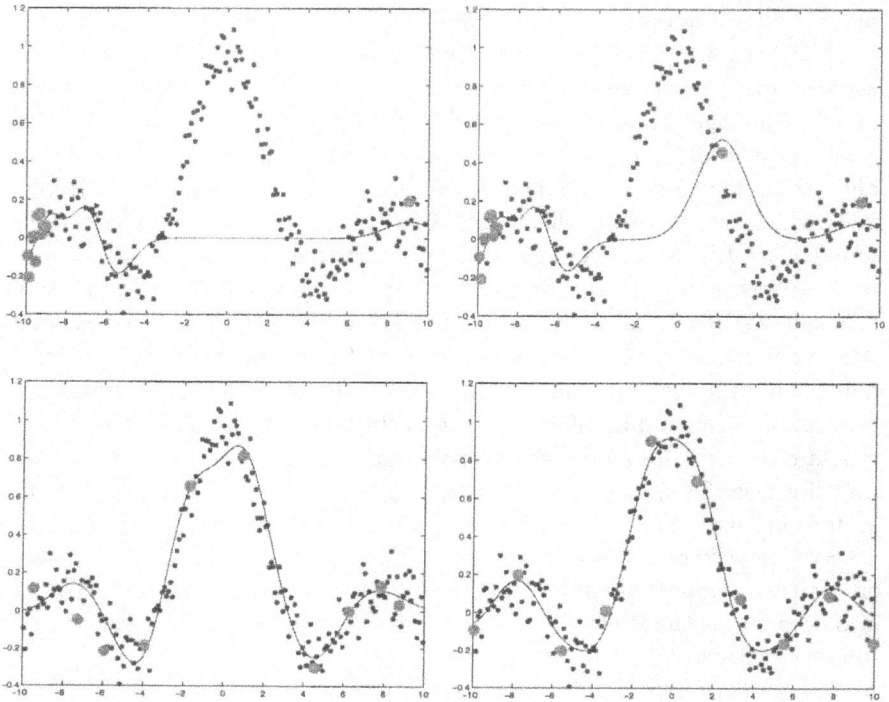

Fig. 6.4 *Fixed size LS-SVM with $M = 10$ and $N = 200$ on a noisy sinc function. The support vectors are actively selected by means of entropy criterion (6.17). A bad initial choice of support vectors (grey dots) is intentionally taken, which after a number of iterations leads to a uniform distribution of support vectors over the x-axis. The figures show different stages of this self-organizing process.*

(Fig. 6.10). The following method has been proposed by Cawley [37] for LS-SVMs.

The complete LS-SVM model based on N training data for function estimation takes the form

$$f(x) = \sum_{k=1}^{N} \alpha_k K(x, x_k) + b \qquad (6.18)$$

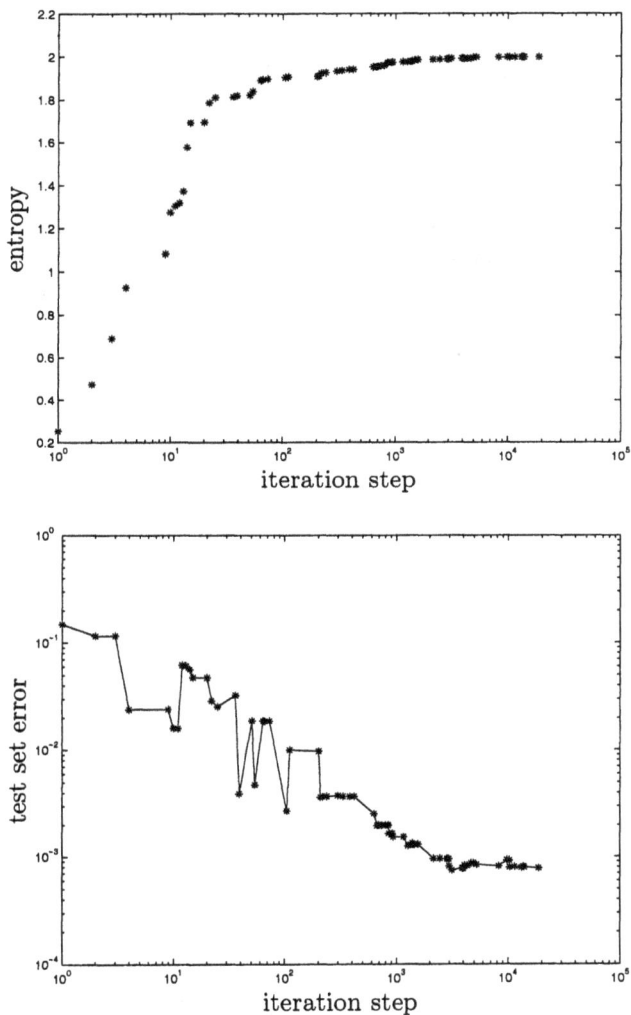

Fig. 6.5 *(Continued) (Top) entropy with respect to iteration steps; (Bottom) Corresponding error on an independent test set for the noisy sinc function of the previous figure.*

where α, b are the solution

$$\begin{bmatrix} \Omega + I/\gamma & 1_v \\ 1_v^T & 0 \end{bmatrix} \begin{bmatrix} \alpha \\ b \end{bmatrix} = \begin{bmatrix} y \\ 0 \end{bmatrix}. \tag{6.19}$$

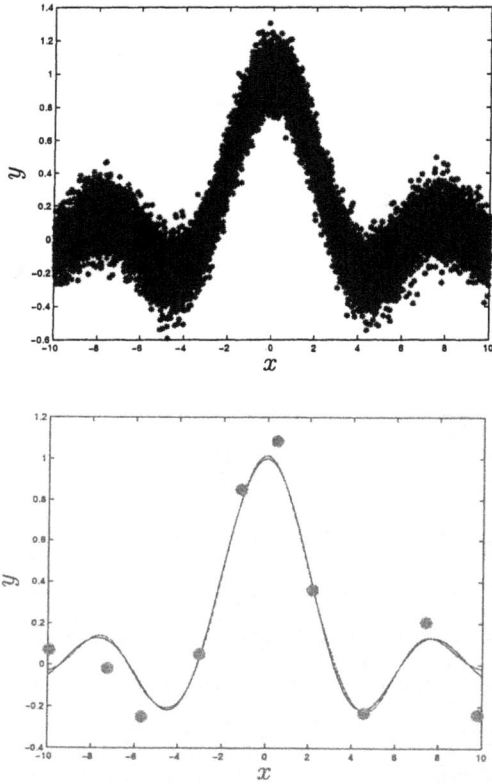

Fig. 6.6   *Fixed size LS-SVM for a noisy sinc function with $M = 10$ and $N = 20000$.
(Top) given training set; (Bottom) true sinc function (solid line), estimate of Fixed Size
LS-SVM (dashed line), and support vectors (grey dots).*

For the solution vector in the primal weight space one has

$$w = \sum_{k=1}^{N} \alpha_k \varphi(x_k). \qquad (6.20)$$

In other words, w is represented by means of a basis $\{\varphi(x_k)\}_{k=1}^{N}$ consisting
of N vectors. In order to find a sparse representation one expresses w in
terms of fewer basis vectors $\{\varphi(\xi_r)\}_{r=1}^{M}$. These vectors ξ_r all belong to the

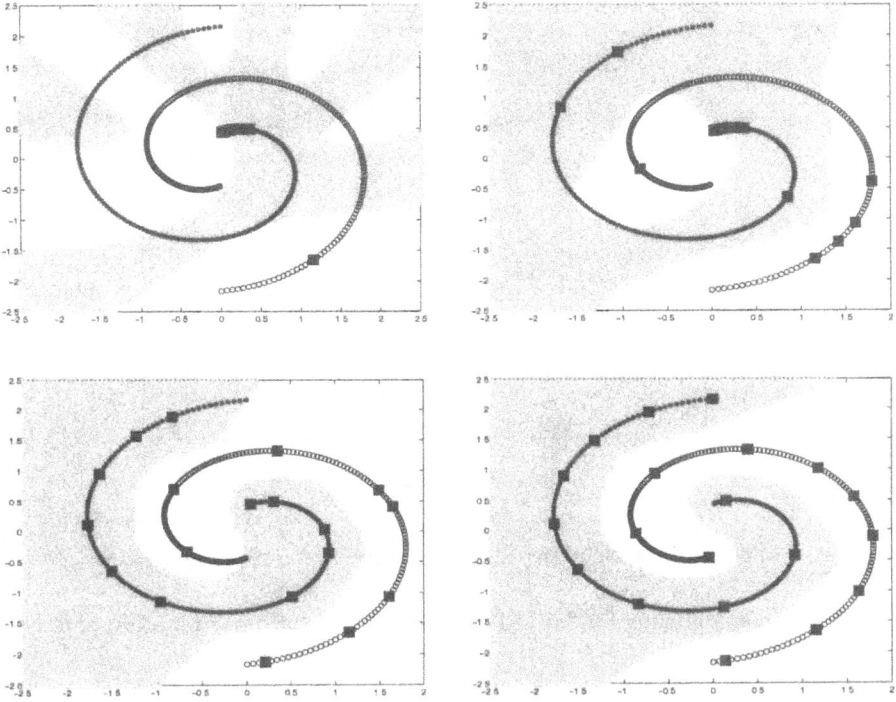

Fig. 6.7   *Fixed size LS-SVM with $M = 20$ and $N = 400$ for a double spiral classification problem. The support vectors are actively selected by means of an entropy criterion. The figures show different stages of this self-organizing process of support vector selection.*

set of training data $\{x_k\}_{k=1}^{N}$. One expresses

$$w = \sum_{r=1}^{M} \beta_r \varphi(\xi_r) \qquad (6.21)$$

where typically $M \ll N$. Such an approach has also been proposed by Schölkopf *et al.* in [204]. In terms of this reduced set the model takes the form

$$\tilde{f}(x) = \sum_{r=1}^{M} \beta_r K(\xi_r, x) + b. \qquad (6.22)$$

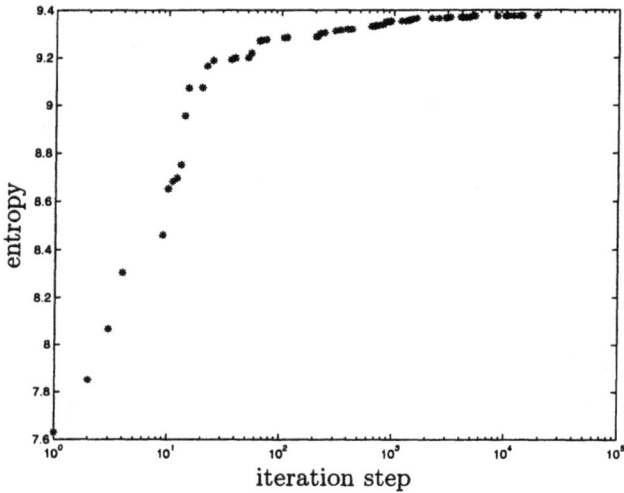

Fig. 6.8 *(Continued) corresponding plot for the entropy during the iteration process.*

In [37] the following objective function is taken for determination of β, b:

$$
\begin{aligned}
\min_{\beta,b} J(\beta, b) &= \frac{1}{2} w^T w + \gamma \sum_{k=1}^{N} \left(y_k - \tilde{f}(x_k) \right)^2 \\
&= \frac{1}{2} \sum_{r=1}^{M} \sum_{s=1}^{M} \beta_r \beta_s K(\xi_r, \xi_s) + \gamma \sum_{k=1}^{N} \left(y_k - \sum_{r=1}^{M} \beta_r K(\xi_r, x_k) - b \right)^2 .
\end{aligned}
$$

$$(6.23)$$

This is a parametric optimization problem in β, b which leads to solving a much smaller linear system of dimension $(M + 1) \times (M + 1)$ instead of $(N + 1) \times (N + 1)$. However, one needs to know the vectors ξ_r before one can actually compute β, b.

The following method was proposed by Baudat & Anouar in [17] for finding the reduced set of basis vectors $\{\varphi(\xi_r)\}_{r=1}^{M}$ and has been further applied in [37]. For each given data point x_k one can consider the ratio

$$
\delta_k = \frac{\|\varphi(x_k) - \varphi_S(x_k)\|_2^2}{\|\varphi(x_k)\|_2^2}
$$

$$(6.24)$$

where $\{\varphi(x_k)\}_{k \in S}$ is the basis constituted by a subset S of selected training data points. In this way one characterizes the quality of the approximation

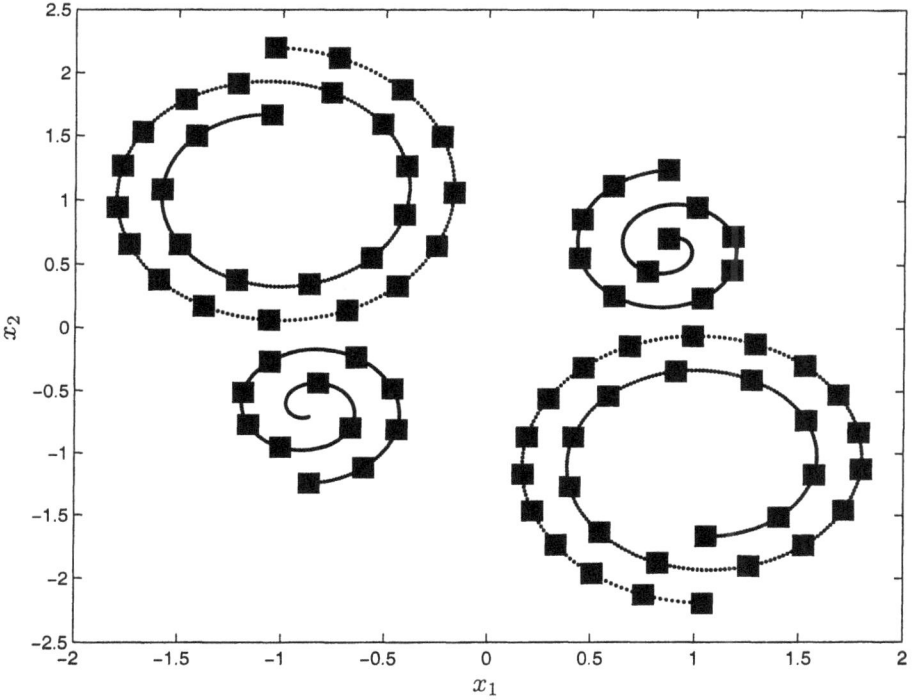

Fig. 6.9   *Illustration of the selection of support vectors for a multi-spiral problem with 1600 training data points and a fixed size LS-SVM (M = 80) with RBF kernel (σ = 0.25). The entropy criterion is used to select the support vectors (depicted by the squares). No class labels are shown for the training data in this unsupervised learning process.*

in the feature space for the reduced set. By application of the kernel trick one obtains

$$\delta_k = 1 - \frac{\Omega_{\mathcal{S}k}^T \Omega_{\mathcal{S}\mathcal{S}}^{-1} \Omega_{\mathcal{S}k}}{\Omega_{kk}} \qquad (6.25)$$

where $\Omega_{\mathcal{S}\mathcal{S}}$ denotes the $M \times M$ submatrix of the kernel matrix Ω of size $N \times N$ and $\Omega_{\mathcal{S}k}$ denotes the k-th column of the matrix $\Omega_{\mathcal{S}\mathcal{S}}$. A related criterion has also been considered in [226]. In order to minimize the error of the approximation in the feature space over all training data N, one

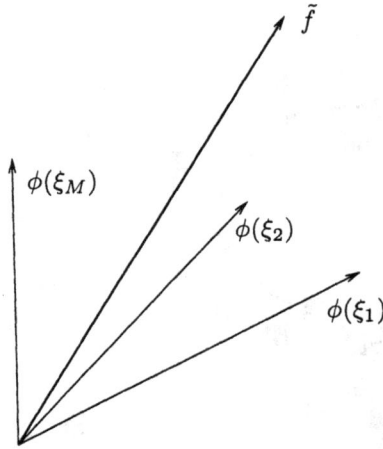

Fig. 6.10 *Expressing an estimate for the model $\tilde{f}$ in terms of a reduced set of basis vectors $\varphi(\xi_1), \varphi(\xi_2), ..., \varphi(\xi_M)$ instead of $\varphi(x_1), \varphi(x_2), ..., \varphi(x_N)$ with $M \ll N$. A suitable selection of the vectors ξ_r is made as a subset of the given training data set.*

considers the cost function

$$\max_{\mathcal{S}} \mathcal{J}(\mathcal{S}) = \frac{1}{N} \sum_{l=1}^{N} \frac{\Omega_{\mathcal{S}l}^T \Omega_{\mathcal{S}\mathcal{S}}^{-1} \Omega_{\mathcal{S}l}}{\Omega_l} \qquad (6.26)$$

where $\Omega_{\mathcal{S}l} = [K(x_1, x_l); ...; K(x_M, x_l)]$ meaning that one lets the vectors ξ_r correspond to the subset $\mathcal{S}$ of the given training set and selects those vectors that optimize this feature space criterion.

As a result, one aims at finding a suitable set of basis vectors that span the model. In this criterion one sees that all chosen vectors will be independent because $\Omega_{\mathcal{S}\mathcal{S}}$ should be invertible (hence full rank). A similar idea of finding a reduced set that spans the model has been proposed for Gaussian processes by Csató & Opper in [53].

6.4   Combining submodels

6.4.1   *Committee network approach*

In the first Chapter we discussed committee methods for combining results from estimated models which is well-known in the area of neural networks.

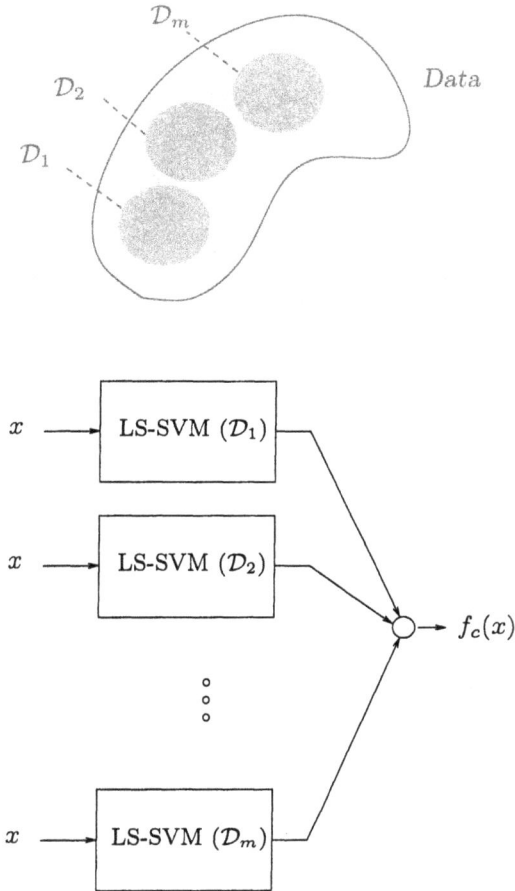

Fig. 6.11 *Mining large data sets using a committee network of LS-SVMs where the individual LS-SVMs are trained on subsets of the entire training data set.*

The models could be MLPs but also different kind of networks. The committee of models aims at realizing the principle of *"the whole is more than the sum of its parts"*, in the sense that the committee as a whole can perform better than each of the individual models in terms of the bias-variance trade-off [12; 23; 179].

We can use this method also to decompose the problem of training LS-SVMs into smaller problems consisting of less training data points and train

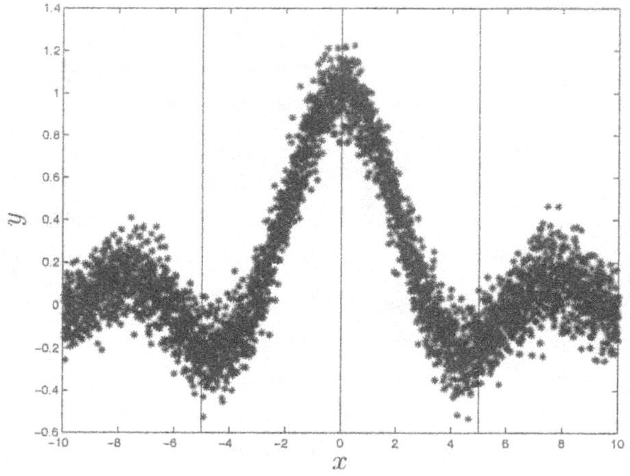

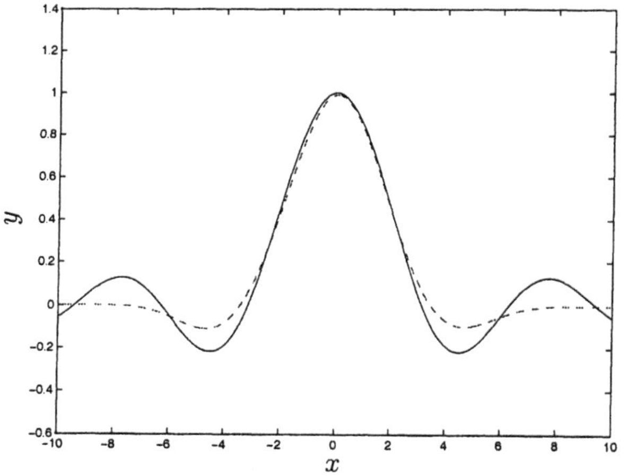

Fig. 6.12   *Illustration of a committee network approach to the estimation of a noisy sinc function with 4000 given data points. (Top) In this illustration the given training data are divided into four intervals. For each of the four parts a separate LS-SVM has been trained on 1000 data points. The results of the four LS-SVMs are combined into an overall model by means of a committee network. (Bottom) true sinc function (solid line), estimate of the committee of LS-SVMs (dashed line). Better results are obtained by randomly selecting training data and assigning them to the subgroups for this example.*

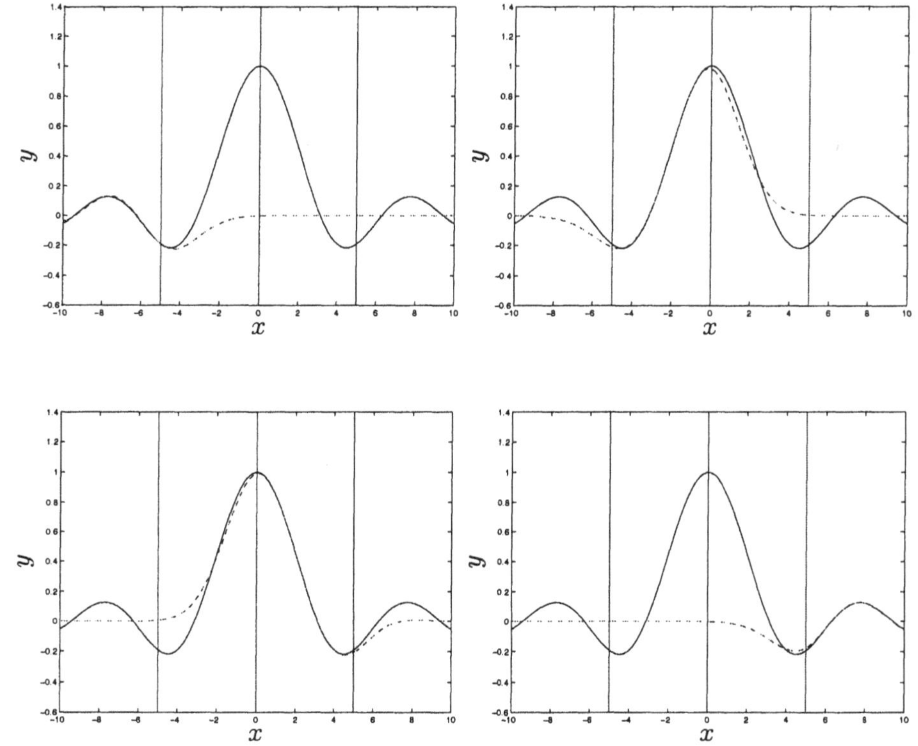

Fig. 6.13 *Results of the individual LS-SVMs for the example of the previous Figure, trained on each of the subintervals: true sinc function (solid line), individual LS-SVM estimates trained on one of the four intervals (dashed line).*

individual LS-SVMs on each of the subproblems. Because the committee is a linear combination of the submodels, the functional form of the LS-SVM committee stays the same as for each of the individual submodels, i.e. a weighted sum of kernel functions evaluated at given data points. In Fig. 6.11 and Fig. 6.12 this approach is shown and illustrated on the training of a noisy sinc function with 4000 given training data points. The training data were generated in the interval $[-10, 10]$. This data set was divided into four parts corresponding to the intervals $[-10, -5]$, $[-5, 0]$, $[0, 5]$ and $[5, 10]$ for the purpose of visualization. Individual LS-SVMs have been trained for these intervals and the results are combined into one single committee

network model. The results of individual LS-SVMs and the committee of
LS-SVMs are shown in Fig. 6.12 and Fig. 6.13, respectively. While for the
purpose of illustration the intervals are chosen to be non-overlapping, in
practice one obtains better results by randomly selecting the data points in
$[-10, 10]$ and assigning it to a training set of one of the individual LS-SVM
models. In the example the same RBF kernel with $\sigma = \sqrt{5}$ and $\gamma = 1$ was
taken for the individual LS-SVMs.

The committee network that consists of the m submodels takes the form

$$
\begin{aligned}
f_c(x) &= \sum_{i=1}^{m} \beta_i f_i(x) \\
&= h(x) + \sum_{i=1}^{m} \beta_i \epsilon_i(x)
\end{aligned}
\tag{6.27}
$$

where $\sum_{i=1}^{m} \beta_i = 1$, $h(x)$ is the true function to be estimated and $\epsilon_i(x) = f_i(x) - h(x)$ where

$$
f_i(x) = \sum_{k=1}^{N_i} \alpha_k^{(i)} K^{(i)}(x, x_k^{(i)}) + b^{(i)}
\tag{6.28}
$$

is the i-th LS-SVM model trained on the data $\{x_k^{(i)}, y_k^{(i)}\}_{k=1}^{N_i}$ with resulting
support values $\alpha_k^{(i)}$, bias term $b^{(i)}$ and kernel $K^{(i)}(\cdot, \cdot)$ for the i-th submodel
and $i = 1, \ldots, m$ with m the number of LS-SVM submodels.

As explained in [12; 23; 179] one considers the covariance matrix

$$
C_{ij} = \mathcal{E}[\epsilon_i(x)\epsilon_j(x)]
\tag{6.29}
$$

where in practice one works with a finite-sample approximation

$$
C_{ij} = \frac{1}{N} \sum_{k=1}^{N} [f_i(x_k) - y_k][f_j(x_k) - y_k]
\tag{6.30}
$$

and the N data are a representative subset of the overall training data set (or the whole training data set itself). The committee error equals

$$
\begin{aligned}
J_c &= \mathcal{E}[\{f_c(x) - h(x)\}^2] \\
&= \mathcal{E}[(\sum_{i=1}^{m} \beta_i \epsilon_i)(\sum_{j=1}^{m} \beta_j \epsilon_j)] \\
&\simeq \sum_{i=1}^{m} \sum_{j=1}^{m} \beta_i \beta_j C_{ij} = \beta^T C \beta.
\end{aligned}
\tag{6.31}
$$

An optimal choice of β follows then from

$$
\min_{\beta} \frac{1}{2} \beta^T C \beta \text{ such that } \sum_{i=1}^{m} \beta_i = 1.
\tag{6.32}
$$

From the Lagrangian with Lagrange multiplier λ

$$
\mathcal{L}(\beta, \lambda) = \frac{1}{2} \beta^T C \beta - \lambda (\sum_{i=1}^{m} \beta_i - 1)
$$

one obtains the conditions for optimality:

$$
\begin{cases}
\frac{\partial \mathcal{L}}{\partial \beta} &= C\beta - \lambda 1_v = 0 \\
\frac{\partial \mathcal{L}}{\partial \lambda} &= 1_v^T \beta - 1 = 0
\end{cases}
\tag{6.33}
$$

with optimal solution

$$
\beta = \frac{C^{-1} 1_v}{1_v^T C^{-1} 1_v}
\tag{6.34}
$$

and corresponding committee error

$$
J_c = 1/(1_v^T C^{-1} 1_v)
\tag{6.35}
$$

with $1_v = [1; 1; ...; 1]$. In case of an ill-conditioned matrix C one can apply additional regularization to the C matrix or impose that $\beta_i \geq 0$ to avoid large negative and positive weights.

Committees have been successfully applied to SVM and Gaussian processes problems e.g. by Tresp [257; 258] and Collobert *et al.* [44]. The committee method is also related to boosting [75; 86]. A method of coupled local minimizers proposed in [248] may be considered as an extension to the committee network method towards solving general differentiable optimization problems. Individual optimizers correspond then to individual LS-SVMs which are interacting through state synchronization constraints, can optimize in a cooperative way a group cost function for the ensemble and in this way realize collective intelligence.

6.4.2 *Multilayer networks of LS-SVMs*

In a committee network one linearly combines the individual LS-SVM models. A natural extension could be to combine the networks in a nonlinear fashion. This might be done in several ways. One option is to take an MLP which has the outputs of the individual LS-SVMs as input to the network. The LS-SVMs are first trained on subsets of data $\mathcal{D}_1, \mathcal{D}_2, ..., \mathcal{D}_m$ and then serve in fact as a first nonlinear preprocessing layer after which an output layer is trained that is represented by an MLP network. Instead of an MLP one may take any other static nonlinear model, hence also LS-SVMs could in principle be taken as a second layer. However, towards large data sets it will be more advantageous to use a parametric modelling approach for the second layer.

A parametric approach to the training of kernel based models was also taken e.g. by Tipping [253] in the relevance vector machine. A multilayer network of LS-SVMs where the combination of the models is determined by solving a parametric optimization problem (e.g. by Bayesian learning), can be considered as a nonlinear extension of this. The *relevance* of the LS-SVM submodels is determined by the estimation of the second layer.

When taking an MLP in the second layer, the model is described by

$$g(z) = w^T \tanh(Vz + d) \qquad (6.36)$$

with

$$z_i(x) = \sum_{k=1}^{N_i} \alpha_k^{(i)} K^{(i)}(x, x_k^{(i)}) + b^{(i)}, \quad i = 1, \ldots, m \qquad (6.37)$$

where m denotes the number of individual LS-SVM models whose outputs z_i are the input to a MLP with output weight vector $w \in \mathbb{R}^{n_h}$, hidden layer matrix $V \in \mathbb{R}^{n_h \times m}$ and bias vector $d \in \mathbb{R}^{n_h}$ where n_h denotes the number of hidden units. One can take multiple outputs as well. The coefficients $\alpha_k^{(i)}, b^{(i)}$ for $i = 1, ..., m$ are the solutions to a number of m linear systems for each of the individual LS-SVMs trained on data sets $\mathcal{D}_i$ (Fig. 6.14).

In Fig. 6.15 and Fig. 6.16 the approach is illustrated on the estimation of a noisy sinc function, both for $m = 4$ and $m = 40$ combined LS-SVMs trained with RBF kernel $\sigma = \sqrt{5}$ and regularization constant $\gamma = 1$. The

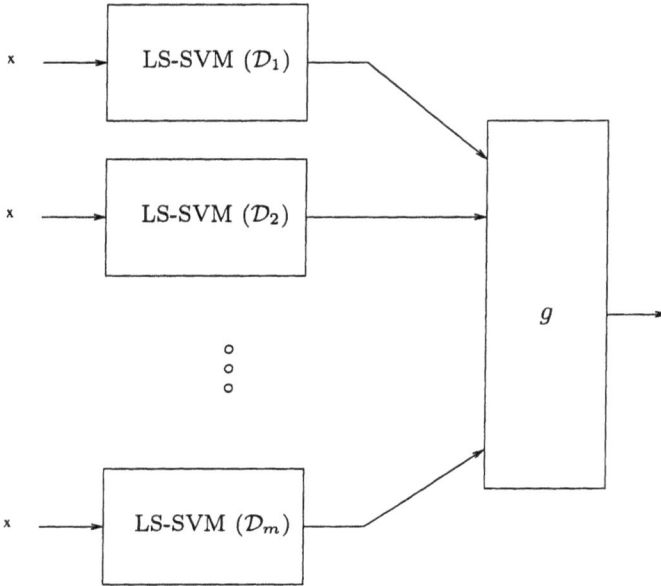

Fig. 6.14 *Nonlinear combinations of trained LS-SVM models. First LS-SVM models are trained on subsets of data $\mathcal{D}_1, ..., \mathcal{D}_m$. Then a nonlinear combination of the LS-SVM network outputs is taken. This leads to a multilayer network where the nonlinear function $g(\cdot)$ can for example be represented by a multilayer perceptron or by a second LS-SVM layer which is trained then with the outcomes of the LS-SVM models as input data. Alternatively, one may also view the LS-SVM models as a linear or nonlinear preprocessing of the input data.*

outputs of these LS-SVMs are taken as input to an MLP with one hidden layer and 6 hidden units. The MLP is trained by Bayesian learning. The results are compared with taking

$$g(z) = w^T z + d \tag{6.38}$$

which corresponds also to one single neuron with linear activation function. Both are trained by the function trainbr in Matlab's neural networks toolbox. In Fig. 6.15 the zero mean random Gaussian noise with standard deviation 0.1 was imposed on the true sinc function for generating the training data, while in Fig. 6.16 a standard deviation of 3 was taken.

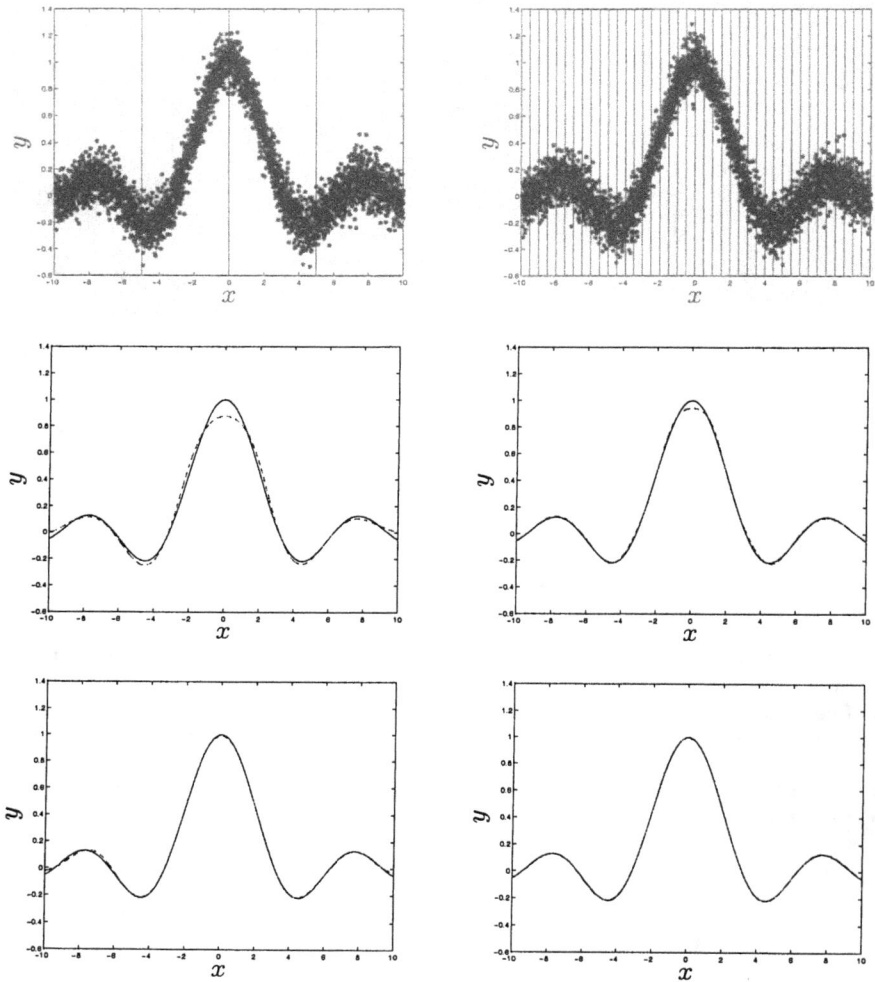

Fig. 6.15   Linear and nonlinear combinations of LS-SVMs with (Left) $m = 4$ LS-SVMs and (Right) $m = 40$ LS-SVMs for a noisy sinc function estimation problem (standard dev. of Gaussian noise is 0.1): (Top) 4000 given training data points, the intervals with vertical lines indicate the training data used for the individual LS-SVMs; (Middle) linear combination of LS-SVMs with single neuron trained by Bayesian learning; (Bottom) nonlinear combination of LS-SVMs with MLP trained by Bayesian learning.

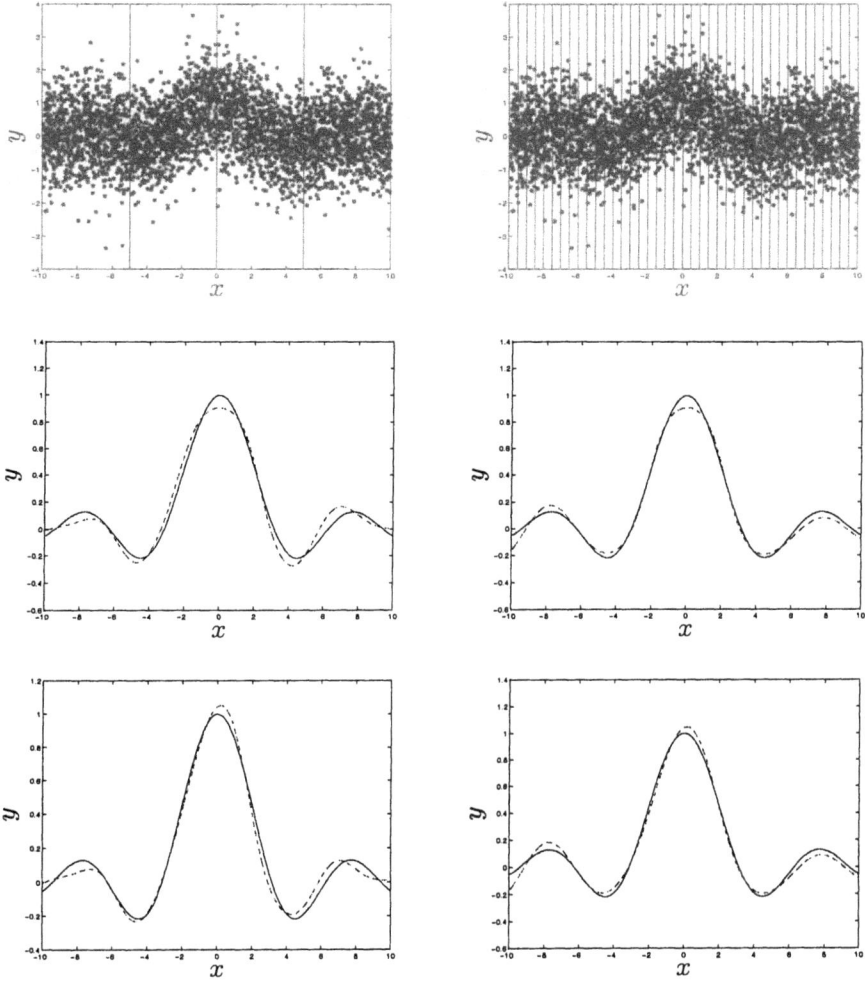

Fig. 6.16 *Similar results as for the previous Figure but with standard dev. of Gaussian noise equal to 3.*

Chapter 7

LS-SVM for Unsupervised Learning

In the previous Chapters we discussed the use of LS-SVMs for problems of supervised learning, including classification and function estimation. In this Chapter we show how support vector machine alike formulations can also be given to the well-known method of principal component analysis (PCA), which is a frequently used technique in unsupervised learning. The new formulation can be extended in a straightforward way to the nonlinear case by applying the kernel trick, and leads to what is presently called kernel PCA. In a similar fashion, new formulations and kernel versions to canonical correlation analysis (CCA) are obtained.

7.1 Support Vector Machines and linear PCA analysis

In the first Chapter we already explained the basic idea of linear and nonlinear PCA methods. The scope now is to first formulate linear PCA analysis within the LS-SVM classifier context and, second, to extend this formulation to a high dimensional feature space with application of the kernel trick. The kernel PCA method was originally introduced by Schölkopf *et al.* in [203]. A difference is that we now formulate an optimization problem with primal-dual interpretations where the dual problem can be related to kernel PCA. This derivation is in the style of LS-SVM primal-dual formulations and interpretations.

7.1.1   *Classical principal component analysis formulation*

In general, there exist different formulations in order to characterize PCA problems [128]. One formulation is to consider a given set of *zero mean* data $\{x_k\}_{k=1}^N$ (only input space data) as a cloud of points for which one tries to find projected variables $w^T x$ with maximal variance. This means

$$
\begin{aligned}
\max_{w} \text{Var}(w^T x) &= \text{Cov}(w^T x, w^T x) \simeq \frac{1}{N} \sum_{k=1}^{N} (w^T x_k)^2 \\
&= w^T C w
\end{aligned}
\tag{7.1}
$$

where $C = (1/N) \sum_{k=1}^N x_k x_k^T$ by definition. One optimizes this objective function under the constraint that $w^T w = 1$. This gives the constrained optimization problem

$$
\mathcal{L}(w; \lambda) = \frac{1}{2} w^T C w - \lambda (w^T w - 1)
\tag{7.2}
$$

with Lagrange multiplier λ. The solution follows from $\partial \mathcal{L}/\partial w = 0$, $\partial \mathcal{L}/\partial \lambda = 0$ and gives the eigenvalue problem

$$
C w = \lambda w.
\tag{7.3}
$$

The matrix C is symmetric and positive semidefinite. The eigenvector w corresponding to the largest eigenvalue determines the projected variable having maximal variance. Efficient and reliable numerical methods are discussed e.g. in [98].

7.1.2   *Support vector machine formulation to linear PCA*

In order to establish now the link with LS-SVM methods, let us reformulate the problem as follows

$$
\max_{w} \sum_{k=1}^{N} [0 - w^T x_k]^2
\tag{7.4}
$$

where 0 is considered as a single target value. The interpretation can be made in a similar fashion as for the links between LS-SVM classifiers and Fisher discriminant analysis. The projected variable to a target space here is $z = w^T x$. While for Fisher discriminant analysis one considers two target values $+1$ and -1 that represent the two classes, in the PCA analysis case a zero target value is considered. Hence, one has in fact a one class modelling problem, but with a different objective function in mind. For Fisher discriminant analysis one aims at minimizing the within scatter around the targets, while for PCA analysis one is interested in finding the direction(s) for which the variance is maximal.

This interpretation of the problem leads to the following primal optimization problem

$$
\left[\ \boxed{\text{P}}:\quad \max_{w,e} J_{\text{P}}(w,e) = \gamma\frac{1}{2}\sum_{k=1}^{N} e_k^2 - \frac{1}{2}w^T w \right.
$$
$$
\left. \text{such that}\quad e_k = w^T x_k,\ k = 1,...,N. \right]
\tag{7.5}
$$

This formulation states that one considers the difference between $w^T x_k$ (the projected data points to the target space) and the value 0 as error variables. The projected variables correspond to what one calls the *score* variables. These error variables are maximized for the given N data points while keeping the norm of w small by the regularization term. The value γ is a positive real constant. The Lagrangian becomes

$$
\mathcal{L}(w,e;\alpha) = \gamma\frac{1}{2}\sum_{k=1}^{N} e_k^2 - \frac{1}{2}w^T w - \sum_{k=1}^{N} \alpha_k \left(e_k - w^T x_k\right)
\tag{7.6}
$$

with conditions for optimality given by

$$
\begin{cases}
\frac{\partial \mathcal{L}}{\partial w} = 0 & \rightarrow\quad w = \displaystyle\sum_{k=1}^{N} \alpha_k x_k \\[2mm]
\frac{\partial \mathcal{L}}{\partial e_k} = 0 & \rightarrow\quad \alpha_k = \gamma e_k, \qquad\quad k = 1,...,N \\[2mm]
\frac{\partial \mathcal{L}}{\partial \alpha_k} = 0 & \rightarrow\quad e_k - w^T x_k = 0, \quad k = 1,...,N.
\end{cases}
\tag{7.7}
$$

By elimination of the variables e, w one obtains

$$
\frac{1}{\gamma}\alpha_k - \sum_{l=1}^{N} \alpha_l x_l^T x_k = 0 , \quad k = 1,...,N.
\tag{7.8}
$$

By defining $\lambda = 1/\gamma$ one has the following dual symmetric eigenvalue problem

$$
\begin{bmatrix} \boxed{D} : & \text{solve in } \alpha : \\ & \begin{bmatrix} x_1^T x_1 & \cdots & x_1^T x_N \\ \vdots & & \vdots \\ x_N^T x_1 & \cdots & x_N^T x_N \end{bmatrix} \begin{bmatrix} \alpha_1 \\ \vdots \\ \alpha_N \end{bmatrix} = \lambda \begin{bmatrix} \alpha_1 \\ \vdots \\ \alpha_N \end{bmatrix} \end{bmatrix} \tag{7.9}
$$

which is the dual interpretation of (7.3). When one writes this as

$$
\Omega \alpha = \lambda \alpha \tag{7.10}
$$

where

$$
\Omega_{kl} = x_k^T x_l, \quad k, l = 1, ..., N. \tag{7.11}
$$

One easily sees that this matrix is the Gram matrix for a linear kernel $K(x_k, x_l) = x_k^T x_l$. The vector of dual variables $\alpha = [\alpha_1; ...; \alpha_N]$ is an eigenvector of the problem and λ is the corresponding eigenvalue. In order to obtain the maximal variance one selects the eigenvector corresponding to the largest eigenvalue.

The score variables become

$$
z(x) = w^T x = \sum_{l=1}^{N} \alpha_l x_l^T x \tag{7.12}
$$

where α is the eigenvector corresponding to the largest eigenvalue. Note that all eigenvalues are positive and real because the matrix is symmetric and positive semidefinite. One has in fact N local minima as solution to the problem for which one selects the solution of interest. The optimal solution is the eigenvector corresponding to the largest eigenvalue because in that case

$$
\sum_{k=1}^{N} (w^T x_k)^2 = \sum_{k=1}^{N} e_k^2 = \sum_{k=1}^{N} \frac{1}{\gamma^2} \alpha_k^2 = \lambda_{max}^2, \tag{7.13}
$$

where $\sum_{k=1}^{N} \alpha_k^2 = 1$ for the normalized eigenvector. For the different score variables one selects the eigenvectors corresponding to the different eigenvalues. The score variables are decorrelated from each other due to the fact that the α eigenvectors are orthonormal. According to [128], one can also additionally stress within the constraints of the formulation that the w vectors related to subsequent scores are orthogonal to each other.

PCA analysis is usually applied to centered data. Therefore one better considers the problem

$$\max_{w} \sum_{k=1}^{N} [w^T (x_k - \hat{\mu}_x)]^2 \tag{7.14}$$

where $\hat{\mu}_x = (1/N) \sum_{k=1}^{N} x_k$. The same derivations can be made and one finally obtains a centered Gram matrix as a result in the eigenvalue problem. One also sees that solving the problem in w is typically advantageous for large data sets, while for fewer given data in huge dimensional input spaces one better solves the dual problem. This point was also addressed in the context of (LS)-SVM classifiers. The approach of taking the eigenvalue decomposition of the centered Gram matrix is also done in *principal co-ordinate analysis* [99; 128].

7.1.3 *Including a bias term*

While in PCA analysis one usually centers the data, the LS-SVM interpretation to PCA analysis also offers the opportunity to analyse the use of a bias term in a straightforward way. The score variables are

$$z(x) = w^T x + b \tag{7.15}$$

and one aims at optimizing the following objective

$$\max_{w,b} \sum_{k=1}^{N} [0 - (w^T x_k + b)]^2. \tag{7.16}$$

Therefore, one formulates the primal optimization problem

$$\left[\boxed{\text{P}} : \quad \max_{w,b,e} J_P(w,e) = \quad \gamma \frac{1}{2} \sum_{k=1}^{N} e_k^2 - \frac{1}{2} w^T w \\ \text{such that} \quad e_k = w^T x_k + b, \quad k = 1, ..., N \right] \tag{7.17}$$

with Lagrangian

$$\mathcal{L}(w, b, e; \alpha) = \gamma \frac{1}{2} \sum_{k=1}^{N} e_k^2 - \frac{1}{2} w^T w - \sum_{k=1}^{N} \alpha_k \left(e_k - w^T x_k - b \right) \tag{7.18}$$

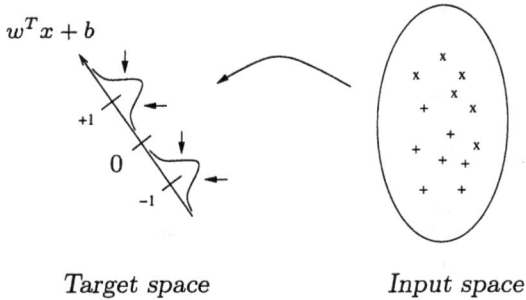

Minimize within class scatter

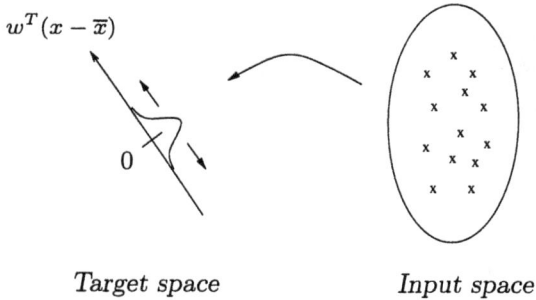

Find direction with maximal variance

Fig. 7.1   *Both Fisher discriminant analysis (FDA) (supervised learning) and PCA anal-*
ysis (unsupervised learning) can be derived from the viewpoint of LS-SVMs as a con-
strained optimization problem formulated in the primal space and solved in the dual
space of Lagrange multipliers. In FDA the within class scatter is minimized around
targets +1 and −1. PCA analysis can be interpreted as maximizing the variance around
target 0, i.e. as a one-class target zero modelling problem.

giving as conditions for optimality

$$
\begin{cases}
\frac{\partial \mathcal{L}}{\partial w} = 0 & \rightarrow \quad w = \sum_{k=1}^{N} \alpha_k x_k \\[2mm]
\frac{\partial \mathcal{L}}{\partial e_k} = 0 & \rightarrow \quad \alpha_k = \gamma e_k, \qquad\qquad k = 1, ..., N \\[2mm]
\frac{\partial \mathcal{L}}{\partial b} = 0 & \rightarrow \quad \sum_{k=1}^{N} \alpha_k = 0 \\[2mm]
\frac{\partial \mathcal{L}}{\partial \alpha_k} = 0 & \rightarrow \quad e_k - w^T x_k - b = 0, \quad k = 1, ..., N.
\end{cases}
\tag{7.19}
$$

Applying $\sum_{k=1}^{N} \alpha_k = 0$ the last condition delivers an expression for the bias term

$$
b = -\frac{1}{N} \sum_{k=1}^{N} \sum_{l=1}^{N} \alpha_l x_l^T x_k. \tag{7.20}
$$

By defining $\lambda = 1/\gamma$ one obtains the dual problem

$\boxed{D}$: \qquad solve in α :

$$
\begin{bmatrix}
(x_1 - \hat{\mu}_x)^T (x_1 - \hat{\mu}_x) & \cdots & (x_1 - \hat{\mu}_x)^T (x_N - \hat{\mu}_x) \\
\vdots & & \vdots \\
(x_N - \hat{\mu}_x)^T (x_1 - \hat{\mu}_x) & \cdots & (x_N - \hat{\mu}_x)^T (x_N - \hat{\mu}_x)
\end{bmatrix}
\begin{bmatrix} \alpha_1 \\ \vdots \\ \alpha_N \end{bmatrix}
= \lambda
\begin{bmatrix} \alpha_1 \\ \vdots \\ \alpha_N \end{bmatrix}
\tag{7.21}
$$

which is an eigenvalue decomposition of the centered Gram matrix

$$
\Omega_c \alpha = \lambda \alpha \tag{7.22}
$$

with $\Omega_c = M_c \Omega M_c$ where $M_c = I - 1_v 1_v^T / N$ and $\Omega_{kl} = x_k^T x_l$ for $k, l = 1, ..., N$. This eigenvalue problem follows from

$$
\frac{1}{\gamma} \alpha_k - \sum_{l=1}^{N} \alpha_l x_l^T x_k + \frac{1}{N} \sum_{k=1}^{N} \sum_{l=1}^{N} \alpha_l x_l^T x_k = 0
$$

by taking into account the fact that $\sum_{k=1}^{N} \alpha_k = 0$. One also sees that considering a bias term in the problem formulation automatically leads to a centering of the matrix.

The score variables equal

$$z(x) = w^T x + b = \sum_{l=1}^{N} \alpha_l x_l^T x + b \qquad (7.23)$$

where α is the eigenvector corresponding to the largest eigenvalue and

$$\sum_{k=1}^{N} (w^T x_k + b)^2 = \sum_{k=1}^{N} e_k^2 = \sum_{k=1}^{N} \frac{1}{\gamma^2} \alpha_k^2 = \lambda_{max}^2. \qquad (7.24)$$

7.1.4   *The reconstruction problem*

It was already mentioned in the first Chapter that PCA analysis can be related to a reconstruction problem with information bottleneck. One aims at minimizing the reconstruction error

$$\min \sum_{k=1}^{N} \|x_k - \tilde{x}_k\|_2^2 \qquad (7.25)$$

where $\tilde{x}_k$ are variables reconstructed from the score variables (Fig. 7.2). Let us denote the data matrix and the matrix with selected score variables as $X = [x_1 x_2 ... x_N] \in \mathbb{R}^{n \times N}$ and $Z = [z_1 z_2 ... z_N] \in \mathbb{R}^{n_s \times N}$, respectively, where n_s denotes the number of selected variables which determines the dimensionality reduction.

In the context of linear PCA analysis one considers a linear mapping from the scores to the reconstructed variables. In order to be able to handle also the bias term formulation case one can take

$$\tilde{x} = Vz + \delta \qquad (7.26)$$

and minimize

$$\min_{V,\delta} \sum_{k=1}^{N} \|x_k - (Vz_k + \delta)\|_2^2. \qquad (7.27)$$

In matrix form this leads to a least squares solution

$$[V\ \delta] = X \begin{bmatrix} Z \\ 1_v^T \end{bmatrix}^T \left(\begin{bmatrix} Z \\ 1_v^T \end{bmatrix} \begin{bmatrix} Z \\ 1_v^T \end{bmatrix}^T \right)^{-1} \qquad (7.28)$$

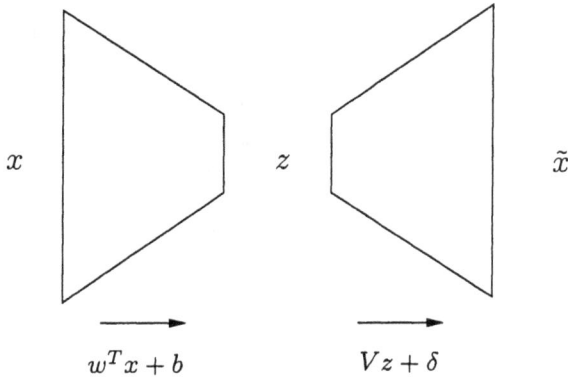

Fig. 7.2 *Reconstruction problem interpretation of linear PCA analysis related to the LS-SVM formulation with bias term. The inputs x are mapped to the score variables z to a lower dimension (dimensionality reduction step) and reconstructed into x̃.*

for the overdetermined problem

$$[V \; \delta] \begin{bmatrix} Z \\ 1_v^T \end{bmatrix} = X. \tag{7.29}$$

In Fig. 7.3 an illustrative example is given of linear PCA analysis with bias term in the problem formulation. Shown are the score variables and reconstructed variables. The two components are reconstructed by $\tilde{x}^{(i)} = V_i z^{(i)} + \delta_i$ for $i \in \{1, 2\}$ where $z^{(i)} \in \mathbb{R}$ are one-dimensional variables and

$$[V_i \; \delta_i] = X \begin{bmatrix} Z^{(i)} \\ 1_v^T \end{bmatrix}^T \left(\begin{bmatrix} Z^{(i)} \\ 1_v^T \end{bmatrix} \begin{bmatrix} Z^{(i)} \\ 1_v^T \end{bmatrix}^T \right)^{-1} \tag{7.30}$$

with $Z^{(i)} \in \mathbb{R}^{1 \times N}$ containing the scores related to the first and second largest eigenvalues, respectively.

In this cost function one usually considers the error on the given (training) data set. Of course issues of generalization are also relevant at this point. In [128] the use of cross-validation for PCA analysis has been discussed.

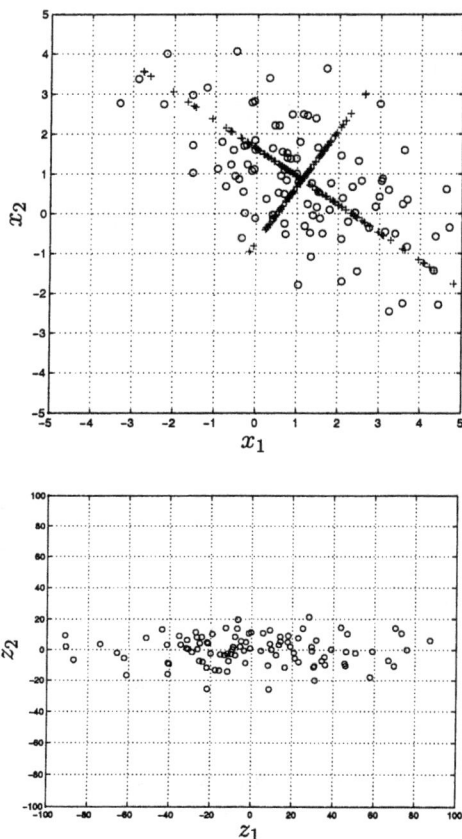

Fig. 7.3   Illustration of PCA analysis with bias term in the problem formulation for given data points depicted as 'o': (Top) reconstructed variables $\tilde{x}_k^{(1)}$ and $\tilde{x}_k^{(2)}$ based upon the scores $z_k^{(1)}$ and $z_k^{(2)}$ depicted as '+'; (Bottom) the score variables $z_k^{(1)}$ and $z_k^{(2)}$ which are decorrelated.

7.2   An LS-SVM approach to kernel PCA

Let us now extend the LS-SVM interpretation of linear PCA analysis to a high dimensional feature space and apply the kernel trick, as illustrated in Fig. 7.4.

Our objective is the following

$$\max_{w} \sum_{k=1}^{N} [0 - w^T(\varphi(x_k) - \hat{\mu}_\varphi)]^2 \tag{7.31}$$

with notation $\hat{\mu}_\varphi = (1/N)\sum_{k=1}^{N} \varphi(x_k)$ and $\varphi(\cdot) : \mathbb{R}^n \to \mathbb{R}^{n_h}$ the mapping to a high dimensional feature space which might be infinite dimensional. We take here the centering approach instead of using a bias term in the formulation. The following optimization problem is formulated now in the primal weight space

$$\left[\; \boxed{P}: \quad \max_{w,e} J_P(w,e) = \gamma\frac{1}{2}\sum_{k=1}^{N} e_k^2 - \frac{1}{2}w^T w \right. \\ \left. \text{such that} \quad e_k = w^T(\varphi(x_k) - \hat{\mu}_\varphi), \; k = 1, ..., N. \;\right] \tag{7.32}$$

This gives the Lagrangian

$$\mathcal{L}(w,e;\alpha) = \gamma\frac{1}{2}\sum_{k=1}^{N} e_k^2 - \frac{1}{2}w^T w - \sum_{k=1}^{N}\alpha_k\left(e_k - w^T(\varphi(x_k) - \hat{\mu}_\varphi)\right) \tag{7.33}$$

with conditions for optimality

$$\begin{cases} \frac{\partial \mathcal{L}}{\partial w} = 0 & \to \quad w = \sum_{k=1}^{N}\alpha_k(\varphi(x_k) - \hat{\mu}_\varphi) \\[2mm] \frac{\partial \mathcal{L}}{\partial e_k} = 0 & \to \quad \alpha_k = \gamma e_k, \qquad\qquad\qquad k = 1, ..., N \\[2mm] \frac{\partial \mathcal{L}}{\partial \alpha_k} = 0 & \to \quad e_k - w^T(\varphi(x_k) - \hat{\mu}_\varphi) = 0, \quad k = 1, ..., N. \end{cases} \tag{7.34}$$

By elimination of the variables e, w one obtains

$$\frac{1}{\gamma}\alpha_k - \sum_{l=1}^{N}\alpha_l(\varphi(x_l) - \hat{\mu}_\varphi)^T(\varphi(x_k) - \hat{\mu}_\varphi) = 0 \;, \; k = 1, ..., N. \tag{7.35}$$

Defining $\lambda = 1/\gamma$ one obtains the following dual problem

$$\left[\; \boxed{D}: \quad \text{solve in } \alpha: \\[4mm] \Omega_c\alpha = \lambda\alpha \;\right] \tag{7.36}$$

with

$$
\Omega_c = \begin{bmatrix} (\varphi(x_1) - \hat{\mu}_\varphi)^T (\varphi(x_1) - \hat{\mu}_\varphi) & \cdots & (\varphi(x_1) - \hat{\mu}_\varphi)^T (\varphi(x_N) - \hat{\mu}_\varphi) \\ \vdots & & \vdots \\ (\varphi(x_N) - \hat{\mu}_\varphi)^T (\varphi(x_1) - \hat{\mu}_\varphi) & \cdots & (\varphi(x_N) - \hat{\mu}_\varphi)^T (\varphi(x_N) - \hat{\mu}_\varphi) \end{bmatrix}.
$$
(7.37)

One has the following elements for the centered kernel matrix

$$
\Omega_{c,kl} = (\varphi(x_k) - \hat{\mu}_\varphi)^T (\varphi(x_l) - \hat{\mu}_\varphi), \quad k, l = 1, ..., N.
$$
(7.38)

For the centered kernel matrix one can apply the kernel trick as follows for given points x_k, x_l:

$$
(\varphi(x_k) - \hat{\mu}_\varphi)^T (\varphi(x_l) - \hat{\mu}_\varphi)
$$

$$
= \left(\varphi(x_k) - \frac{1}{N} \sum_{r=1}^{N} \varphi(x_r) \right)^T \left(\varphi(x_l) - \frac{1}{N} \sum_{r=1}^{N} \varphi(x_r) \right)
$$

$$
= \varphi(x_k)^T \varphi(x_l) - \varphi(x_k)^T \frac{1}{N} \sum_{r=1}^{N} \varphi(x_r) - \varphi(x_l)^T \frac{1}{N} \sum_{r=1}^{N} \varphi(x_r) +
$$

$$
\frac{1}{N^2} \sum_{r=1}^{N} \sum_{s=1}^{N} \varphi(x_r)^T \varphi(x_s)
$$

$$
= K(x_k, x_l) - \frac{1}{N} \sum_{r=1}^{N} K(x_k, x_r) - \frac{1}{N} \sum_{r=1}^{N} K(x_l, x_r) + \frac{1}{N^2} \sum_{r=1}^{N} \sum_{s=1}^{N} K(x_r, x_s).
$$
(7.39)

This solution is equivalent with kernel PCA as proposed by Schölkopf *et al.* in [203]. The centered kernel matrix can be computed as $\Omega_c = M_c \Omega M_c$ with $\Omega_{kl} = K(x_k, x_l)$ with M_c the centering matrix. This issue of centering is also of importance in methods of principal co-ordinate analysis [128].

The optimal solution to the formulated problem is obtained by selecting the eigenvector corresponding to the largest eigenvalue. The projected variables become

$$
\begin{aligned}
z(x) &= w^T \left(\varphi(x) - \hat{\mu}_\varphi \right) \\
&= \sum_{l=1}^{N} \alpha_l \left(\varphi(x_l) - \hat{\mu}_\varphi \right)^T \left(\varphi(x) - \hat{\mu}_\varphi \right) \\
&= \sum_{l=1}^{N} \alpha_l \left(K(x_l, x) - \frac{1}{N} \sum_{r=1}^{N} K(x_r, x) - \frac{1}{N} \sum_{r=1}^{N} K(x_r, x_l) + \right. \\
&\qquad \left. \frac{1}{N^2} \sum_{r=1}^{N} \sum_{s=1}^{N} K(x_r, x_s) \right).
\end{aligned}
\tag{7.40}
$$

One may choose here any positive definite kernel satisfying the Mercer condition, with the RBF kernel as a typical choice.

For the nonlinear PCA case the number of score variables n_s can be larger than the dimension of the input space n. One selects then as few score variables as possible and minimize the reconstruction error (Fig. 7.5). In this form of nonlinear PCA the mappings are nonlinear. The mapping from the score variables to the reconstructed input variables is done as

$$
\tilde{x} = h(z)
\tag{7.41}
$$

such that one minimizes the reconstruction error

$$
\min \sum_{k=1}^{N} \|x_k - h(z_k)\|_2^2.
\tag{7.42}
$$

This form of nonlinear PCA analysis is common in the area of neural networks [23]. A different reconstruction method has been discussed by Schölkopf *et al.* in [204]. In Fig. 7.6 an illustrative example is given of kernel PCA with RBF kernel applied to a noisy sine function problem. The intrinsic dimensionality of the problem is 1. Based upon the second score variable a good reconstruction with *denoising* of the given data can be made. For the nonlinear mapping $h(\cdot)$ an MLP with one hidden layer has been taken which was trained by Bayesian learning. The eigenvalues of the kernel matrix are shown in Fig. 7.7.

$$\boxed{\textbf{LS-SVM interpretation to Kernel PCA}}$$

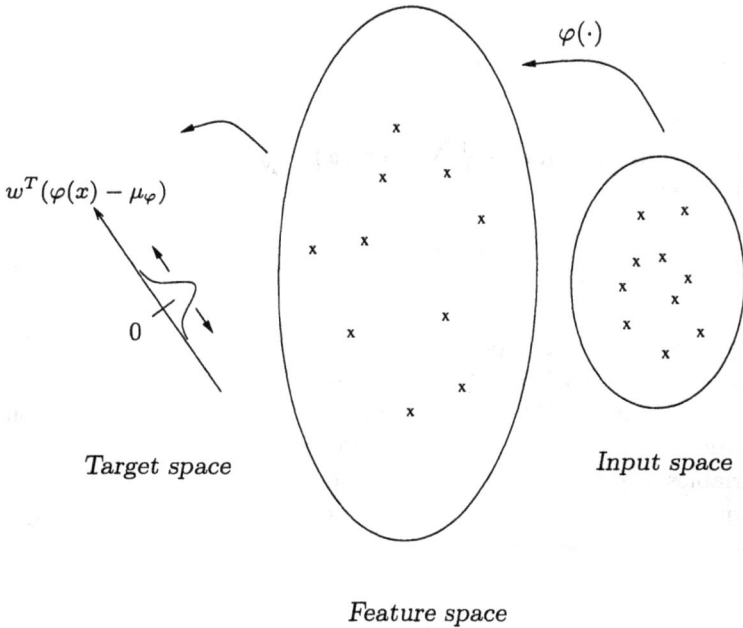

Feature space

Find direction with maximal variance

Fig. 7.4   *LS-SVM approach to kernel principal component analysis: the input data are mapped to a high dimensional feature space and next to the score variables. The score variables are interpreted as error variables in a one-class modelling problem with target zero for which one aims at having maximal variance.*

7.3   Links with density estimation

A link between kernel PCA and orthogonal series density estimation has been established by Girolami [94]. A probability density function that is square integrable can be represented by a convergent orthogonal series

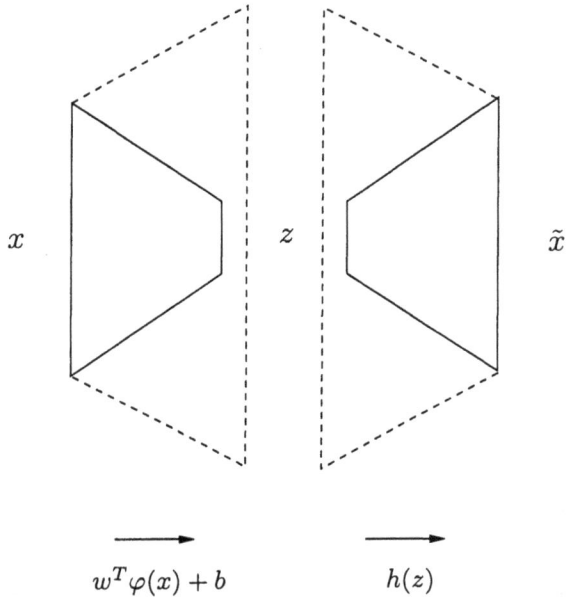

$$w^T\varphi(x) + b \qquad\qquad h(z)$$

Fig. 7.5 *Reconstruction problem interpretation of kernel PCA analysis related to the LS-SVM formulation. The inputs x are mapped to the score variables z. This number can be larger than the dimension of the input space, but one selects as few as possible and minimizes the reconstruction error. In this form of nonlinear PCA the mappings are nonlinear.*

expansion such that

$$p(x) = \sum_{i=1}^{\infty} c_i \varphi_i(x) \qquad\qquad (7.43)$$

with $x \in \mathbb{R}^n$ and $\{\varphi_i(x)\}_{i=1}^{\infty}$ an orthonormal set of functions [120]. For an orthonormal series expansion in a Hilbert space with density function $p(x)$ the expansion coefficients equal [136]

$$c_i = \int \varphi_i(x) p(x) dx. \qquad\qquad (7.44)$$

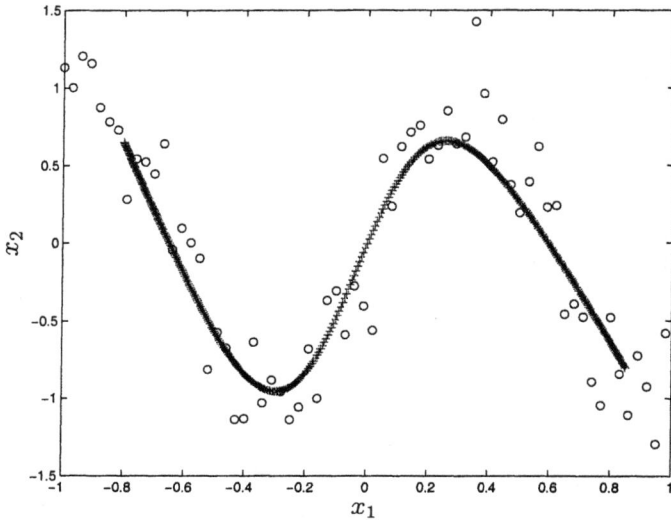

Fig. 7.6   *Illustration of kernel PCA to noisy sine function data depicted by 'o' in a two-dimensional input space. The reconstructed variables $\tilde{x}_k$ are shown as '+' and are reconstructed based upon one single score variable. This shows that the method is capable of discovering the intrinsic dimensionality equal to one of the noisy sine function line in the input space and denoise the noisy sine function.*

Towards illustrating the link with kernel PCA it is needed to consider the truncated series

$$\hat{p}_M(x) = \sum_{i=1}^{M} \hat{c}_i \varphi_i(x) \tag{7.45}$$

consisting of M terms with estimated coefficients

$$\hat{c}_i = \frac{1}{N} \sum_{k=1}^{N} \varphi_i(x_k) \tag{7.46}$$

which gives

$$\hat{p}_M(x) = \frac{1}{N} \sum_{i=1}^{M} \sum_{k=1}^{N} \varphi_i(x_k) \varphi_i(x). \tag{7.47}$$

Although it can be shown that this method is asymptotically unbiased it may produce negative point values which is an important drawback of the

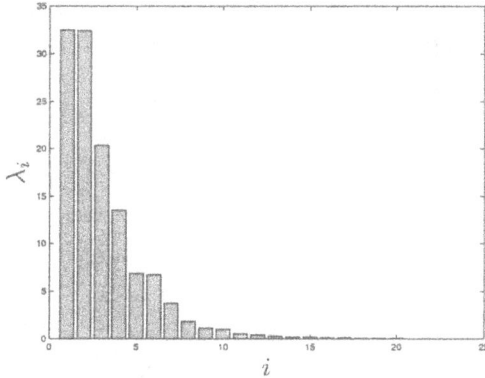

Fig. 7.7 *Eigenvalues of the centered kernel matrix (scree graph) related to the previous Figure of the noisy sine function.*

method.

One can then take the scores resulting from kernel PCA as basis function for a density estimator. Therefore, one considers the eigenvalue decomposition of the centered kernel matrix

$$\Omega_c U = U \tilde{\Lambda} \tag{7.48}$$

where $\tilde{\Lambda} = \mathrm{diag}([\tilde{\lambda}_1; ...; \tilde{\lambda}_N])$ contains the eigenvalues and $U = [u_1...u_N] \in \mathbb{R}^{N \times N}$ the corresponding eigenvectors. This can be used in order to estimate the eigenfunctions $\phi_i(x)$ and eigenvalues λ_i for the integral equation (Karhunen-Loeve expansion)

$$\int K(x, x')\phi_i(x)p(x)dx = \lambda_i\phi_i(x') \tag{7.49}$$

with the following estimates

$$
\begin{aligned}
\hat{\lambda}_i &= \tfrac{1}{N}\tilde{\lambda}_i \\
\hat{\phi}_i(x_k) &= \sqrt{N}u_{ki} \\
\hat{\phi}_i(x') &= \tfrac{\sqrt{N}}{\tilde{\lambda}_i}\sum_{k=1}^{N} u_{ki}K(x_k, x')
\end{aligned}
\tag{7.50}
$$

where u_{ki} denotes the ki-th entry of the matrix U. This approach is a form of Nyström method [63; 294]. Using the eigenvectors as finite sample estimates of the corresponding eigenfunctions, the truncated estimate of the probability density function at point x' is given by

$$
\begin{aligned}
\hat{p}_M(x') &= \frac{1}{N} 1_v^T \sum_{i=1}^{M} \sqrt{\tilde{\lambda}_i} u_i \sum_{k=1}^{N} \frac{1}{\sqrt{\tilde{\lambda}_i}} u_{ki} K(x_k, x') \\
&= \frac{1}{N} 1_v^T U_M U_M^T \theta(x')
\end{aligned}
\tag{7.51}
$$

where $\theta(x') = [K(x', x_1); K(x', x_2); ...; K(x', x_N)]$, $1_v = [1; 1; ...; 1]$ and $U_M \in \mathbb{R}^{N \times M}$ is the matrix with eigenvectors of Ω_c consisting of the eigenvectors corresponding to the M largest eigenvalues. For the case of $M = N$ this reduces to the well-known Parzen window density estimator $p(x') = \frac{1}{N} 1_v^T \theta(x')$. Note that in order to obtain a density the integral of the function over the entire space should be equal to one. It is assumed here that normalizations are done for the kernel K to achieve this.

Concerning the number of terms to be taken in the expansion many criteria have been investigated in literature [67]. In [94] the following has been investigated with cutoff value for determination of the value of M:

$$
(\frac{1}{N} 1_v^T u_i)^2 > \frac{2N}{1 + N}.
\tag{7.52}
$$

An estimate for the overall integrated square truncation error $\sum_{i=M+1}^{\infty} c_i^2$ is given by

$$
c_i^2 \simeq \tilde{\lambda}_i (\frac{1}{N} 1_v^T u_i)^2.
\tag{7.53}
$$

This can also be related to the quadratic Renyi entropy

$$
H_R = -\log \int p(x)^2 dx
\tag{7.54}
$$

which forms a measure of distribution compactness [87] and has recently been used within the context of information theoretic learning [187]. In [94] Girolami shows that

$$
\int \hat{p}(x)^2 dx = \sum_{i=1}^{N} \tilde{\lambda}_i (\frac{1}{N} 1_v^T u_i)^2.
\tag{7.55}
$$

Large contributions to the entropy come from components that have small values of $\tilde{\lambda}_i (\frac{1}{N} 1_v^T u_i)^2$ and are related to elements with little or no structure, caused by observation noise or diffuse regions in the data. Large values of $\tilde{\lambda}_i(\frac{1}{N}1_v^T u_i)^2$ on the other hand indicate regions of high density or compactness. A more general result according to [94] is

$$\int \hat{p}(x)^2 dx = \frac{1}{N^2}1_v^T \Omega 1_v \qquad (7.56)$$

in the sense that this result is not restricted to RBF kernels.

7.4   Kernel CCA

7.4.1   *Classical canonical correlation analysis formulation*

The problem of canonical correlation analysis (CCA) is related to the PCA analysis problem [128]. In CCA analysis (originally studied by Hotelling in 1936 [116]) one is interested in finding maximal correlation between projected variables $z_x = w^T x$ and $z_y = v^T y$ where $x \in \mathbb{R}^{n_x}, y \in \mathbb{R}^{n_y}$ denote given random vectors with zero mean. Linear CCA analysis has been applied in subspace algorithms for system identification [275], with links to system theory, information theory and signal processing [61].

The objective function is to maximize the correlation coefficient

$$
\begin{aligned}
\max_{w,v} \rho &= \frac{\mathcal{E}[z_x z_y]}{\sqrt{\mathcal{E}[z_x z_x]}\sqrt{\mathcal{E}[z_y z_y]}} \\
&= \frac{w^T C_{xy} v}{\sqrt{w^T C_{xx} w}\sqrt{v^T C_{yy} v}}
\end{aligned}
\qquad (7.57)
$$

with $C_{xx} = \mathcal{E}[xx^T]$, $C_{yy} = \mathcal{E}[yy^T]$, $C_{xy} = \mathcal{E}[xy^T]$. This is usually formulated as the constrained optimization problem

$$
\begin{aligned}
\max_{w,v} &\quad w^T C_{xy} v \\
\text{such that} &\quad w^T C_{xx} w = 1 \\
&\quad v^T C_{yy} v = 1
\end{aligned}
\qquad (7.58)
$$

which leads to a generalized eigenvalue problem. The solution follows from the Lagrangian

$$\mathcal{L}(w,v;\eta,\nu) = w^T C_{xy} v - \eta\frac{1}{2}(w^T C_{xx} w - 1) - \nu\frac{1}{2}(v^T C_{yy} v - 1) \qquad (7.59)$$

with Lagrange multipliers η, ν, which gives

$$
\begin{cases}
C_{xy}v = \eta\, C_{xx}w \\
C_{yx}w = \nu\, C_{yy}v.
\end{cases}
\tag{7.60}
$$

7.4.2   Support vector machine formulation to linear CCA

In a similar fashion as for PCA analysis one can develop a support vector machine type formulation. This is done by considering the primal problem

$$
\boxed{P}: \quad \max_{w,v,e,r} \quad \gamma \sum_{k=1}^{N} e_k r_k - \nu_1 \frac{1}{2} \sum_{k=1}^{N} e_k^2 - \nu_2 \frac{1}{2} \sum_{k=1}^{N} r_k^2 - \frac{1}{2} w^T w - \frac{1}{2} v^T v
$$

$$
\text{such that } e_k = w^T x_k, \qquad\qquad k = 1, ..., N
$$

$$
r_k = v^T y_k \qquad\qquad k = 1, ..., N
$$

$$
\tag{7.61}
$$

with Lagrangian

$$
\mathcal{L}(w,v,e,r;\alpha,\beta) = \gamma \sum_{k=1}^{N} e_k r_k - \nu_1 \frac{1}{2} \sum_{k=1}^{N} e_k^2 - \nu_2 \frac{1}{2} \sum_{k=1}^{N} r_k^2 - \frac{1}{2} w^T w -
$$

$$
\frac{1}{2} v^T v - \sum_{k=1}^{N} \alpha_k [e_k - w^T x_k] - \sum_{k=1}^{N} \beta_k [r_k - v^T y_k]
$$

$$
\tag{7.62}
$$

where α_k, β_k are Lagrange multipliers. The objective function in the primal problem does not have the same expression as the correlation coefficient but takes the contribution of the numerator with a plus sign and the contributions of the denominator with a minus sign. However, the cost function can be considered as a generalization of the objective $\min_{w,v} \sum_k \|w^T x_k - v^T y_k\|_2^2$ which is known in the area of CCA analysis [96]. Additional unknown error variables e, r are taken which are equal to the score variables z_x, z_y. By considering the expressions of the score variables as constraints one is able to find a meaningful dual problem for which one is able to create a kernel version.

The conditions for optimality are

$$
\begin{cases}
\frac{\partial \mathcal{L}}{\partial w} = 0 & \rightarrow \quad w = \sum_{k=1}^{N} \alpha_k x_k \\[2mm]
\frac{\partial \mathcal{L}}{\partial v} = 0 & \rightarrow \quad v = \sum_{k=1}^{N} \beta_k y_k \\[2mm]
\frac{\partial \mathcal{L}}{\partial e_k} = 0 & \rightarrow \quad \gamma v^T y_k = \nu_1 w^T x_k + \alpha_k, \quad k = 1, ..., N \\[2mm]
\frac{\partial \mathcal{L}}{\partial r_k} = 0 & \rightarrow \quad \gamma w^T x_k = \nu_2 v^T y_k + \beta_k, \quad k = 1, ..., N \\[2mm]
\frac{\partial \mathcal{L}}{\partial \alpha_k} = 0 & \rightarrow \quad e_k = w^T x_k, \quad\quad\quad\quad k = 1, ..., N \\[2mm]
\frac{\partial \mathcal{L}}{\partial \beta_k} = 0 & \rightarrow \quad r_k = v^T y_k, \quad\quad\quad\quad k = 1, ..., N
\end{cases}
\tag{7.63}
$$

which results in the following dual problem after defining $\lambda = 1/\gamma$

$$
\boxed{D}: \qquad \text{solve in } \alpha, \beta :
$$

$$
\begin{bmatrix}
 & 0 & & y_1^T y_1 & \cdots & y_1^T y_N \\
 & & & \vdots & & \vdots \\
 & & & y_N^T y_1 & \cdots & y_N^T y_N \\
\hline
x_1^T x_1 & \cdots & x_1^T x_N & & & \\
\vdots & & \vdots & & 0 & \\
x_N^T x_1 & \cdots & x_N^T x_N & & &
\end{bmatrix}
\begin{bmatrix}
\alpha_1 \\ \vdots \\ \alpha_N \\ \beta_1 \\ \vdots \\ \beta_N
\end{bmatrix}
$$

$$
= \lambda
\begin{bmatrix}
\nu_1 x_1^T x_1 + 1 & \cdots & \nu_1 x_1^T x_N & & & \\
\vdots & & \vdots & & 0 & \\
\nu_1 x_N^T x_1 & \cdots & \nu_1 x_N^T x_N + 1 & & & \\
\hline
 & & & \nu_2 y_1^T y_1 + 1 & \cdots & \nu_2 y_1^T y_N \\
 & 0 & & \vdots & & \vdots \\
 & & & \nu_2 y_N^T y_1 & \cdots & \nu_2 y_N^T y_N + 1
\end{bmatrix}
\begin{bmatrix}
\alpha_1 \\ \vdots \\ \alpha_N \\ \beta_1 \\ \vdots \\ \beta_N
\end{bmatrix}.
$$

$$
\tag{7.64}
$$

This is a generalized eigenvalue problem to be solved with eigenvectors $[\alpha; \beta]$ and corresponding eigenvalues λ. One can then select the value of λ and the corresponding eigenvectors such that the correlation coefficient is maximized

$$\max \rho(\lambda). \tag{7.65}$$

The resulting score variables are

$$
\begin{aligned}
z_{x_k} &= e_k = \sum_{l=1}^{N} \alpha_l x_l^T x_k \\
z_{y_k} &= r_k = \sum_{l=1}^{N} \beta_l y_l^T y_k.
\end{aligned}
\tag{7.66}
$$

The eigenvalues λ will be both positive and negative for the CCA problem. Also note that one has $\rho \in [-1, 1]$.

7.4.3   *Extension to kernel CCA*

Let us now extend this formulation for linear CCA to a nonlinear version by mapping the input space to a high dimensional feature space. The score variables are now

$$
\begin{aligned}
z_x &= w^T (\varphi_1(x) - \hat{\mu}_{\varphi_1}) \\
z_y &= v^T (\varphi_2(y) - \hat{\mu}_{\varphi_2})
\end{aligned}
\tag{7.67}
$$

where $\varphi_1(\cdot) : \mathbb{R}^{n_x} \to \mathbb{R}^{n_{h_x}}$ and $\varphi_2(\cdot) : \mathbb{R}^{n_y} \to \mathbb{R}^{n_{h_y}}$ are mappings (which can be chosen to be different) to high dimensional feature spaces and $\hat{\mu}_{\varphi_1} = (1/N) \sum_{k=1}^{N} \varphi_1(x_k)$, $\hat{\mu}_{\varphi_2} = (1/N) \sum_{k=1}^{N} \varphi_2(y_k)$. It is important to take centering into account for the nonlinear CCA problem.

One starts from the primal problem

$$
\boxed{\text{P}}: \quad \max_{w,v,e,r} \quad \gamma \sum_{k=1}^{N} e_k r_k - \nu_1 \frac{1}{2} \sum_{k=1}^{N} e_k^2 - \nu_2 \frac{1}{2} \sum_{k=1}^{N} r_k^2 - \frac{1}{2} w^T w - \frac{1}{2} v^T v
$$

$$
\begin{aligned}
\text{such that} \quad e_k &= w^T (\varphi_1(x_k) - \hat{\mu}_{\varphi_1}), && k = 1, ..., N \\
r_k &= v^T (\varphi_2(y_k) - \hat{\mu}_{\varphi_2}), && k = 1, ..., N
\end{aligned}
\tag{7.68}
$$

with Lagrangian

$$\mathcal{L}(w, v, e, r; \alpha, \beta) = \gamma \sum_{k=1}^{N} e_k r_k - \nu_1 \frac{1}{2} \sum_{k=1}^{N} e_k^2 - \nu_2 \frac{1}{2} \sum_{k=1}^{N} r_k^2 - \frac{1}{2} w^T w -$$

$$\frac{1}{2} v^T v - \sum_{k=1}^{N} \alpha_k [e_k - w^T (\varphi_1(x_k) - \hat{\mu}_{\varphi_1})] - \sum_{k=1}^{N} \beta_k [r_k - v^T (\varphi_2(y_k) - \hat{\mu}_{\varphi_2})]$$

$$(7.69)$$

where α_k, β_k are Lagrange multipliers. Note that w and v might be infinite dimensional now. For this reason it is necessary to derive the dual problem.

The conditions for optimality are

$$\begin{cases} \frac{\partial \mathcal{L}}{\partial w} = 0 & \rightarrow \quad w = \sum_{k=1}^{N} \alpha_k (\varphi_1(x_k) - \hat{\mu}_{\varphi_1}) \\[2mm] \frac{\partial \mathcal{L}}{\partial v} = 0 & \rightarrow \quad v = \sum_{k=1}^{N} \beta_k (\varphi_2(y_k) - \hat{\mu}_{\varphi_2}) \\[2mm] \frac{\partial \mathcal{L}}{\partial e_k} = 0 & \rightarrow \quad \gamma v^T (\varphi_2(y_k) - \hat{\mu}_{\varphi_2}) = \nu_1 w^T (\varphi_1(x_k) - \hat{\mu}_{\varphi_1}) + \alpha_k \\ & \qquad\qquad\qquad\qquad\qquad\qquad\qquad\quad k = 1, ..., N \\[2mm] \frac{\partial \mathcal{L}}{\partial r_k} = 0 & \rightarrow \quad \gamma w^T (\varphi_1(x_k) - \hat{\mu}_{\varphi_1}) = \nu_2 v^T (\varphi_2(y_k) - \hat{\mu}_{\varphi_2}) + \beta_k \\ & \qquad\qquad\qquad\qquad\qquad\qquad\qquad\quad k = 1, ..., N \\[2mm] \frac{\partial \mathcal{L}}{\partial \alpha_k} = 0 & \rightarrow \quad e_k = w^T (\varphi_1(x_k) - \hat{\mu}_{\varphi_1}) \\ & \qquad\qquad\qquad\qquad\qquad\qquad\qquad\quad k = 1, ..., N \\[2mm] \frac{\partial \mathcal{L}}{\partial \beta_k} = 0 & \rightarrow \quad r_k = v^T (\varphi_2(y_k) - \hat{\mu}_{\varphi_2}) \\ & \qquad\qquad\qquad\qquad\qquad\qquad\qquad\quad k = 1, ..., N \end{cases}$$

$$(7.70)$$

which results in the following dual problem after defining $\lambda = 1/\gamma$

$$\begin{bmatrix} \boxed{D}: & \text{solve in } \alpha, \beta : \\[4mm] & \begin{bmatrix} 0 & \Omega_{c,2} \\ \Omega_{c,1} & 0 \end{bmatrix} \begin{bmatrix} \alpha \\ \beta \end{bmatrix} \\[6mm] & = \lambda \begin{bmatrix} \nu_1 \Omega_{c,1} + I & 0 \\ 0 & \nu_2 \Omega_{c,2} + I \end{bmatrix} \begin{bmatrix} \alpha \\ \beta \end{bmatrix} \end{bmatrix}$$

$$(7.71)$$

where

$$\begin{aligned} \Omega_{c,1_{kl}} &= (\varphi_1(x_k) - \hat{\mu}_{\varphi_1})^T (\varphi_1(x_l) - \hat{\mu}_{\varphi_1}) \\ \Omega_{c,2_{kl}} &= (\varphi_2(y_k) - \hat{\mu}_{\varphi_2})^T (\varphi_2(y_l) - \hat{\mu}_{\varphi_2}) \end{aligned} \qquad (7.72)$$

are the elements of the centered Gram matrices for $k, l = 1, ..., N$. In practice these matrices can be computed by $\Omega_{c,1} = M_c \Omega_1 M_c$, $\Omega_{c,2} = M_c \Omega_2 M_c$ with centering matrix M_c. The eigenvalues and eigenvectors that give an optimal correlation coefficient value are selected. The resulting score variables can be computed by applying the kernel trick with kernels $K_1(x_k, x_l) = \varphi_1(x_k)^T \varphi_1(x_l)$, $K_2(y_k, y_l) = \varphi_2(y_k)^T \varphi_2(y_l)$.

Using the CCA method one is in search of finding interesting relations between variables. One may apply this for example for input selection. At this point it is important to make a good choice of the tuning parameters and of the kernels and their tuning parameters. One may use an additional validation set to ensure meaningful generalization of the method. Further extensions towards independent component analysis (ICA) in relation to multiway parafac-candecomp models may be studied in view of rank-one approximation to high order tensors and a generalized Rayleigh quotient as defined in [302]. The CCA formulation can also be further related to partial least squares (PLS), for which kernel versions have been studied in [195].

Chapter 8

LS-SVM for Recurrent Networks and Control

The problems discussed in the previous Chapters were either supervised or unsupervised but on the other hand always static in the sense that there are no recursive equations involved in these formulations. In this Chapter we show how LS-SVM methods can be extended to problems with dynamics, more specifically an extension is made from feedforward to recurrent LS-SVMs and the use of LS-SVMs in the context of optimal control. Although the resulting problems are non-convex one can still apply the kernel trick after formulating the primal problem and taking conditions for optimality from the Lagrangian.

8.1 Recurrent Least Squares Support Vector Machines

8.1.1 *From feedforward to recurrent LS-SVM formulations*

The methods for nonlinear function estimation by LS-SVM as discussed in previous Chapters can be used in a dynamical systems context only in combination with NARX models

$$\hat{y}_k = f(y_{k-1}, y_{k-2}, ..., y_{k-p}, u_{k-1}, u_{k-2}, ..., u_{k-p}) \qquad (8.1)$$

where $\hat{y}$ denotes the estimated output, y the true output and f a smooth nonlinear mapping. This model structure is intended for modelling a dynamical system with input $u \in \mathbb{R}$ and output $y \in \mathbb{R}$ with discrete time index k. The value p denotes the model order. On the other hand, the model is essentially feedforward despite the fact that it is used to model a dynamical system. When using static LS-SVM models the given training data set consists of given input data vectors $\{[y_{k-1}; y_{k-2}; ...; y_{k-p}; u_{k-1}; u_{k-2}; ...; u_{k-p}]\}_{k=p+1}^{p+N}$ and corresponding given target output values $\{y_k\}_{k=p+1}^{p+N}$. Although this model structure is employed to model dynamics of a system, it is inherently feedforward due to the fact that the equation is not recursive.

A *recurrent model* [230; 231] on the other hand in input-output description takes the form

$$\hat{y}_k = f(\hat{y}_{k-1}, \hat{y}_{k-2}, ..., \hat{y}_{k-p}, u_{k-1}, u_{k-2}, ..., u_{k-p}) \qquad (8.2)$$

which is a recursive equation in the estimated output. This is a nonlinear output error model (NOE). Let us further study now the autonomous case of this recurrent model

$$\hat{y}_k = f(\hat{y}_{k-1}, \hat{y}_{k-2}, ..., \hat{y}_{k-p}) \qquad (8.3)$$

and explain how LS-SVM methodology can be used in the context of such model structures. We will do this again by stating the problem in the primal weight space as a constrained optimization problem. Next we construct the Lagrangian, take conditions for optimality and solve the dual problem in the Lagrange multipliers.

The following model and optimization problem were formulated by Suykens & Vandewalle in [240]. The recurrent model in the primal weight space is given by

$$\hat{y}_k = w^T \varphi(\begin{bmatrix} \hat{y}_{k-1} \\ \hat{y}_{k-2} \\ \vdots \\ \hat{y}_{k-p} \end{bmatrix}) + b, \qquad (8.4)$$

where $\varphi(\cdot)$ is the mapping to the high dimensional feature space. The initial condition for this model is assumed to be given. Starting from the initial condition one can simulate the system (8.4) in time to generate a sequence of values $\{\hat{y}\}$, while the model (8.1) is simply a one step ahead predictor model. The model (8.4) is expressed then in terms of the true measured output values and a set of unknown error variables:

$$y_k - e_k = w^T \varphi(y_{k-1\_k-p} - e_{k-1\_k-p}) + b \qquad (8.5)$$

where $e_k = y_k - \hat{y}_k$, $y_{k-1\_k-p} = [y_{k-1}; y_{k-2}; ...; y_{k-p}]$, $e_{k-1\_k-p} = [e_{k-1}; e_{k-2}; ...; e_{k-p}]$ by definition. The output weight vector and bias term are denoted by $w \in \mathbb{R}^{n_h}$ and $b \in \mathbb{R}$. Note that recurrent neural networks are classically trained by dynamic backpropagation or backpropagation through time [170; 289], which is more complicated than a backpropagation method for training of feedforward neural networks.

The recurrent LS-SVM approach starts from the following constrained optimization problem in primal weight space [240]

$$
\boxed{P}: \quad \min_{w,b,e} J_P(w,e) = \frac{1}{2}w^T w + \gamma \frac{1}{2} \sum_{k=p+1}^{p+N} e_k^2
$$

$$
\text{such that} \quad y_k - e_k = w^T \varphi(y_{k-1\_k-p} - e_{k-1\_k-p}) + b,
$$
$$
k = 1, ..., N. \tag{8.6}
$$

At this point it is important to note that the constraints are *nonlinear* in the unknown e variables, while in the static LS-SVM case the constraints are linear in the unknowns w, b, e. Hence, the problem will become non-convex in general with a non-unique solution. The Lagrangian for the problem is

$$
\mathcal{L}(w,b,e;\alpha) = \mathcal{J}(w,b,e) + \sum_{k=p+1}^{p+N} \alpha_{k-p}[y_k - e_k - w^T \varphi(y_{k-1\_k-p} - e_{k-1\_k-p}) - b].
$$
$$
\tag{8.7}
$$

The conditions for optimality are given by

$$
\begin{cases}
\dfrac{\partial \mathcal{L}}{\partial w} &=\ w - \sum_{k=p+1}^{p+N} \alpha_{k-p}\,\varphi(y_{k-1\_k-p} - e_{k-1\_k-p}) = 0 \\[2em]
\dfrac{\partial \mathcal{L}}{\partial b} &=\ \sum_{k=p+1}^{p+N} \alpha_{k-p} = 0 \\[2em]
\dfrac{\partial \mathcal{L}}{\partial e_k} &=\ \gamma e_k - \alpha_{k-p} - \sum_{i=1}^{p} \alpha_{k-p+i} \dfrac{\partial}{\partial e_{k-i}} [w^T \varphi(y_{k-1\_k-p} \\
& \quad\ -e_{k-1\_k-p})] = 0, \qquad\qquad\qquad k = p+1, ..., N \\[2em]
\dfrac{\partial \mathcal{L}}{\partial \alpha_{k-p}} &=\ y_k - e_k - w^T \varphi(y_{k-1\_k-p} - e_{k-1\_k-p}) - b = 0, \\
& \qquad\qquad\qquad\qquad\qquad\qquad k = p+1, ..., p+N. \tag{8.8}
\end{cases}
$$

It is difficult to eliminate the unknowns e_k from this set of nonlinear equations. However, it is straightforward to eliminate w (which can be infinite dimensional) and apply the kernel trick by $K(y_{k-1\_k-p}-e_{k-1\_k-p}, y_{l-1\_l-p}-e_{l-1\_l-p}) = \varphi(y_{k-1\_k-p}-e_{k-1\_k-p})^T \varphi(y_{l-1\_l-p}-e_{l-1\_l-p})$.

This leads to the following set of nonlinear equations:

$\boxed{\text{PD}}$:   solve set of nonlinear equations in α, e, b :

$$
\begin{cases}
\displaystyle\sum_{k=p+1}^{p+N} \alpha_{k-p} = 0 \\[2mm]
\gamma e_k - \alpha_{k-p} - \displaystyle\sum_{i=1}^{p} \alpha_{k-p+i} \frac{\partial}{\partial e_{k-i}} \Big(\sum_{l=p+1}^{p+N} \alpha_{l-p} K(y_{k-1\_k-p}- \\[1mm]
\qquad e_{k-1\_k-p}, y_{l-1\_l-p} - e_{l-1\_l-p}) = 0, \qquad k = p+1, ..., N \\[2mm]
y_k - e_k - \displaystyle\sum_{l=p+1}^{p+N} \alpha_{l-p} K(y_{k-1\_k-p} - e_{k-1\_k-p}, y_{l-1\_l-p} \\[1mm]
\qquad - e_{l-1\_l-p}) - b = 0, \qquad\qquad k = p+1, ..., p+N.
\end{cases}
$$
$$(8.9)$$

8.1.2 A simplification to the problem

Finding a solution to the equations (8.9) is computationally heavy. A simplification can be made by letting the regularization constant $\gamma \to \infty$. This leads to the problem:

$\boxed{\text{PD}}$: $\displaystyle\min_{e,b,\alpha} J_{\text{PD}}(e) = \frac{1}{2} \sum_{k=p+1}^{p+N} e_k^2$

such that

$$
y_k - e_k = \sum_{l=p+1}^{p+N} \alpha_{l-p} K(y_{l-1\_l-p}-
$$
$$
e_{l-1\_l-p}, y_{k-1\_k-p} - e_{k-1\_k-p}) + b, \qquad k = p+1, ..., N+p
$$

$$
\sum_{k=p+1}^{p+N} \alpha_{k-p} = 0.
$$

$$(8.10)$$

This constrained nonlinear optimization problem can be solved e.g. by sequential quadratic programming (SQP) [82]. For large data sets special methods for large scale nonlinear optimization exist and can be applied [174]. Because the results are based on the limit case $\gamma \to \infty$ the regularization term $\frac{1}{2} w^T w$ is partially neglected (however the form of the solution still takes into account the regularization term) there is a danger for overfitting by minimizing the training set error. Therefore, instead of minimizing the cost function to its local minimum one may apply early stopping then, which is another form of regularization. Note that standard VC bounds on the generalization error cannot be applied here due to the fact that the data are not i.i.d. because the system is dynamic instead of static.

Instead of solving this problem in the unknowns e, b, α one may further eliminate e and solve

$$\boxed{\text{D}}: \quad \min_{\alpha, b} J_{\text{D}}(\alpha, b) = \frac{1}{2} \sum_{k=p+1}^{p+N} (y_k - \hat{y}_k(\alpha))^2$$

such that

$$\hat{y}_k = \sum_{l=p+1}^{p+N} \alpha_{l-p} K(\hat{y}_{l-1\_l-p}, \hat{y}_{k-1\_k-p}) + b, \quad k = p+1, ..., N+p$$

$$\sum_{k=p+1}^{p+N} \alpha_{k-p} = 0,$$

$$(8.11)$$

where the recurrent simulation model can be used

$$\hat{y}_k = \sum_{l=p+1}^{p+N} \alpha_{l-p} K\left(\hat{y}_{l-1\_l-p}, \begin{bmatrix} \hat{y}_{k-1} \\ \hat{y}_{k-2} \\ \vdots \\ \hat{y}_{k-p} \end{bmatrix}\right) + b \qquad (8.12)$$

with given initial condition $\hat{y}_i = y_i$ for $i = 1, 2, ..., p$. For RBF kernels one employs

$$K(\hat{y}_{k-1\_k-p}, \hat{y}_{l-1\_l-p}) = \exp(-\nu \|\hat{y}_{k-1\_k-p} - \hat{y}_{l-1\_l-p}\|_2^2) \qquad (8.13)$$

where ν is a positive real constant. This simplified case is equivalent to parameterizing the recurrent model structure (8.2) by a radial basis function network and minimizing the fitting error and requiring that the average error on the training data is zero.

8.1.3 *Example: trajectory learning of chaotic systems*

We illustrate the training of recurrent LS-SVMs now on an example of trajectory learning of chaotic systems, according to [240]. A well-known paradigm for chaos in electrical circuits is Chua's circuit [43]. It is described by the following equations:

$$\begin{cases} \dot{x}_1 &= a\left[x_2 - g(x_1)\right] \\ \dot{x}_2 &= x_1 - x_2 + x_3 \\ \dot{x}_3 &= -bx_2 \end{cases} \qquad (8.14)$$

with piecewise linear characteristic

$$g(x_1) = m_1 x_1 + \frac{1}{2}(m_0 - m_1)\left(|x_1 + 1| - |x_1 - 1|\right). \qquad (8.15)$$

A double scroll strange attractor is obtained by taking $a = 9$, $b = 14.286$, $m_0 = -1/7$, $m_1 = 2/7$. A trajectory has been generated with initial condition $[0.1; 0; -0.1]$ for the state vector with three state variables $[x_1; x_2; x_3]$ by using a Runge-Kutta integration rule (ode23 in Matlab).

In Fig. 8.1 the first $N = 300$ data points were used for trajectory learning of the double scroll by a recurrent LS-SVM with RBF kernel. For the model structure $p = 12$ has been taken and $\nu = 1$ for the RBF kernel. In order to solve the constrained nonlinear optimization problem (8.11), sequential quadratic programming (SQP) has been applied (*constr* in Matlab with a specification of the number of equality constraints). The model (8.12) has been simulated in C by using Matlab's cmex facility. In all simulations, for the initial unknown parameter vector α_k, e_k in (8.10) have been chosen randomly according to a Gaussian distribution with zero mean and standard deviation 0.1 and $b = 0$. Although no further optimization of the model structure has been done, it has been observed that small values of p (that are chosen in accordance with Takens' embedding theorem) are slowing down the training process. In Fig. 8.1 the result after training of the recurrent LS-SVM is shown in dashed line. The model is simulated for initial condition $\hat{y}_{1\_12} = y_{1\_12}$. As one can see the training data are well reproduced 300 points ahead starting from this initial condition. The prediction beyond the time horizon of 300 training data points is very good, given the limited

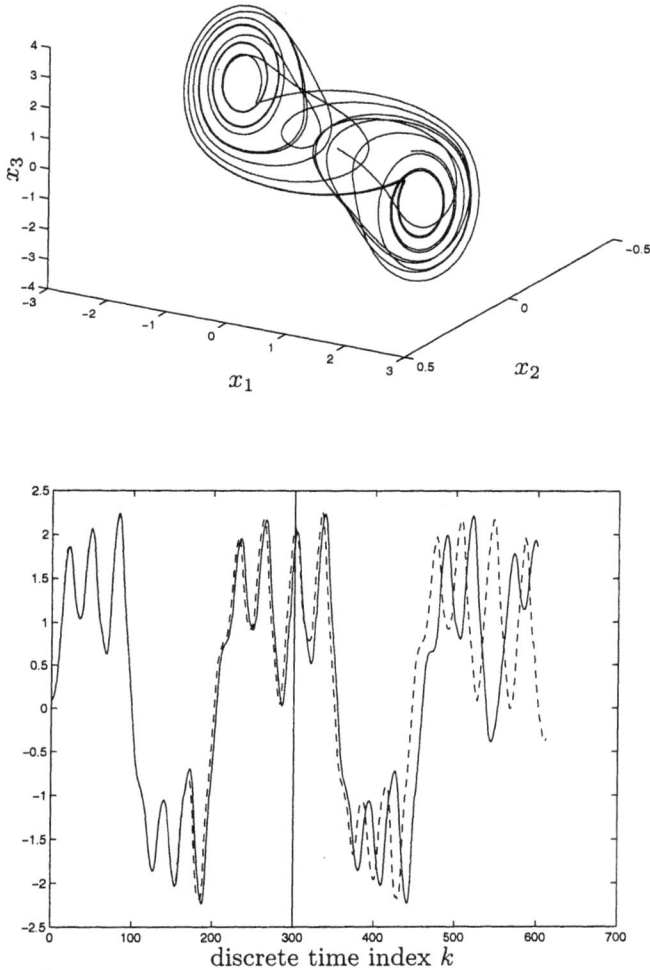

Fig. 8.1 *(Top) Double-scroll attractor; (Bottom) Trajectory learning of the double scroll (full line) by a recurrent LS-SVM with RBF kernel. The simulation result after training (dashed line) on $N = 300$ data points is shown with data points $k = 1$ to 12 i.e. $\hat{y}_{1\_12} = y_{1\_12}$ taken as initial condition. Good qualitative generalization is obtained for the prediction from $k = 301$ to about $k = 500$.*

amount of training data. The two next jumps to the other scroll are quite well predicted. The early stopping in the training procedure was done based upon the training data itself (instead of on an independent validation set). This is because it is non-trivial to reproduce the trajectory in recurrent mode on the training data itself. The overfitting was observed in recurrent mode on these training data. At that point training was stopped. The resulting recurrent LS-SVM model is generalizing well to a double scroll attractor, also for small perturbations on the initial state. One should also note that for the recurrent LS-SVMs it is important to simulate the form (8.12) because when solving the constrained nonlinear optimization problem (8.11) the constraints will usually not hold exactly but only for a certain tolerance. For chaotic systems such small perturbations could cause significant differences between the solution vector from (8.11) and the recurrent simulation model (8.12), while for globally asymptotically stable systems such perturbations are not amplified.

8.2   LS-SVMs and optimal control

Besides applying LS-SVM methods to recurrent models one can also follow a similar methodology towards solving optimal control problems. Both the optimal control problems and support vector methods are closely related to optimization theory. Hence one could try to merge the two formulations. This approach has been originally proposed by Suykens *et al.* in [241]. Let us first discuss a few aspects of the N-stage optimal control problem, next explain how one embeds LS-SVMs in the problem formulation and finally show how to apply the kernel trick and solve the problem in the Lagrange multipliers. In control problems stability and robustness are important issues. These aspects are discussed in the next subsection together with illustrative examples.

8.2.1   *The N-stage optimal control problem*

In the N-stage optimal control problem one aims at solving the following problem [33]:

$$\min \mathcal{J}_N(x_k, u_k) = \rho(x_{N+1}) + \sum_{k=1}^{N} h(x_k, u_k) \qquad (8.16)$$

subject to the system dynamics

$$x_{k+1} = f(x_k, u_k), \quad k = 1, ..., N \quad (x_1 \text{ given}),$$

where $\rho(\cdot)$ and $h(\cdot, \cdot)$ are positive definite functions and N is the finite discrete time horizon. A typical choice is the quadratic cost $h(x_k, u_k) = x_k^T Q x_k + u_k^T R u_k$, $\rho(x_{N+1}) = x_{N+1}^T Q x_{N+1}$ with $Q = Q^T > 0$, $R = R^T > 0$. The functions $\rho(\cdot)$, $h(\cdot, \cdot)$, $f(\cdot, \cdot)$ are assumed to be twice continuously differentiable, $x_k \in \mathbb{R}^n$ denotes the state vector, $u_k \in \mathbb{R}$ is the input of the system with discrete time index k. We consider here a single input system but it can be extended to multi-input systems as well.

In order to find the optimal control law, one constructs the Lagrangian

$$\mathcal{L}_N(x_k, u_k; \lambda_k) = \mathcal{J}_N(x_k, u_k) + \sum_{k=1}^{N} \lambda_k^T [x_{k+1} - f(x_k, u_k)] \quad (8.17)$$

with Lagrange multipliers $\lambda_k \in \mathbb{R}^n$. The conditions for optimality are

$$\left\{ \begin{array}{lll} \frac{\partial \mathcal{L}_N}{\partial x_k} & = & \frac{\partial h}{\partial x_k} + \lambda_{k-1} - (\frac{\partial f}{\partial x_k})^T \lambda_k = 0, \quad k = 2, ..., N \\ & & \hspace{5cm} \text{(adjoint equation)} \\[2mm] \frac{\partial \mathcal{L}_N}{\partial x_{N+1}} & = & \frac{\partial \rho}{\partial x_{N+1}} + \lambda_N = 0 \hspace{2cm} \text{(adjoint final condition)} \\[2mm] \frac{\partial \mathcal{L}_N}{\partial u_k} & = & \frac{\partial h}{\partial u_k} - \lambda_k^T \frac{\partial f}{\partial u_k} = 0, \hspace{1.5cm} k = 1, ..., N \\ & & \hspace{5cm} \text{(variational condition)} \\[2mm] \frac{\partial \mathcal{L}_N}{\partial \lambda_k} & = & x_{k+1} - f(x_k, u_k) = 0, \hspace{1.5cm} k = 1, ..., N \\ & & \hspace{5cm} \text{(system dynamics).} \end{array} \right.$$

$$(8.18)$$

For the case of a quadratic cost function subject to linear system dynamics with infinite time horizon, the optimal control law can be represented by full static state feedback control (which is the well-known LQR (Linear Quadratic Regulator) problem). However, in general, the optimal control law cannot be represented by state feedback as the optimal control law may also depend on the co-state λ_k. Nevertheless, one is often interested in finding a suboptimal control strategy of this form for example in state feedback form $u_k = g(x_k)$. The nonlinear function $g(\cdot)$ can be parameterized by a linear function resulting in a linear state feedback controller. An alternative which has been studied in the area of neural networks is to parameterize it by means of a multilayer perceptron, which results in a powerful nonlinear state feedback law [230]. It has been shown for example in [230; 232] how difficult control problems such as swinging up an inverted and double inverted pendulum for a cart-pole system with local stabilization at the endpoint can be performed well by simple MLP neural networks.

For MLPs it is straightforward to parameterize the control law by the architecture. However, when one intends to do the same using SVMs, this is more complicated due to the nonparametric nature.

8.2.2   *LS-SVMs as controllers within the N-stage optimal control problem*

Formulation with error variables

We discuss now the approach outlined in [241], for introducing LS-SVMs within the N-stage optimal control problem formulation. The fact that LS-SVMs work with equality constraints and sum squared error simplifies the formulation in comparison with standard SVMs. One starts again by formulating the primal problem, then constructs the Lagrangian and takes the conditions for optimality. The equations for the Lagrange multipliers related to the LS-SVM control law and the Lagrange multipliers related to the system dynamics will become coupled.

The problem formulation starts from

$$
\boxed{P}: \quad \min \quad J_P(x_k, u_k, w, e_k) = J_N(x_k, u_k) + \frac{1}{2}w^T w + \gamma \frac{1}{2}\sum_{k=1}^{N} e_k^2
$$

subject to the system dynamics

$$
x_{k+1} = f(x_k, u_k), \quad k = 1, ..., N \quad (x_1 \text{ given})
$$

and the control law

$$
u_k = w^T \varphi(x_k) + e_k, \quad k = 1, ..., N
$$

$$(8.19)$$

where J_N is given by (8.16), $\varphi(\cdot) : \mathbb{R}^n \to \mathbb{R}^{n_h}$ with n_h the dimension of the high dimensional feature space, which can be infinite dimensional. The control law is restricted to a nonlinear state feedback which is assumed to be representable by a LS-SVM. A promising aspect of using SVM here is that they are suitable for learning and generalization in large dimensional input spaces (in this case the state space), while other methods of fuzzy control or control by RBF neural networks suffer from the curse of dimensionality as explained e.g. by Sanner & Slotine in [197]. The actual control signal applied to the plant is assumed to be $w^T \varphi(x_k)$. In the linear control case one has $\varphi(x_k) = x_k$ with $n_h = n$. In the sequel we will employ RBF kernels.

The Lagrangian takes the form

$$\mathcal{L}(x_k, u_k, w, e_k; \lambda_k, \alpha_k) = \mathcal{J}_N(x_k, u_k) + \frac{1}{2}w^T w + \gamma \frac{1}{2}\sum_{k=1}^{N} e_k^2 +$$
$$\sum_{k=1}^{N} \lambda_k^T [x_{k+1} - f(x_k, u_k)] + \sum_{k=1}^{N} \alpha_k [u_k - w^T \varphi(x_k) - e_k]. \tag{8.20}$$

The conditions for optimality are given by

$$
\begin{cases}
\frac{\partial \mathcal{L}}{\partial x_k} = \frac{\partial h}{\partial x_k} + \lambda_{k-1} - (\frac{\partial f}{\partial x_k})^T \lambda_k - \\
\qquad \alpha_k \frac{\partial}{\partial x_k}[w^T \varphi(x_k)] = 0, \qquad k = 2, ..., N \\
\qquad\qquad\qquad\qquad\qquad\qquad \text{(adjoint equation)} \\[2mm]
\frac{\partial \mathcal{L}}{\partial x_{N+1}} = \frac{\partial \rho}{\partial x_{N+1}} + \lambda_N = 0 \\
\qquad\qquad\qquad\qquad\qquad\qquad \text{(adjoint final condition)} \\[2mm]
\frac{\partial \mathcal{L}}{\partial u_k} = \frac{\partial h}{\partial u_k} - \lambda_k^T \frac{\partial f}{\partial u_k} + \alpha_k = 0, \qquad k = 1, ..., N \\
\qquad\qquad\qquad\qquad\qquad\qquad \text{(variational condition)} \\[2mm]
\frac{\partial \mathcal{L}}{\partial w} = w - \sum_{k=1}^{N} \alpha_k \varphi(x_k) = 0 \\
\qquad\qquad\qquad\qquad\qquad\qquad \text{(support vectors)} \\[2mm]
\frac{\partial \mathcal{L}}{\partial e_k} = \gamma e_k - \alpha_k = 0 \qquad\qquad k = 1, ..., N \\
\qquad\qquad\qquad\qquad\qquad\qquad \text{(support values)} \\[2mm]
\frac{\partial \mathcal{L}}{\partial \lambda_k} = x_{k+1} - f(x_k, u_k) = 0, \qquad k = 1, ..., N \\
\qquad\qquad\qquad\qquad\qquad\qquad \text{(system dynamics)} \\[2mm]
\frac{\partial \mathcal{L}}{\partial \alpha_k} = u_k - w^T \varphi(x_k) - e_k = 0, \qquad k = 1, ..., N \\
\qquad\qquad\qquad\qquad\qquad\qquad \text{(SVM control).}
\end{cases}
\tag{8.21}
$$

From the variational condition one can see that the Lagrange multipliers which are related to the system dynamics interact with the support values of the LS-SVM controller. The obtained set of nonlinear equations is of the form

$$F_1(x_k, x_{N+1}, u_k, w, e_k, \lambda_k, \alpha_k) = 0 \qquad (8.22)$$

for $k = 1, ..., N$ with x_1 given. Although the resulting problem is non-convex one can still apply the kernel trick and eliminate the unknown w vector. After elimination of w, a set of nonlinear equations of the form

$$F_2(x_k, x_{N+1}, u_k, \lambda_k, \alpha_k) = 0 \qquad (8.23)$$

for $k = 1, ..., N$ with x_1 given is obtained. The solution is characterized as follows:

$$
\left[
\begin{array}{l}
\boxed{\text{PD}} : \quad \text{solve set of nonlinear equations in } x_k, x_{N+1}, u_k, \lambda_k, \alpha_k : \\[2mm]
\left\{
\begin{array}{ll}
\dfrac{\partial h}{\partial x_k} + \lambda_{k-1} - \left(\dfrac{\partial f}{\partial x_k}\right)^T \lambda_k - \alpha_k \displaystyle\sum_{l=1}^{N} \alpha_l \dfrac{\partial K(x_k, x_l)}{\partial x_k} = 0, & \\[2mm]
& k = 2, ..., N \\[4mm]
\dfrac{\partial \rho}{\partial x_{N+1}} + \lambda_N = 0 & \\[4mm]
\dfrac{\partial h}{\partial u_k} - \lambda_k^T \dfrac{\partial f}{\partial u_k} + \alpha_k = 0, & k = 1, ..., N \\[4mm]
x_{k+1} - f(x_k, u_k) = 0, & k = 1, ..., N \\[4mm]
u_k - \displaystyle\sum_{l=1}^{N} \alpha_l K(x_l, x_k) - \alpha_k/\gamma = 0, & k = 1, ..., N.
\end{array}
\right.
\end{array}
\right]
$$

$$(8.24)$$

The actual control signal applied to the plant becomes

$$u(x) = \sum_{l=1}^{N} \alpha_l K(x_l, x) \qquad (8.25)$$

where $\{x_l\}_{l=1}^{N}$, $\{\alpha_l\}_{l=1}^{N}$ are obtained from solving the set of nonlinear equations and x_k is the actual state vector at discrete time k. The data $\{x_l\}_{l=1}^{N}$ are used as support vector data for the control signal and are the solution to the set of nonlinear equations, while in problems of static nonlinear function estimation the support vectors correspond to given training data.

Formulation without error variables

In the previous formulation, error variables were taken in order to allow a difference between the optimal control signal u_k and the output of the LS-SVM. It is possible to simplify this and leave out the variables e_k from the problem formulation. As a result the problem has less unknowns.

According to [241] the simplified problem formulation is

$$\min_{x_k, w} \; \mathcal{J}_N(x_k, u_k(w, x_k)) + \frac{1}{2} w^T w \tag{8.26}$$

subject to

$$\begin{cases} x_{k+1} &= f(x_k, u_k(w, x_k)), & k = 1, ..., N \;\; (x_1 \text{ given}) \\[2mm] u_k &= w^T \varphi(x_k) & k = 1, ..., N. \end{cases}$$

This gives the Lagrangian

$$\mathcal{L}(x_k, w; \lambda_k) = \mathcal{J}_N(x_k, u_k(w, x_k)) + \frac{1}{2} w^T w + \sum_{k=1}^{N} \lambda_k^T [x_{k+1} - f(x_k, w^T \varphi(x_k))].$$
$$\tag{8.27}$$

The conditions for optimality are

$$\begin{cases} \frac{\partial \mathcal{L}}{\partial x_k} &= \frac{\partial h}{\partial x_k} + \lambda_{k-1} - (\frac{\partial f}{\partial x_k})^T \lambda_k - (\frac{\partial f}{\partial u_k} \frac{\partial u_k}{\partial x_k})^T \lambda_k = 0, \quad k = 2, ..., N \\[3mm] \frac{\partial \mathcal{L}}{\partial x_{N+1}} &= \frac{\partial \rho}{\partial x_{N+1}} + \lambda_N = 0 \\[3mm] \frac{\partial \mathcal{L}}{\partial w} &= w - \sum_{k=1}^{N} \lambda_k^T \frac{\partial f}{\partial u_k} \, \varphi(x_k) = 0 \\[3mm] \frac{\partial \mathcal{L}}{\partial \lambda_k} &= x_{k+1} - f(x_k, w^T \varphi(x_k)) = 0, & k = 1, ..., N. \end{cases}$$
$$\tag{8.28}$$

The support values are directly related here to the Lagrange multipliers λ_k. The resulting set of nonlinear equations is of the form

$$F_3(x_k, x_{N+1}, w, \lambda_k) = 0. \tag{8.29}$$

The kernel trick can again be applied, finally yielding a set of equations of the form

$$F_4(x_k, x_{N+1}, \lambda_k) = 0 \tag{8.30}$$

after elimination of w.

8.2.3 *Alternative formulation and stability issues*

An alternative formulation, directly expressed in the unknowns x_k, α_k can be made by considering the optimization problem

$$\min_{x_k,\alpha_k} J_N(x_k, x_k^r, u_k) + \lambda \sum_{k=1}^{N} \alpha_k^2 \tag{8.31}$$

subject to

$$\begin{cases} x_{k+1} &= f(x_k, u_k), \quad x_1 \text{ given} \\ u_k &= \sum_{l=1}^{N} \alpha_l K([x_l; x_l^r], [x_k; x_k^r]) \end{cases}$$

which does not start from the primal weight space but rather specifies the form of the control signal in a dual space. In this formulation a reference state vector x_k^r is considered. A regularization term $\sum_k \alpha_k^2$ is included with λ a positive real constant.

In order to impose local stability at a fixed equilibrium point x^{eq}, one locally linearizes the autonomous closed-loop simulation model by evaluating $\frac{\partial f}{\partial \hat{x}_k}$ at x^{eq}. LS-SVM control with a local stability constraint can then be formulated as:

$$\min_{x_k,\alpha_k,Q} J_N(x_k, x^{eq}, u_k) + \lambda \sum_{k=1}^{N} \alpha_k^2 \tag{8.32}$$

subject to

$$\begin{cases} x_{k+1} = f(x_k, \sum_{l=1}^{N} \alpha_l K([x_l; x^{eq}], [x_k; x^{eq}])), \quad x_1 \text{ given} \\ A^T P A - P < 0 \end{cases}$$

where $P = Q^T Q$, $A = \frac{\partial f}{\partial \hat{x}_k} |_{\hat{x}_k = x^{eq}}$. The last equation is a matrix inequality which expresses that the matrix $A^T P A - P$ is negative definite (maximal eigenvalue negative) meaning that A should be stable [29].

Imposing global asymptotic stability can be done in a somewhat similar way as done in NLq neural control theory [230; 233] or one may impose robust local stability by enlarging the basin of attraction around the equilibrium point according to [212; 284]. Within NLq theory sufficient conditions for global asymptotic stability and input/output stability with finite L_2-gain have been derived, which can also be employed in order to impose robust local stability of the origin for the closed-loop system. Based upon NLq stability criteria, dynamic backpropagation has been modified by imposing matrix inequality constraints. One may proceed in a similar fashion in the case of LS-SVM control laws based upon [212; 284].

When applying an optimization method, the constraints will only hold with a certain tolerance. These small numerical errors may lead to differences between the state vectors as solution to the optimization problem and the simulation of the closed-loop system with LS-SVM control, especially in the control of unstable systems. For control of stable systems this problem will be less critical. A simulation of the closed-loop system with LS-SVM control

$$\hat{x}_{k+1} = f(\hat{x}_k, \sum_{l=1}^{N} \alpha_l \, K([x_l; x_l^r], [\hat{x}_k; x_k^r])) \quad \hat{x}_1 = x_1 \text{ given} \qquad (8.33)$$

is always needed in order to validate the results, where $\{x_l\}_{l=1}^{N}$ and $\{\alpha_l\}_{l=1}^{N}$ are obtained as a solution to the optimization problem.

8.2.4 *Illustrative examples*

A simple state vector tracking example

Here the LS-SVM optimal control method is illustrated on an example reported by Narendra & Mukhopadhyay in [241]. Given the nonlinear system

$$\begin{cases} x_{1,k+1} &= 0.1x_{1,k} + 2\frac{u_k + x_{2,k}}{1 + (u_k + x_{2,k})^2} \\ \\ x_{2,k+1} &= 0.1x_{2,k} + u_k(2 + \frac{u_k^2}{1 + x_{1,k}^2 + x_{2,k}^2}) \end{cases} \qquad (8.34)$$

discrete time index k

discrete time index k

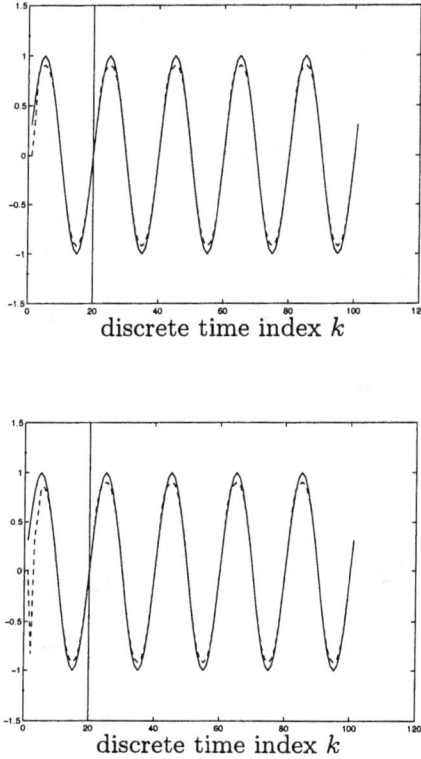

Fig. 8.2   *(Top) Optimal control by LS-SVM with RBF kernel and application of Mercer's theorem. (Full line) first state variable to be tracked; (Dashed line) actual state variable by closed-loop simulation of the system. The data $k = 1, ..., 20$ were used for training of the controller with the origin as initial state. (Bottom) simulation result for another randomly chosen initial state. The LS-SVM controller shows a good generalization performance with respect to other initial states and beyond the time horizon.*

with state variables x_1, x_2 and input u. Consider a state vector tracking problem with

$$
\begin{aligned}
h(x_k, u_k) &= (x_k - x_k^r)^T Q (x_k - x_k^r) + u_k^T R u_k \\
\rho(x_{N+1}) &= (x_{N+1} - x_{N+1}^r)^T Q (x_{N+1} - x_{N+1}^r)
\end{aligned}
\tag{8.35}
$$

where x_k^r is the reference trajectory to be tracked. Suppose one aims at tracking the first state variable and choose $Q = \mathrm{diag}([1; 0.001])$, $R = 1$, $x_k^r = [\sin(2\pi k/20); \cos(2\pi k/20)]$ with $k = 1, ..., N$ and $N = 20$. The given initial state vector is $[0; 0]$. As control law is taken $w^T \varphi([x_k; x_k^r])$. The derivations are similar to the ones derived for the LS-SVM state feedback control law. In Fig. 8.2 simulation results are shown for the solution obtained from (8.24). The set of nonlinear equations was solved using Matlab's optimization toolbox (function *leastsq*) with unknowns $x_k, u_k, \lambda_k, \alpha_k, \sigma$ (variables e_k eliminated) for $\gamma = 100$ taking an LS-SVM controller with RBF kernel. The unknowns were randomly initialized with zero mean and standard deviation 0.3. The plots show the simulation results for the closed-loop system. The controller is generalizing well towards other initial conditions than the origin (for which it has been trained) and beyond the time horizon of $N = 20$.

Swinging up an inverted pendulum cart-pole system

Control by LS-SVMs is illustrated now on the problem of swinging up an inverted pendulum with local stabilization at the target point. It was shown in [230; 232] that this can also be successfully done by simple MLP architectures.

A nonlinear state space model for the inverted pendulum system (Fig. 8.3) is given by

$$\dot{x} = F(x) + G(x)u \qquad (8.36)$$

with

$$F(x) = \begin{bmatrix} x_2 \\ \frac{\frac{4}{3} m l x_4^2 \sin x_3 - \frac{mg}{2} \sin(2x_3)}{\frac{4}{3} m_t - m \cos^2 x_3} \\ x_4 \\ \frac{m_t g \sin x_3 - \frac{ml}{2} x_4^2 \sin(2x_3)}{l(\frac{4}{3} m_t - m \cos^2 x_3)} \end{bmatrix}, \; G(x) = \begin{bmatrix} 0 \\ \frac{4}{3} \frac{1}{\frac{4}{3} m_t - m \cos^2 x_3} \\ 0 \\ -\frac{\cos x_3}{l(\frac{4}{3} m_t - m \cos^2 x_3)} \end{bmatrix}.$$

The state variables x_1, x_2, x_3, x_4 are the position and velocity of the cart, angle of the pole with the vertical and rate of change of the angle, respectively. The input signal u is the force applied to the cart's center of mass. The symbols m, m_t, l, g denote respectively mass of the pole, total mass of cart and pole, half pole length and the acceleration due to gravity. Here $m = 0.1$, $m_t = 1.1$, $l = 0.5$ is taken. Remark that in the autonomous case $x = [0; 0; 0; 0]$ (pole up) and $x = [0; 0; \pi; 0]$ (pole down) are equilibrium points. The linearized system around the target equilibrium point (the

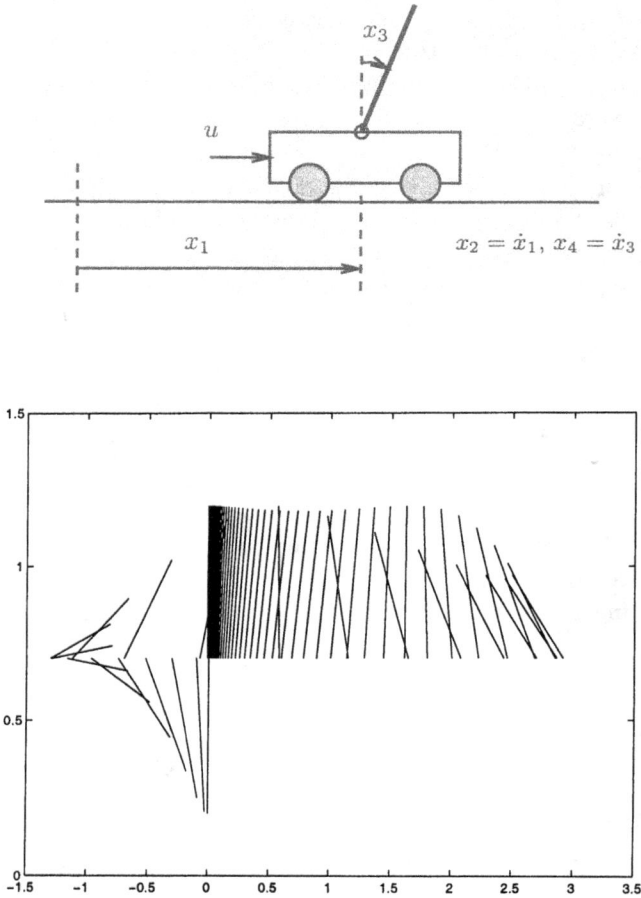

Fig. 8.3   Swinging up an inverted pendulum by an LS-SVM controller with local stabilization in the upright position. Around the target point the controller is behaving as an LQR controller. (Top) inverted pendulum system; (Bottom) simulation result which visualizes the several pole positions with respect to time.

origin) is given by

$$\dot{x} = Ax + Bu \qquad (8.37)$$

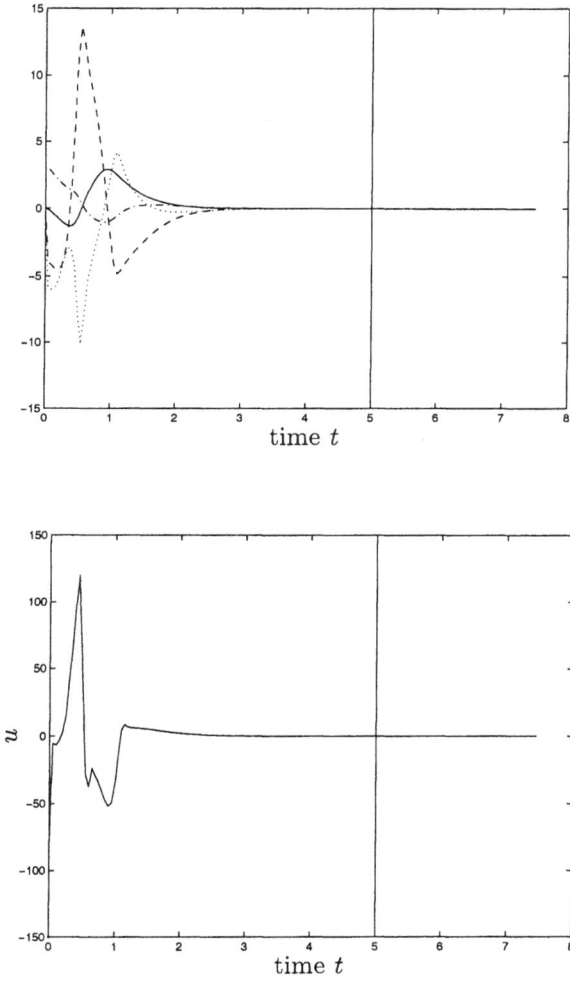

Fig. 8.4 *(Continued) (Top) state variables with respect to time of the closed-loop simulation model with LS-SVM controller: $x_1(t)$ (-); $x_2(t)$ (- -); $x_3(t)$ (-.); $x_4(t)$ (:). The state vector data before the vertical line were used in the training process as support vector data with an RBF kernel; (Bottom) control signal u_k.*

with

$$A = \begin{bmatrix} 0 & 1 & 0 & 0 \\ 0 & 0 & -\frac{mg}{\frac{4}{3}m_t-m} & 0 \\ 0 & 0 & 0 & 1 \\ 0 & 0 & \frac{m_t g}{l(\frac{4}{3}m_t-m)} & 0 \end{bmatrix}, B = \begin{bmatrix} 0 \\ \frac{4}{3}\frac{1}{\frac{4}{3}m_t-m} \\ 0 \\ -\frac{1}{l(\frac{4}{3}m_t-m)} \end{bmatrix}. \tag{8.38}$$

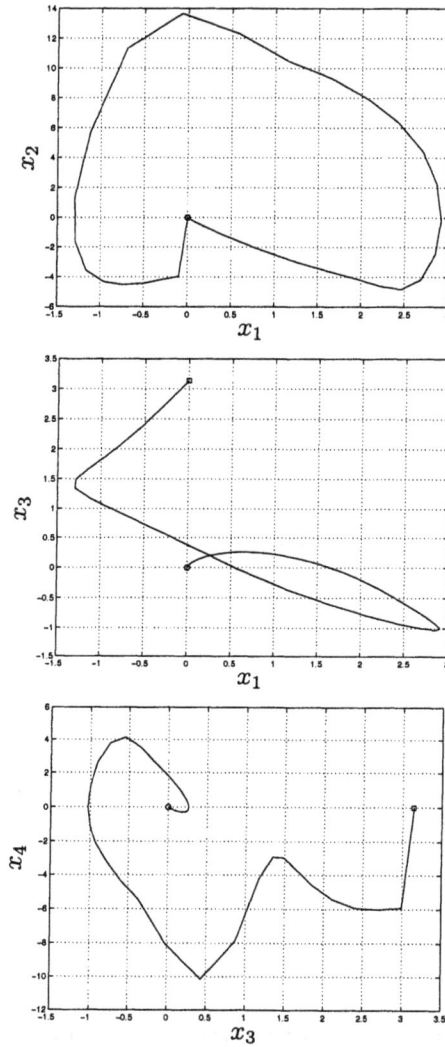

Fig. 8.5 *(Continued) (Top) $(x_1(t), x_2(t))$; (Middle) $(x_1(t), x_3(t))$; (Bottom) $(x_3(t), x_4(t))$. The initial state is marked by a square and the target state by o.*

According to [232] the strategy for the neural controller was to first design a Linear Quadratic Regulator (LQR) controller [28; 85] based upon the linearized model (eventually, also robust linear controllers might be designed, based on H_∞ control or μ theory, if additional robustness with respect to noise and uncertainties would be needed). This was done in order to impose a set of constraints on the choice of the interconnection weights of a multi-layer perceptron controller. The MLP has additional degrees of freedom in comparison with a linear mapping. These additional degrees of freedom are used in order to swing up the pole with local stabilization at the endpoint.

In order to solve the swinging up problem using LS-SVM control, a somewhat similar approach is followed. Based on the continuous time model an LQR controller is designed e.g. for $Q = I$, $R = 0.01$ in the LQR cost function $\int_0^\infty (x^T Q x + u^T R u) dt$ (Matlab command lqr). Then the continuous time model has been discretized by using a fourth order Runge-Kutta integration rule with constant step $h_s = 0.05$, resulting in a discrete time model of the form

$$x_{k+1} = \mathcal{F}(x_k, u_k) \tag{8.39}$$

where u_k is assumed to be constant in the time intervals $[kh_s, (k+1)h_s)$ (zero order hold). The control law is chosen as

$$u_k = (L_{\text{lqr}} - L_\Delta)x_k + u_k^{\text{svm}} \tag{8.40}$$

with

$$u_k^{\text{svm}} = \sum_{l=1}^{N} \alpha_l K(x_l, x_k) \tag{8.41}$$

where L_{lqr} is the resulting feedback matrix from LQR design and

$$L_\Delta = \frac{\partial u_k^{\text{svm}}}{\partial x_k} \Big|_{x_k=0} \tag{8.42}$$

is a modification to the LS-SVM control law such that, locally at the origin, the control law u_k is acting as a LQR controller, in a continuous time sense. The combination between a continuous time LQR result and the discrete time SVM control law may look surprising at first sight, but here it is a convenient way to design an LS-SVM controller for the given continuous time model. An approach according to (8.24) would be more complicated here due to the Runge-Kutta integration rule. An RBF kernel function is taken for the LS-SVM part.

The resulting closed-loop simulation model is given by

$$\hat{x}_{k+1} = \mathcal{F}(\hat{x}_k, (L_{\mathrm{lqr}} - L_\Delta)\hat{x}_k + \sum_{l=1}^{N} \alpha_l K(x_l, \hat{x}_k)) \qquad (8.43)$$

with $\hat{x}_1 = x_1 = 0$ given and $\{x_l\}_{l=1}^{N}$, $\{\alpha_l\}_{l=1}^{N}$ are the solution to the constrained optimization problem (8.32). As cost function has been taken

$$\mathcal{J}_N = \sum_{k=N-5}^{N} \|x_k\|_2^2 \qquad (8.44)$$

with $x_k^r = 0$, $\lambda = 0$ and time horizon $N = 100$ (5 seconds). The constrained optimization problem has been solved using Matlab's optimization toolbox by SQP (Sequential Quadratic Programming) (function *constr*). The values x_k and α_k were initialized by taking small random values. The parameter of the RBF kernel was taken as additional unknown in the optimization problem. In order to emphasize the equations of the constraints, these have been multiplied by a factor 1000. This is needed due to the fact that the system to be controlled is unstable and small differences between x_k (as solution to (8.24)) and $\hat{x}_k$ (in the simulation model (8.33)) may cause large differences otherwise. Simulation results of the closed-loop simulation model are shown in Figs. 8.3-8.5.

Ball and beam control example

Here we discuss the LS-SVM control method on a ball and beam system as described in [241]. The continuous time system description of the ball and beam system (Fig. 8.6) is given by

$$F(x) = \begin{bmatrix} x_2 \\ B(x_1 x_4^2 - G\sin x_3) \\ x_4 \\ 0 \end{bmatrix}, \quad G(x) = \begin{bmatrix} 0 \\ 0 \\ 0 \\ 1 \end{bmatrix} \qquad (8.45)$$

where $x = [x_1; x_2; x_3; x_4] = [r; \dot{r}; \theta; \dot{\theta}]$ with r the ball position and θ the beam angle and $B = M/(J_b/R_b^2 + M)$ where M, J_b, R_b are the mass, moment of inertia and radius of the ball, respectively. For the control input one has $\tau = 2Mr\dot{r}\dot{\theta} + MGr\cos\theta + (Mr^2 + J + J_b)u$ where τ, G, J denote the torque applied to the beam, the acceleration of gravity and the moment of inertia

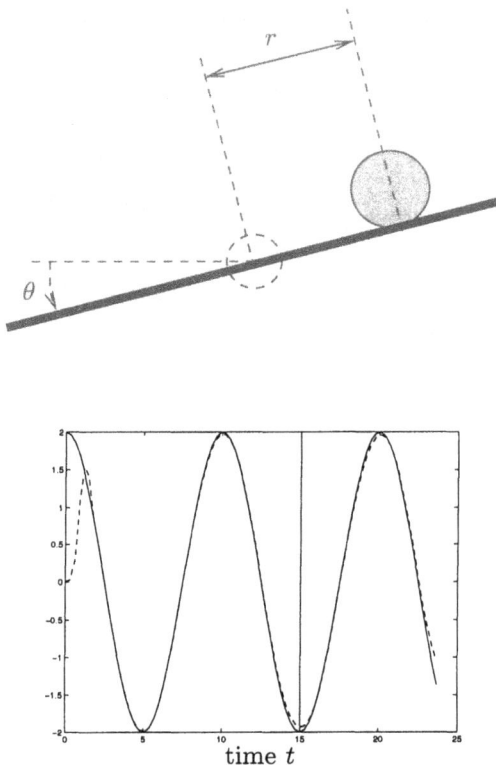

Fig. 8.6 *Tracking control of a ball and beam system by a LS-SVM controller: (Top) ball and beam system; (Bottom) reference input (-) and position of the ball (- -). The state vector data before the vertical line were used in the training process as support vector data with an RBF kernel. Shown is the closed-loop simulation result.*

of the beam, respectively. As control objective we consider tracking of the ball position for a reference input $d(t) = 2\cos(\pi t/5)$.

As for the inverted pendulum example, an LQR design is combined with the LS-SVM control law. The continuous time state space description has been linearized around the origin, for which a LQR controller is designed with $Q = I$, $R = 1$ in the LQR cost function. Then the continuous time model is discretized by using a fourth order Runge-Kutta integration rule with constant step $h_s = 0.3$ resulting in a discrete time model. The first component of the reference state vector $x_{1,k}^r$ is derived from $d(t)$ according to this step size. The control law is taken as

$$u_k = (L_{\text{lqr}} - L_\Delta) \begin{bmatrix} x_k \\ x_{1,k}^r \end{bmatrix} + u_k^{\text{svm}} \tag{8.46}$$

with

$$u_k^{\text{svm}} = \sum_{l=1}^{N} \alpha_l K(\begin{bmatrix} x_l \\ x_{1,l}^r \end{bmatrix}, \begin{bmatrix} x_k \\ x_{1,k}^r \end{bmatrix}) \tag{8.47}$$

where L_{lqr} is the resulting feedback matrix from LQR (for the autonomous closed-loop system) and

$$L_\Delta = \frac{\partial u_k^{\text{svm}}}{\partial x_k} |_{x_k=0} . \tag{8.48}$$

Note that the LS-SVM part also depends on the reference $x_{1,k}^r$. A zero initial state has been taken. SQP was applied for constrained nonlinear optimization with similar initialization as for the inverted pendulum example. Simulation results of the closed-loop simulation model are shown in Fig. 8.6.

Appendix A

A.1 Optimization theory

For several topics in nonlinear programming, constrained optimization, convex optimization and interior point algorithms one may consult e.g. [20; 82; 86; 147; 154; 174; 190; 261]. The following introductory material is based on [82].

Constrained optimization problems, Lagrangian

Consider the constrained optimization problem

$$
\begin{aligned}
\text{minimize} \quad & f(x) & & x \in \mathbb{R}^n \\
\text{subject to} \quad & c_i(x) = 0, & & i \in \mathcal{C}_E \\
& c_i(x) \geq 0, & & i \in \mathcal{C}_I
\end{aligned}
$$

where $f(x)$ is the objective function, c_i with $i = 1, 2, ..., p$ are the constraint functions, $\mathcal{C}_E$ is the index set of equality constraints in the problem and $\mathcal{C}_I$ the set of inequality constraints. Any point that satisfies all the constraints is called a feasible point and the set of all such points is referred to as the feasible region. Active (or binding) constraints at any point x' are defined by the index set $\mathcal{A}(x') = \{i : c_i(x') = 0\}$ so that any constraint is active at x' if x' is on the boundary of its feasible region. One introduces the Lagrangian function as

$$
\mathcal{L}(x, \lambda) = f(x) - \sum_i \lambda_i c_i(x).
$$

KKT conditions

One has the following first order (i.e. first derivative) necessary conditions. If x^* is a local minimizer of the above constrained optimization problem and a certain regularity assumption holds at x^*, then there exist Lagrange multipliers λ^* such that x^*, λ^* satisfy the following system:

$$\nabla_x \mathcal{L}(x, \lambda) = 0$$
$$c_i(x) = 0 \qquad i \in \mathcal{C}_E$$
$$c_i(x) \geq 0 \qquad i \in \mathcal{C}_I$$
$$\lambda_i \geq 0 \qquad i \in \mathcal{C}_I$$
$$\lambda_i c_i(x) = 0 \qquad \forall i$$

called the Karush-Kuhn-Tucker (KKT) conditions. A point x^* which satisfies the conditions is sometimes referred to as a KKT point. The final condition $\lambda_i^* c_i^* = 0$ is referred to as the complementarity condition and states that λ_i^* and c_i^* cannot be both non-zero, or equivalently that inactive constraints have a zero multiplier. If there is no i such that $\lambda_i^* = c_i^* = 0$ then strict complementarity holds. Strongly active means $\lambda_i^* > 0$, $c_i^* = 0$, weakly active $\lambda_i^* = c_i^* = 0$ and inactive $\lambda_i^* = 0$, $c_i^* > 0$. Second order conditions can be expressed in terms of the Hessian matrix with respect to x of the Lagrangian function $\nabla_x^2 \mathcal{L}(x^*, \lambda^*) = \nabla^2 f(x^*) - \sum_i \lambda_i^* \nabla^2 c_i(x^*)$.

Convex sets and convex functions

For problems of convex optimization it is often possible to state alternative (dual) problems from which the solution to the original (primal) problem can be obtained. These dual problems involve as dual variables the Lagrange multipliers. A convex set $\mathcal{S}$ in $\mathbb{R}^n$ is defined by the property that for any two points $x_1 \neq x_2 \in \mathcal{S}$ it follows that $x \in \mathcal{S}$ with

$$x = (1 - \theta)x_1 + \theta x_2, \quad \forall \theta \in [0, 1].$$

and more generally

$$x = \sum_{i=1}^{m} \theta_i x_i, \qquad \sum_{i=1}^{m} \theta_i = 1, \qquad \theta_i \geq 0$$

where $x_1, x_2, ..., x_m \in S$. This convex set is called the convex hull of the set of points. The intersection of convex sets is also a convex set.

A convex function $f(x)$ is defined by the condition that for any $x_1, x_2 \in S$ it follows that

$$f((1-\theta)x_1 + \theta x_2) \le (1-\theta)f(x_1) + \theta f(x_2), \quad \forall \theta \in [0,1].$$

The function is called strictly convex if the strict inequality holds here. For a convex function $f(x)$ the epigraph is a convex set, where the epigraph is the set of points in $\mathbb{R} \times \mathbb{R}^n$ that lies on or above the graph of $f(x)$. If $f_i(x)$ for $i = 1, 2, ..., m$ are convex functions on a convex set S and if $\lambda_i \ge 0$, then $\sum_i \lambda_i f_i(x)$ is a convex function on S.

Convex programming, dual problem

The problem of minimizing a convex function on a convex set S is said to be a convex programming problem.

$$\text{minimize} \quad f(x)$$
$$\text{subject to} \quad x \in S = \{x | c_i(x) \ge 0, \ i = 1, 2, ..., m\}$$

where $f(x)$ is a convex function on S and the functions $c_i(x)$ for $i = 1, 2, ..., m$ are concave on $\mathbb{R}^n$ (a concave function is defined as $-f(x)$ convex). Note that if $c(x)$ is a concave function then the set $S(k) = \{x | c(x) \ge k\}$ is convex. Every local solution x^* to a convex programming problem is a global solution and the set of global solutions S is convex. If $f(x)$ is also strictly convex on S then any global solution is unique. If x^* solves the convex programming primal problem above and if the objective function and constraints are differentiable and a certain regularity assumption holds, then x^*, λ^* solves the dual problem

$$\text{maximize}_{x,\lambda} \quad \mathcal{L}(x, \lambda)$$
$$\text{subject to} \quad \nabla_x \mathcal{L}(x, \lambda) = 0, \ \lambda \ge 0.$$

Furthermore the minimum primal and maximum dual function values are equal.

Quadratic programming problem

A quadratic programming problem is of the form

$$\text{minimize} \quad q(x) = \frac{1}{2}x^T G x + g^T x$$

$$\text{subject to} \quad a_i^T x = b_i, \quad i \in C_E$$

$$a_i^T x \geq b_i, \quad i \in C_I.$$

If the Hessian matrix G is positive semidefinite, the solution x^* is global. If G is positive definite the solution x^* is global and unique. When the Hessian G is indefinite other local solutions than the global solution may exist.

A.2   Linear algebra

For an introduction to linear algebra and elements of matrix analysis one may consult [98; 114; 162; 229]. The following elements have been discussed in [98; 162].

Independence, subspaces

Let $a_i \in \mathbb{R}^m$ for $i = 1, ..., n$ be a set of real valued vectors. This set is linearly independent if $\sum_{j=1}^{n} \alpha_j a_j = 0$ implies that vector $\alpha = 0$. A subspace spanned by $\{a_1, ..., a_n\}$ is span$\{a_1, ..., a_n\} = \{\sum_{j=1}^{n} \beta_j a_j : \beta_j \in \mathbb{R}\}$. If $\{a_1, ..., a_n\}$ is linearly independent and $b \in$ span$\{a_1, ..., a_n\}$ then b is a unique linear combination of the a_j.

Important subspaces associated with $A \in \mathbb{R}^{m \times n}$ (a real valued matrix of size $m \times n$) are the range of A defined by range$(A) = \{y \in \mathbb{R}^m : y = Ax$ for some $x \in \mathbb{R}^n\}$ and the null space defined by null$(A) = \{x \in \mathbb{R}^n : Ax = 0\}$. For $A = [a_1 ... a_n]$ one has range$(A) =$ span$\{a_1, ..., a_n\}$. The rank of a matrix A is defined by rank$(A) = \dim(\text{range}(A))$. One has rank$(A) =$ rank(A^T). We say that $A \in \mathbb{R}^{m \times n}$ is rank deficient if rank$(A) < \min\{m, n\}$. If $A \in \mathbb{R}^{m \times n}$ then $\dim(\text{null}(A)) + \text{rank}(A) = n$.

Sherman-Morrison-Woodbury formula

If A and X are in $\mathbb{R}^{n \times n}$ and satisfy $AX = I$, then X is the inverse of A, denoted by A^{-1}. If A^{-1} exists, then A is nonsingular, otherwise A is singular. One has $(AB)^{-1} = B^{-1}A^{-1}$, $(AB)^T = B^T A^T$. The Sherman-Morrison-Woodbury formula give an expression for the inverse of $(A+UV^T)$ where $A \in \mathbb{R}^{n \times n}$ and $U, V \in \mathbb{R}^{n \times k}$:

$$(A + UV^T)^{-1} = A^{-1} - A^{-1}U(I + V^T A^{-1}U)^{-1}V^T A^{-1}$$

and in general the matrix inversion lemma

$$(A + BCD)^{-1} = A^{-1} - A^{-1}B(DA^{-1}B + C^{-1})^{-1}DA^{-1}$$

with matrices $A \in \mathbb{R}^{n \times n}$, $B \in \mathbb{R}^{n \times m}$, $C \in \mathbb{R}^{m \times m}$, $D \in \mathbb{R}^{m \times n}$. A special case is

$$(A + uu^T)^{-1} = A^{-1} - \frac{A^{-1}uu^T A^{-1}}{1 + u^T A^{-1}u}$$

with $u \in \mathbb{R}^n$.

Block inverse

The inverse of a block matrix is given by

$$\begin{bmatrix} A_{11} & A_{12} \\ A_{21} & A_{22} \end{bmatrix}^{-1} = \begin{bmatrix} F_{11}^{-1} & -A_{11}^{-1}A_{12}F_{22}^{-1} \\ -F_{22}^{-1}A_{21}A_{11}^{-1} & F_{22}^{-1} \end{bmatrix}$$

$$= \begin{bmatrix} A_{11}^{-1} + A_{11}^{-1}A_{12}F_{22}^{-1}A_{21}A_{11}^{-1} & -F_{11}^{-1}A_{12}A_{22}^{-1} \\ -A_{22}^{-1}A_{21}F_{11}^{-1} & A_{22}^{-1} + A_{22}^{-1}A_{21}F_{11}^{-1}A_{12}A_{22}^{-1} \end{bmatrix}$$

with Schur complements

$$F_{11} = A_{11} - A_{12}A_{22}^{-1}A_{21}$$
$$F_{22} = A_{22} - A_{21}A_{11}^{-1}A_{12}.$$

Note that in the case that A_{12}, A_{21} are vectors and A_{22} a scalar:

$$\begin{bmatrix} A & b \\ b^T & c \end{bmatrix}^{-1} = \begin{bmatrix} A^{-1} + \frac{1}{d}A^{-1}bb^T A^{-1} & -\frac{1}{d}A^{-1}b \\ -\frac{1}{d}b^T A^{-1} & \frac{1}{d} \end{bmatrix}$$

where the Schur complement $d = c - b^T A^{-1} b$ is a scalar.

Determinant, trace, quadratic form, positive definite

The determinant of $A \in \mathbb{R}^{n \times n}$ is defined in terms of order $n-1$ determinants $\det(A) = \sum_{j=1}^{n} (-1)^{j+1} a_{1j} \det(A_{1j})$ where A_{1j} denotes an $(n-1) \times (n-1)$ matrix obtained by deleting the first row and j-th column of A. One has $A = (a) \in \mathbb{R}^{1 \times 1}$ with $\det(A) = a$ and the following properties

$$
\begin{aligned}
\det(AB) &= \det(A)\det(B) & A, B \in \mathbb{R}^{n \times n} \\
\det(A^{-1}) &= 1/\det(A) & A \in \mathbb{R}^{n \times n} \\
\det(A^T) &= \det(A) & A \in \mathbb{R}^{n \times n} \\
\det(cA) &= c^n \det(A) & c \in \mathbb{R}, A \in \mathbb{R}^{n \times n}
\end{aligned}
$$

and $\det(A) \neq 0$ if and only if A is nonsingular. With respect to the eigenvalue decomposition

$$ AX = X\Lambda $$

for matrix $A \in \mathbb{R}^{n \times n}$ with eigenvector matrix X and the corresponding diagonal matrix $\Lambda = \mathrm{diag}([\lambda_1; ...; \lambda_n])$ with eigenvalues λ_i one has the properties $\det(A) = \prod_{i=1}^{n} \lambda_i$ and $\mathrm{trace}(A) = \sum_{i=1}^{n} a_{ii} = \sum_{i=1}^{n} \lambda_i$. One has that $\mathrm{trace}(AB) = \mathrm{trace}(BA)$ and $y^T x = \mathrm{trace}(xy^T)$ for $x, y \in \mathbb{R}^n$. A symmetric matrix has real eigenvalues. A matrix $A \in \mathbb{R}^n$ is positive semidefinite if A is symmetric and $x^T A x \geq 0$ for all $x \in \mathbb{R}^n$ and positive definite if A is symmetric and $x^T A x > 0$ for all $x \neq 0$. The notation $A > 0$ for a symmetric matrix $A = A^T$ means that A is positive definite (all eigenvalues of A positive) and $A < 0$ means that A is negative definite (all eigenvalues of A negative). Note that for a quadratic form with A non-symmetric one can write $x^T A x = x^T [(A + A^T)/2] x$ where $(A + A^T)/2$ is the symmetric part of matrix A. A matrix A that satisfies $A^2 = A$ (by definition $A^2 = AA$) is called idempotent. An example is the projection matrix $H(H^T H)^{-1} H$ where $H \in \mathbb{R}^{m \times n}$ with $m > n$.

Differentiation

With respect to differentiation of matrices one has the following properties. Suppose α is a scalar and that $A(\alpha)$ is an $m \times n$ matrix with ij-th entry

$a_{ij}(\alpha)$. If $a_{ij}(\alpha)$ is a differentiable function of α for all i, j (and similarly for a matrix B with compatible dimension) then one has

$$
\begin{aligned}
\dot{A}(\alpha) &= \tfrac{d}{d\alpha} A(\alpha) = \dot{a}_{ij}(\alpha) \\
\tfrac{d}{d\alpha}[A(\alpha)B(\alpha)] &= [\tfrac{d}{d\alpha}A(\alpha)]B(\alpha) + A(\alpha)[\tfrac{d}{d\alpha}B(\alpha)] \\
\tfrac{d}{d\alpha}[A(\alpha)^{-1}] &= -A(\alpha)^{-1}[\tfrac{d}{d\alpha}A(\alpha)]A(\alpha)^{-1}.
\end{aligned}
$$

Vector norms

A vector norm on $\mathbb{R}^n$ is a function $f : \mathbb{R}^n \to \mathbb{R}$ that satisfies the following properties

$$
\begin{aligned}
f(x) &\geq 0 & x &\in \mathbb{R}^n, & (f(x) = 0 \text{ iff } x = 0) \\
f(x + y) &\leq f(x) + f(y) & x, y &\in \mathbb{R}^n \\
f(\alpha x) &= |\alpha| f(x) & \alpha &\in \mathbb{R}, x \in \mathbb{R}^n
\end{aligned}
$$

and is denoted by $f(x) = \|x\|$. A class of vector norms called p-norms is defined by

$$
\|x\|_p = (|x_1|^p + \ldots + |x_n|^p)^{1/p}, \qquad p \geq 1
$$

with as special cases

$$
\begin{aligned}
\|x\|_1 &= |x_1| + \ldots + |x_n| \\
\|x\|_2 &= (x^T x)^{1/2} \\
\|x\|_\infty &= \max_{1 \leq i \leq n} |x_i|.
\end{aligned}
$$

Often if a norm $\|x\|$ is not further specified one means the 2-norm (Euclidean norm). For the p-norm the Holder inequality holds:

$$
|x^T y| \leq \|x\|_p \|y\|_q \text{ for } \frac{1}{p} + \frac{1}{q} = 1
$$

with as special case the Cauchy-Schwartz inequality

$$
|x^T y| \leq \|x\|_2 \|y\|_2.
$$

The angle θ between the vectors x and y is characterized by

$$
\cos \theta = \frac{x^T y}{\|x\|_2 \|y\|_2}.
$$

Furthermore if $\|\cdot\|_\alpha$ and $\|\cdot\|_\beta$ are norms on $\mathbb{R}^n$ then there are positive constants c_1 and c_2 such that

$$c_1\|x\|_\alpha \leq \|x\|_\beta \leq c_2\|x\|_\alpha$$

for all $x \in \mathbb{R}^n$. For example one has

$$
\begin{aligned}
\|x\|_2 &\leq \|x\|_1 \leq \sqrt{n}\|x\|_2 \\
\|x\|_\infty &\leq \|x\|_2 \leq \sqrt{n}\|x\|_\infty \\
\|x\|_\infty &\leq \|x\|_1 \leq n\|x\|_\infty.
\end{aligned}
$$

Bilinear functional, metric, inner product, Hilbert space

Let us consider an n-dimensional vector space X over the field of real numbers. If f is a bilinear functional on a real vector space X, then $f : X \times X \to \mathbb{R}$ and

$$
\begin{aligned}
f(\alpha x_1 + \beta x_2, y) &= \alpha f(x_1, y) + \beta f(x_2, y) \\
f(x, \alpha y_1 + \beta y_2) &= \alpha f(x, y_1) + \beta f(x, y_2)
\end{aligned}
$$

for all $\alpha, \beta \in \mathbb{R}$ and for all $x, x_1, x_2, y, y_1, y_2 \in X$. As a consequence, more generally

$$f\Big(\sum_{j=1}^n \alpha_j x_j, \sum_{k=1}^m \beta_k y_k\Big) = \sum_{j=1}^n \sum_{k=1}^m \alpha_j \beta_k f(x_j, y_k).$$

A bilinear functional f on a real vector space X is said to be an inner product on X if (i) f is symmetric (i.e. $f(x, y) = f(y, x)$) and (ii) f is strictly positive. A real finite-dimensional vector space on which an inner product is defined is called a Euclidean space. For Euclidean norms the inner product between vectors x, y is defined as $x^T y$. Let $\rho(x, y)$ denote the distance between two vectors x and y of X. A function $\rho(x, y)$ that has the following properties for all $x, y, z \in X$ is then called a metric:

      (i) $\rho(x, y) = \rho(y, x)$

     (ii) $\rho(x, y) \geq 0$ and $\rho(x, y) = 0$ iff $x = y$

    (iii) $\rho(x, y) \leq \rho(x, z) + \rho(z, y).$

A Euclidean space is one example of a metric space. A function defined on $X \times X$ into $\mathbb{R}$ denoted by $\langle x, y \rangle$ for $x, y \in X$ is called an inner product if

   (i) $\langle x, x \rangle > 0 \;\; \forall x \neq 0 \;\; (\langle x, x \rangle = 0 \; \text{if} \; x = 0)$

   (ii) $\langle x, y \rangle = \langle y, x \rangle \;\; \forall x, y \in X$

   (iii) $\langle \alpha x + \beta y, z \rangle = \alpha \langle x, z \rangle + \beta \langle y, z \rangle \;\; \forall x, y, z \in X, \;\; \forall \alpha, \beta \in \mathbb{R}.$

One calls a real (or complex) space X on which the inner product is defined a real (or complex) inner product space. A complete inner product space is called a Hilbert space.

Matrix norms

For matrices $f : \mathbb{R}^{m \times n} \to \mathbb{R}$ is a matrix norm if the following three properties hold

$$
\begin{array}{lll}
f(A) \geq 0 & A \in \mathbb{R}^{m \times n}, & (f(A) = 0 \text{ iff } A = 0) \\
f(A + B) \leq f(A) + f(B) & A, B \in \mathbb{R}^{m \times n} & \\
f(\alpha A) = |\alpha| f(A) & \alpha \in \mathbb{R}, A \in \mathbb{R}^{m \times n} &
\end{array}
$$

and is denoted by $f(A) = \|A\|$. Examples are the Frobenius norm

$$
\|A\|_F = \sqrt{\sum_{i=1}^{m} \sum_{j=1}^{n} |a_{ij}|^2}
$$

and the p-norms

$$
\|A\|_p = \sup_{x \neq 0} \frac{\|Ax\|_p}{\|x\|_p} = \max_{\|x\|_p = 1} \|Ax\|_p.
$$

One has

$$
\begin{array}{rcl}
\|A\|_1 & = & \max_{1 \leq j \leq n} \sum_{i=1}^{m} |a_{ij}| \\
\|A\|_2 & = & \max_j [\lambda_j (A^T A)]^{1/2} \\
\|A\|_\infty & = & \max_{1 \leq i \leq m} \sum_{j=1}^{n} |a_{ij}|.
\end{array}
$$

One can verify that

$$
\|AB\|_p \leq \|A\|_p \|B\|_p.
$$

Some properties of matrix norms are

$$\|A\|_2 \quad \leq \quad \|A\|_F \quad \leq \quad \sqrt{n}\,\|A\|_2$$

$$\max_{ij}|a_{ij}| \quad \leq \quad \|A\|_2 \quad \leq \quad \sqrt{mn}\,\max_{i,j}|a_{ij}|$$

$$\tfrac{1}{\sqrt{n}}\|A\|_\infty \quad \leq \quad \|A\|_2 \quad \leq \quad \sqrt{m}\,\|A\|_\infty$$

$$\tfrac{1}{\sqrt{m}}\|A\|_1 \quad \leq \quad \|A\|_2 \quad \leq \quad \sqrt{n}\,\|A\|_1.$$

Singular value decomposition

A set of vectors $\{x_1, \ldots, x_p\}$ in $\mathbb{R}^m$ is orthogonal if $x_i^T x_j = 0$ whenever $i \neq j$ and orthonormal if $x_i^T x_j = \delta_{ij}$. A matrix $Q \in \mathbb{R}^{m \times m}$ is said to be orthogonal if $Q^T Q = I$. The 2-norm is invariant under orthogonal transformations: if $Q^T Q = I$ then $\|Qx\|_2^2 = x^T Q^T Q x = x^T x = \|x\|_2^2$. For orthogonal matrices Q, Z with appropriate dimensions one has $\|QAZ\|_F = \|A\|_F$ and $\|QAZ\|_2 = \|A\|_2$. For $A \in \mathbb{R}^{m \times n}$ there exist orthogonal matrices $U = [u_1 ... u_m] \in \mathbb{R}^{m \times m}$ $(U^T U = I_m)$, $V = [v_1 ... v_n] \in \mathbb{R}^{n \times n}$ $(V^T V = I_n)$ such that

$$U^T A V = \Sigma = \mathrm{diag}([\sigma_1; ...; \sigma_p]) \in \mathbb{R}^{m \times n}$$

where $p = \min\{m, n\}$ with $\sigma_1 \geq \sigma_2 \geq ... \geq 0$, called singular values. the vectors u_i, v_i are called left and right singular vectors respectively. The singular values σ_i of A and the eigenvalues λ_i of $A^T A$ are related as follows for an SVD $A = U\Sigma V^T$:

$$A^T A = V\Sigma^2 V^T$$
$$AA^T = U\Sigma^2 U^T.$$

If one has $\sigma_1 \geq ... \geq \sigma_r > \sigma_{r+1} = ... = \sigma_p = 0$, then

$$\mathrm{rank}(A) = r$$
$$\mathrm{null}(A) = \mathrm{span}\{v_{r+1}, ..., v_n\}$$
$$\mathrm{range}(A) = \mathrm{span}\{u_1, ..., u_r\}$$

with dyadic decomposition

$$A = \sum_{i=1}^{r} \sigma_i u_i v_i^T.$$

The link with matrix norms is as follows

$$\|A\|_F^2 = \sigma_1^2 + ... + \sigma_p^2 \quad p = \min\{m, n\}$$

$$\|A\|_2 = \sigma_1 = \sigma_{\max}$$

$$\min_{x \neq 0} \frac{\|Ax\|_2}{\|x\|_2} = \sigma_n \quad m \geq n.$$

Due to rounding errors and noisy data, rank determination may be difficult. One defines then a numerical rank deficiency as $\mathrm{rank}(A, \epsilon) = \min_{\|A-B\| \leq \epsilon} \mathrm{rank}(B)$. If $k < r = \mathrm{rank}(A)$ and $A_k = \sum_{i=1}^{k} \sigma_i u_i v_i^T$, then

$$\min_{\mathrm{rank}(B)=k} \|A - B\|_2 = \|A - A_k\|_2 = \sigma_{k+1}.$$

For the maximal singular value one can also show that

$$\sigma_{\max}(A) = \max_{y \in \mathbb{R}^m, x \in \mathbb{R}^n} \frac{y^T A x}{\|x\|_2 \|y\|_2}.$$

In relation to solving the linear system $Ax = b$ with $A \in \mathbb{R}^{n \times n}$ nonsingular and $b \in \mathbb{R}^n$ the solution can be expressed as

$$x = A^{-1}b = (U\Sigma V^T)^{-1}b = \sum_{i=1}^{n} \frac{u_i^T b}{\sigma_i} v_i$$

with $A = U\Sigma V^T$.

Condition number

For square matrices A one defines the condition number $\kappa(A) = \|A\| \|A^{-1}\|$ (note that it is always ≥ 1) where

$$\kappa_2(A) = \|A\|_2 \|A^{-1}\|_2 = \frac{\sigma_{\max}(A)}{\sigma_{\min}(A)}.$$

Matrices with small condition numbers are called well-conditioned. Given an unperturbed linear system $Ax = b$ and the perturbed problem $(A + \Delta A)y = b + \Delta b$ with $\|\Delta A\| \leq \epsilon\|A\|$ and $\|\Delta b\| \leq \epsilon\|b\|$. If $\epsilon\kappa(A) = r < 1$, then $A + \Delta A$ is nonsingular and

$$\frac{\|y\|}{\|x\|} \leq \frac{1+r}{1-r}$$

and

$$\frac{\|y - x\|}{\|x\|} \leq \frac{2\epsilon}{1 - r}\kappa(A)$$

holds.

A.3 Statistics

There exist many introductory books about statistics, e.g. [8; 62; 77; 124; 125; 127; 191; 213; 215]. We discuss here a few elementary aspects with respect to densities and descriptive statistics.

Random variable, sample space, cumulative distribution function

In general, a random variable X is a function from a sample space Ω^s to the real numbers $\mathbb{R}$ where $X : \Omega^s \to \mathbb{R} : \omega_k \mapsto x_k$ with $\omega_k \in \Omega^s$ and x_k a realization of the random variable X.

The cumulative distribution function (cdf) of a random variable X, denoted by $F(x) : \mathbb{R} \to [0, 1]$ is defined by

$$F(x) = P(X \leq x), \quad \forall x \in \mathbb{R}.$$

The function $F(x)$ is a cdf if and only if the following three conditions hold: (i) $\lim_{x \to -\infty} F(x) = 0$ and $\lim_{x \to \infty} F(x) = 1$; (ii) $F(x)$ is a nondecreasing function of x; (iii) $F(x)$ is right-continuous.

Probability density function

Associated with a random variable X and its cdf $F(x)$ is the probability density function (pdf) in the continuous case or the probability mass function (pmf) in the discrete case. The pmf of a discrete random variable X is given by

$$p(x) = P(X = x), \quad \forall x \in \mathbb{R}$$

and the pdf of a continuous random variable X is the function that satisfies

$$F(x) = \int_{-\infty}^{x} p(t)dt$$

for all $x \in \mathbb{R}$. For any $a < b$ the probability that X falls within the interval (a, b) is the area under the pdf curve over that interval equals

$$P(a < X < b) = \int_{a}^{b} p(x)dx.$$

A function $p(x)$ is a pdf or pmf of a random variable X if and only if (i) $p(x) \geq 0$, $\forall x$; (ii) $\int_{-\infty}^{\infty} p(x)dx = 1$.

Expected value and variance

The two most important parameters used to describe or summarize the properties of a random variable X are the mean $\mathcal{E}[X]$ and the variance $\mathrm{Var}[X]$. The mean or expected value of X, $\mu = \mathcal{E}[X]$ is given by

$$\mu = \mathcal{E}[X] = \sum_{k} x_k p(x_k)$$

for the case that X is discrete (provided that $\sum_k |x_k| p(x_k) < \infty$), and

$$\mu = \mathcal{E}[X] = \int_{-\infty}^{\infty} xp(x)dx$$

for the case that X is continuous (provided that $\int_{-\infty}^{\infty} |x| p(x)dx < \infty$). The variance of X is defined by $\sigma^2 = \mathrm{Var}[X] = \mathcal{E}[(X - \mu)^2]$ which equals

$$\sigma^2 = \mathrm{Var}[X] = \sum_{k} (x_k - \mu)^2 p(x_k)$$

for discrete random variables X, and

$$\sigma^2 = \mathrm{Var}[X] = \int_{-\infty}^{\infty} (x - \mu)^2 p(x)dx.$$

Discrete distribution

A list of common pmf's for discrete distributions is

- Bernoulli (q)

$$P(X = x|q) = q^x(1-q)^{1-x}, \ x = 0, 1; \ 0 \le q \le 1$$

  with mean $\mathcal{E}[X] = q$ and variance $\mathrm{Var}[X] = q(1-q)$.

- Binomial (n, q)

$$P(X = x|n, q) = \binom{n}{x} q^x(1-q)^{n-x}, \ x = 0, 1, 2, ..., n; \ 0 \le q \le 1$$

  with mean $\mathcal{E}[X] = nq$ and variance $\mathrm{Var}[X] = nq(1-q)$.

- Geometric (q)

$$P(X = x|q) = q(1-q)^{x-1}, \ x = 1, 2, ...; \ 0 \le q \le 1$$

  with mean $\mathcal{E}[X] = 1/q$ and variance $\mathrm{Var}[X] = (1-q)/q^2$.

- Poisson (λ)

$$P(X = x|\lambda) = \frac{e^{-\lambda}\lambda^x}{x!}, \ x = 0, 1, 2, ...; \ 0 \le \lambda < \infty$$

  with mean $\mathcal{E}[X] = \lambda$ and variance $\mathrm{Var}[X] = \lambda$. The Poisson frequency function can be used to approximate binomial probabilities for large n and small q.

Continuous distributions

Common pdf's for continuous distributions are

- Beta (α, β)

$$p(x|\alpha, \beta) = \frac{1}{B(\alpha, \beta)} x^{\alpha-1}(1-x)^{\beta-1}, \ 0 \le x \le 1$$

  with mean $\mathcal{E}[X] = \alpha/(\alpha + \beta)$ and variance $\mathrm{Var}[X] = \alpha\beta/((\alpha + \beta)^2(\alpha + \beta + 1))$ and $\alpha > 0, \beta > 0$ where

$$B(\alpha, \beta) = \frac{\Gamma(\alpha)\Gamma(\beta)}{\Gamma(\alpha + \beta)}$$

with Gamma function $\Gamma(\cdot)$.

- t-distribution v

$$p(x|v) = \frac{\Gamma(\frac{v+1}{2})}{\Gamma(\frac{v}{2})\sqrt{v\pi}} \frac{1}{(1 + (\frac{x^2}{v}))^{(v+1)/2}}$$

with $v = 1, 2, \ldots$ the degrees of freedom, mean $\mathcal{E}[X] = 0$ $(v > 1)$ and variance $\text{Var}[X] = v/(v-2)$ $(v > 2)$. For $v = 1$ one has the Cauchy distribution and for $v \to \infty$ the normal distribution.

- Cauchy (θ, σ)

$$p(x|\theta, \sigma) = \frac{1}{\pi\sigma} \frac{1}{1 + \frac{(x-\theta)^2}{\sigma^2}}, \quad -\infty < x < \infty$$

with $-\infty < \theta < \infty$ and $\sigma > 0$. The mean and variance are not defined.

- Gamma (α, β)

$$p(x|\alpha, \beta) = \frac{1}{\Gamma(\alpha)\beta^\alpha} x^{\alpha-1} e^{-x/\beta}, \quad 0 \le x \le \infty$$

and $\alpha, \beta > 0$ with mean $\mathcal{E}[X] = \alpha\beta$ and variance $\text{Var}[X] = \alpha\beta^2$. Some special cases are the exponential distribution $(\alpha = 1)$ and the Chi-squared distribution $(\alpha = v/2, \beta = 2)$.

- Chi-squared χ_v^2

$$p(x|v) = \frac{1}{2^{v/2}\Gamma(v/2)} x^{(v/2)-1} e^{-x/2}, \quad 0 \le x \le \infty$$

with degrees of freedom $v = 1, 2, \ldots$, mean $\mathcal{E}[X] = v$ and variance $\text{Var}[X] = 2v$.

- Exponential (λ)

$$p(x|\lambda) = \frac{1}{\lambda} e^{-x/\lambda}, \quad 0 \le x < \infty$$

and $\lambda > 0$ with mean $\mathcal{E}[X] = \lambda$ and variance $\text{Var}[X] = \lambda^2$.

- Laplace (μ, σ) (or double exponential distribution)

$$p(x|\mu, \sigma) = \frac{1}{2\sigma} e^{-|x-\mu|/\sigma}$$

with mean $\mathcal{E}[X] = \mu$ and variance $\text{Var}[X] = 2\sigma^2$.

- Normal (μ, σ^2) (or Gaussian distribution)

$$p(x|\mu, \sigma^2) = \frac{1}{\sqrt{2\pi}\sigma} e^{-(x-\mu)^2/(2\sigma^2)}$$

with mean $\mathcal{E}[X] = \lambda$ and variance $\text{Var}[X] = \sigma^2$.

- Uniform (a, b)

$$p(x|a, b) = \frac{1}{b - a}, \ a \leq x \leq b$$

with mean $\mathcal{E}[X] = (b + a)/2$ and variance $\text{Var}[X] = (b - a)^2/12$. This is a special case of the Beta distribution for $\alpha = \beta = 1$.

Multivariate Gaussian density

For a vector X of random variables with mean $\mu_x \in \mathbb{R}^n$ and covariance matrix $\Sigma_{xx} = \mathcal{E}[(X - \mu_x)(X - \mu_x)^T]$ one has

$$\mathcal{E}[AX + B] = A\mu_x + B$$
$$\text{Var}[AX + B] = A\Sigma_{xx}A^T.$$

Consider the multivariate Gaussian density

$$p(X|\mu_x, \Sigma_{xx}) = \frac{1}{\sqrt{(2\pi)^n \det\Sigma_{xx}}} \exp\left(-\frac{1}{2}(x - \mu_x)^T \Sigma_{xx}^{-1} (x - \mu_x)\right).$$

Assuming that X is distributed according to this Gaussian density denoted as $X \sim \mathcal{N}(\mu_x, \Sigma_{xx})$, then

$$X \sim \mathcal{N}(\mu_x, \Sigma_{xx}) \ \Rightarrow \ (AX + B) \sim \mathcal{N}(A\mu_x + B, A\Sigma_{xx}A^T)$$
$$X \sim \mathcal{N}(\mu_x, \Sigma_{xx}) \ \Rightarrow \ (X - \mu_x)^T \Sigma_{xx}^{-1} (X - \mu_x) \sim \chi_n^2.$$

Skewness, kurtosis

In descriptive statistics there are various ways to characterize the shape of a distribution. A first index of shape which gives an idea about the asymmetry of a sample around the mean is the skewness which is defined by

$$\text{skewness}(x) = \frac{\frac{1}{N}\sum_{k=1}^{N}(x_k - \bar{x})^3}{s^3}$$

with $\bar{x} = \sum_{k=1}^{N} x_k/N$ the sample mean and $s = (\sum_{k=1}^{N}(x_k - \bar{x})^2/(N-1))^{1/2}$ the standard deviation. Symmetric distributions may deviate from normality by being heavy tailed and/or not as highly peaked as the Gaussian density. This deviation is characterized by the coefficient of kurtosis

$$\text{kurtosis}(x) = \frac{\frac{1}{N}\sum_{k=1}^{N}(x_k - \bar{x})^4}{s^4}.$$

Measures of location

- Mode:
  The mode is the value of the random variable which corresponds to the maximum of the density function.

- Mean:
  The mean can be considered as the mass of the density. The sample mean is given by $\bar{x}$.

- Median:
  The median m of a random variable X satisfies $P(X \leq m) \geq 0.5$ and $P(X \geq m) \geq 0.5$. In the continuous case one has

$$\int_{-\infty}^{m} p(x)dx = \int_{m}^{\infty} p(x)dx = 0.5.$$

- Quantiles:
  The q-th quantile, denoted by x_q is defined by

$$x_q = F^{-}(q) = \inf_{x \in \mathbb{R}}\{x : F(X) \geq q\}, \; \forall q \in (0,1).$$

In the case where the cdf is continuous and strictly increasing, x_q is unique and defined by

$$x_q = F^{-1}(q) \text{ or } F(x_q) = q.$$

Measures of dispersion

- Range:
  The range is defined as the difference between the largest and the smallest value taken by the random variable X.

- Interquartile range:
  This is equal to difference between the lower and the upper quartile:

  $$\text{IQR}(x) = x_{3/4} - x_{1/4}.$$

- The standard deviation s and variance $\bar{x}$.

Boxplot

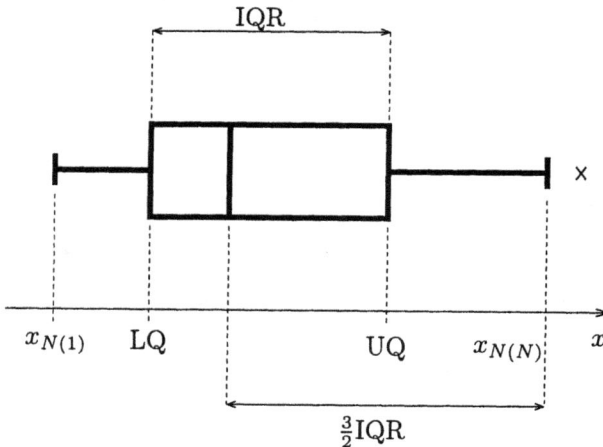

A boxplot graphically displays a measure of location (the median), a measure of dispersion (the interquartile range) and the presence of possible outliers and gives an idea about the symmetry or skewness of the distribution and also gives an indication of possible outliers. A boxplot is a rectangle with length equal to the IQR and separated into two parts by the sample median. At each side of this box, two line segments are drawn describing the distribution tails. The left and right boundaries for the distribution tails are given by

$$x_L = \max\{x_{N(1)}, \text{LQ} - \tfrac{3}{2}\text{IQR}\}$$
$$x_R = \max\{x_{N(N)}, \text{UQ} + \tfrac{3}{2}\text{IQR}\}$$

where $x_{N(i)}$ with $i = 1, ..., N$ are ordered samples for $x_1, ..., x_N$ such that $x_{N(1)} \leq x_{N(2)} \leq ... \leq x_{N(N)}$ and LQ and UQ denote the lower and upper sample quartiles, respectively. An example of a boxplot is shown above. Here '×' denotes an outlier.

Bibliography

[1] Abrahamsen P. (1997) "A review of Gaussian random fields and correlation functions", *Technical Report 917, Norwegian Computing Center, Oslo Norway*.

[2] Abrahamsen P., Omre H. (1994) "Random functions and geological subsurface". *Technical Report, Norwegian Computing Center, Oslo Norway*.

[3] Aha D., Kibler D. (1991) "Instance-based learning algorithms", *Machine Learning*, **6**, 37–66.

[4] Aizerman M.A., Braverman E.M., Rozonoér L.I. (1964) "Theoretical foundations of the potential function method in pattern recognition learning", *Automation and Remote Control*, **25**, 821–837.

[5] Akaike H. (1970) "Statistical predictor identification", *Ann. Inst. Stat. Math.*, **22**, 202–217.

[6] Akaike H. (1974) "A new look at statistical model identification", *IEEE Transactions on Automatic Control*, **19**, 716–723.

[7] Amari S., Wu S. (1999) "Improving support vector machine classifiers by modifying kernel functions", *Neural Networks*, **12**, 783–789.

[8] Anderson T.W. (1966) *An Introduction to Multivariate Analysis*, John Wiley & Sons, New York.

[9] Andrews D.F., Bickel P.J., Hampel F.R., Huber P.J., Rogers W.H., Tukey J.W. (1972) *Robust Estimates of Location: Survey and Advances*, Princeton, NJ: Princeton University Press.

[10] Anguita D., Boni A., Ridella S. (2000) "Evaluating the generalization ability of support vector machines through the bootstrap", *Neural Processing Letters*, **11**(1), 51–58.

[11] Anthony M., Bartlett P. (1999) *Neural Network Learning: Theoretical Foundations*, Cambridge University Press, New York.

[12] Arbib M.A. (Ed.) (1995) *The Handbook of Brain Theory and Neural Networks*, MIT Press, Cambridge MA.

[13] Aronszajn N. (1950) "Theory of reproducing kernels", *Trans. American Mathematical Soc.*, **68**, 337–404.

[14] Baker C.T.H. (1977) *The Numerical Treatment of Integral Equations*, Oxford: Clarendon Press.

[15] Bartlett P.L., Mendelson S. (2002) "Rademacher and Gaussian complexities: risk bounds and structural results", *Journal of Machine Learning Research*, to appear.

[16] Baudat G., Anouar F. (2000) "Generalized discriminant analysis using a kernel approach", *Neural Computation*, **12**, 2385–2404.

[17] Baudat G., Anouar F. (2001) "Kernel-based methods and function approximation", *Proc. of the International Joint Conference on Neural Networks (IJCNN 2001)*, Washington DC, 1244–1249.

[18] Bay S.D. (1999) "Nearest neighbor classification from multiple feature subsets", *Intelligent Data Analysis*, **3**, 191–209.

[19] Bernardo J.M., Smith A.F.M. (1994) *Bayesian Theory*, John Wiley & Sons, New York.

[20] Bertsekas D.P. (1982) *Constrained Optimization and Lagrange Multiplier Methods*, Academic Press, New York.

[21] Bickel P.J. (1965) "On some robust estimates of location", *Ann. Math. Statist.*, **36**, 847–858.

[22] Bickel P.J., Lehman E.L. (1975) "Descriptive statistics for non-parametric models", *Ann. Statist.*, **3**, 1038–1045.

[23] Bishop C.M. (1995) *Neural Networks for Pattern Recognition*, Oxford University Press.

[24] Blake C.L., Merz C.J. (1998) *UCI repository of machine learning databases* [http://www.ics.uci.edu/~mlearn/MLRepository.html]. Irvine, CA: University of California, Dept. of Information and Computer Science.

[25] Blum L., Cucker F., Shub M., Smale S. (1998), *Complexity and Real Computation*, Springer-Verlag, New York.

[26] Bochner S. (1959) *Lectures on Fourier Integrals*, Princeton University Press, New Jersey.

[27] Boser B.E., Guyon I.M., Vapnik V.N. (1992) "A training algorithm for optimal margin classifiers", In *Proceedings of the Fifth Annual ACM Workshop on Computational Learning Theory*, 144–152.

[28] Boyd S., Barratt C. (1991) *Linear Controller Design, Limits of Performance*, Prentice-Hall.

[29] Boyd S., El Ghaoui L., Feron E., Balakrishnan V. (1994) *Linear Matrix Inequalities in System and Control Theory*, SIAM Studies in Applied Mathematics 15.

[30] Bradley P.S., Mangasarian O.L. (1998) "Feature selection via concave minimization and support vector machines", In: J. Shavlik (Ed.), *Machine Learning Proc. of the Fifteenth Int. Conf. (ICML'98)*, Morgan Kaufmann, San Francisco, California, 82–90.

[31] Brown M., Grundy W., Lin D., Cristianini N., Sugnet C, Furey T., Ares M., Haussler D. (2000) "Knowledge-based analysis of microarray gene expression data using support vector machines", *Proceedings of the National*

Academy of Science, **97**(1), 262–267.

[32] Brown P.J. (1977) "Centering and scaling in ridge regression", *Technometrics*, **19**, 35–36.

[33] Bryson A.E., Ho Y.C. (1969) *Applied Optimal Control*, Waltham, MA: Blaisdel.

[34] Bunke O., Droge B. (1984) "Bootstrap and cross-validation estimates of the prediction error for linear regression models", *Ann. Statist*, **12**, 1400–1424.

[35] Burges C.J.C. (1998) "A tutorial on support vector machines for pattern recognition", *Knowledge Discovery and Data Mining*, **2**(2), 121–167.

[36] Burman P. (1989) "A comparative study of ordinary cross-validation, v-fold cross-validation and the repeated learning-testing methods", *Biometrika*, **76**(3), 503–514.

[37] Cawley G.C., Talbot N.L.C. (2002) "Efficient formation of a basis in a kernel induced feature space", *Proc. European Symposium on Artificial Neural Networks (ESANN 2002)*, Brugge Belgium, 1–6.

[38] Chen S.C., Donoho D.L., Saunders M.A. (1998) "Atomic decomposition by basis pursuit", *SIAM J. Scientific Computing*, **20**(1), 33–61.

[39] Chen S., Cowan C., Grant P. (1991) "Orthogonal least squares learning algorithm for radial basis function networks", *IEEE Transactions on Neural Networks*, **2**(2), 302–309.

[40] Chen S., Billings S.A., Grant M. (1992) "Recursive hybrid algorithm for non-linear system identification using radial basis function networks", *International Journal of Control*, **55**(5), 1051–1070.

[41] Cherkassky V., Mulier F. (1998) *Learning from Data: Concepts, Theory and Methods*, John Wiley & Sons, New York.

[42] Cherkassky V., Shao X. (2001) "Signal estimation and denoising using VC-theory", *Neural Networks*, **14**, 37–52.

[43] Chua L.O., Komuro M., Matsumoto T. (1986) "The double scroll family", *IEEE Trans. Circuits and Systems-I*, **33**(11), 1072–1118.

[44] Collobert R., Bengio S., Bengio Y. (2002) "A parallel mixture of SVMs for very large scale problems", *Neural Computation*, **14**(5), 1105–1114.

[45] Conover W.J. (1980) *Practical Nonparametric Statistics* (2nd ed.), John Wiley & Sons, New York.

[46] Cortes C., Vapnik V. (1995) "Support vector networks", *Machine Learning*, **20**, 273–297.

[47] Cover T., Thomas J. (1991) *Elements of Information Theory*, John Wiley & Sons, New York.

[48] Courant R., Hilbert D. (1953) *Methods of Mathematical Physics*, Interscience Publishers, New York.

[49] Craven P., Wahba G. (1979) "Smoothing noisy data with spline functions: estimating the correct degree of smoothing by the method of generalized cross-validation", *Numer. Math.*, **31**, 377–403.

[50] Cressie N. (1993) *Statistics for Spatial Data*, John Wiley & Sons, New York.

[51] Cristianini N., Shawe-Taylor J. (2000) *An Introduction to Support Vector*

Machines, Cambridge University Press.

[52] Cristianini N., Shawe-Taylor J., Elisseeff A., Kandola J. (2001) "On kernel-target alignment", in T. G. Dietterich, S. Becker, and Z. Ghahramani (Eds.) *Advances in Neural Information Processing Systems 14. MIT Press*, Cambridge, MA, 2002.

[53] Csató L., Opper M. (2002) "Sparse on-line Gaussian processes", *Neural Computation*, **14**(3), 641–668.

[54] Cucker F., Smale S. (2002) "On the mathematical foundations of learning theory", *Bulletin of the AMS*, **39**, 1–49.

[55] Daniel C. (1920) "Observations weighted according to order", *Amer. J. Math*, **42**, 222–236.

[56] David H.A. (1998) "Early sample measures of variability", *Statistical Science*, **13**(4), 368–377.

[57] Davison A.C., Hinkley D.V. (1997) *Bootstrap Methods and Their Application*, Cambridge University Press.

[58] Debnath L., Mikusiński P. (1999) *Introduction to Hilbert spaces with Applications* (2nd ed.), Academic Press.

[59] De Brabanter J., Pelckmans K., Suykens J.A.K., Vandewalle J. (2002) "Robust cross-validation score function for non-linear function estimation", *International Conference on Artificial Neural Networks (ICANN 2002)*, Madrid Spain, August 2002, pp. 713–719.

[60] De Brabanter J., Pelckmans K., Suykens J.A.K., Vandewalle J., De Moor B. (2002) "Robust cross-validation score function for LS-SVM non-linear function estimation", *K.U. Leuven ESAT-SISTA Technical Report 2002-94*.

[61] De Cock K. (2002) *Principal Angles in System Theory, Information Theory and Signal Processing*, PhD Thesis, K.U. Leuven Department of Electrical Engineering Leuven Belgium.

[62] De Groot M.H. (1986) *Probability and Statistics* (2nd ed.), Addison-Wesley, Reading, MA.

[63] Delves L.M., Mohamed J.L. (1985) *Computational Methods for Integral Equations*, Cambridge: Cambridge University Press.

[64] Devroye L., Györfi L., Lugosi G. (1996) *A Probabilistic Theory of Pattern Recognition*, NY: Springer.

[65] Dietterich T.G. (1998) "Approximate statistical tests for comparing supervised classification learning algorithms", *Neural Computation*, **10**, 1895–1924.

[66] Dietterich T.G., Bakiri G. (1995) "Solving multiclass learning problems via error-correcting output codes", *Journal of Artificial Intelligence Research*, **2**, 263–286.

[67] Diggle P.J., Hall P. (1986) "The selection of terms in an orthogonal series density estimator", *Journal of the American Statistical Association*, **81**, 230–233.

[68] Domingos P. (1996) "Unifying instance-based and rule-based induction", *Machine Learning*, **24**, 141–168.

[69] Donoho D.L., Huber P.J. (1983) "The notion of breakdown point", in A Festschrift for Erich Lehmann, (Ed. P. Bickel, K. Doksum, J.L. Hodges Jr.), Belmont, CA: Wadsworth.

[70] Doob J.L. (1953) *Stochastic Processes*, John Wiley & Sons, New York.

[71] Duda R.O., Hart P.E., Stork D.G. (2001) *Pattern Classification* (2nd ed.), John Wiley & Sons, New York.

[72] Eubank R.L. (1999) *Nonparametric Regression and Spline Smoothing*, Statistics: textbooks and monographs, Vol. 157, second edition, Marcel Dekker, New York.

[73] Evgeniou T., Pontil M., Poggio T. (2000) "Regularization networks and support vector machines", *Advances in Computational Mathematics,* **13**(1), 1–50.

[74] Evgeniou T., Pontil M., Papageorgiou C., Poggio T. (2000) "Image representations for object detection using kernel classifiers", *Proc. Fourth Asian Conference on Computer Vision (ACCV 2000)*, 687–692.

[75] Evgeniou T., Pontil M., Elisseeff A. (2001) "Leave-one-out error, stability, and generalization of voting combinations of classifiers", *INSEAD Technical Report 2001-21-TM.*

[76] Fayyad U.M., Piatetsky-Shapiro G., Smyth P., Uthurasamy R. (Eds.) (1996) *Advances in Knowledge Discovery and Data Mining,* MIT Press.

[77] Feller W. (1966) *An Introduction to Probability Theory and its Applications,* Vol.I (2nd ed.) and Vol.II, John Wiley & Sons, New York.

[78] Fernholz L.T. (1983) *Von Mises Calculus for Statistical Functionals,* Lecture Notes in Statistics, Springer-Verlag.

[79] Fine S., Scheinberg K. (2001) "Efficient SVM training using low-rank kernel representation", *Journal of Machine Learning Research,* **2**, 243–264.

[80] Fisher R.A. (1936) "The use of multiple measurements in taxonomic problems", *Annals of Eugenics,* **7**, 179–188.

[81] Fisher R.A. (1950) *Contributions to Mathematical Statistics,* W.A. Shewhart (Ed.), John Wiley & Sons, New York.

[82] Fletcher R. (1987) *Practical Methods of Optimization,* Chichester and New York: John Wiley & Sons, New York.

[83] Fosgren A., Sporre G. (2001) "On weighted linear least-squares problems related to interior methods for convex quadratic programming", *SIAM Journal on Matrix Analysis and Applications,* **23**(1), 42–56.

[84] Frank I.E., Friedman, J.H. (1993) "A statistical view of some chemometrics regression tools", *Technometrics,* **35**, 109–135.

[85] Franklin G.F., Powell J.D., Workman M.L. (1990). *Digital Control of Dynamic Systems,* Reading MA: Addison-Wesley.

[86] Freund R.M., Mizuno S. (2000) "Interior point methods: current status and future directions", in *High Performance Optimization,* Frenk H. *et al.* (Eds.), Kluwer Academic Publishers, 441–466.

[87] Friedman J.H., Tukey J.W. (1974) "A projection pursuit algorithm for exploratory data analysis", *IEEE Transactions on Computing,* **23**, 881–

890.

[88] Friedman J. (1989) "Regularized discriminant analysis", *Journal of the American Statistical Association*, **84**, 165–175.

[89] Furey T., Cristianini N., Duffy N., Bednarski D., Schummer M., Haussler D. (2000) "Support vector machine classification and validation of cancer tissue samples using microarray expression", *Bioinformatics*, **16**(10), 906–914.

[90] Gammerman A. (1996) *Machine Learning: Progress and Prospects*, Royal Holloway University of London, Inaugural Lecture Series.

[91] Gardner W.A. (1990) *Introduction to Random Processes*, Second edition, McGraw-Hill New York.

[92] Gauss C.F. (1809) *Theoria Motus Corporum Coelesium in Sectionibus Conicis Solem Ambientieum*, Perthes F., and Besser J.H., Hamburg. Reprinted in Carl Friedrich Gauss Werke, vol. 12, Königlichen Gesellschaft der Wissenschaften zu Göttingen. Translated in Theory of the Motion of the Heavenly Bodies Moving about the Sun in Conic Sections, Little, Brown and Co., Boston, 1857, reprinted by Dover, New York, 1963.

[93] Genton M. (2001) "Classes of kernels for machine learning: a statistics perspective", *Journal of Machine Learning Research*, **2**, 299-312.

[94] Girolami M. (2002) "Orthogonal series density estimation and the kernel eigenvalue problem", *Neural Computation*, **14**(3), 669–688.

[95] Girosi F. (1998) "An equivalence between sparse approximation and support vector machines", *Neural Computation*, **10**(6), 1455–1480.

[96] Gittins R. (1985) *Canonical analysis*, Springer-Verlag.

[97] Golub G.H., Heath M., Wahba G. (1979) "Generalized cross-validation a method for choosing a good ridge regression parameter", *Technometrics*, **21**, 215–223.

[98] Golub G.H., Van Loan C.F. (1989) *Matrix Computations*, Baltimore MD: Johns Hopkins University Press.

[99] Gower J.C. (1966) "Some distance properties of latent root and vector methods used in multivariate analysis", *Biometrika*, **53**, 325–338.

[100] Greenbaum A. (1997) *Iterative Methods for Solving Linear Systems*, SIAM, Philadelphia.

[101] Guyon I., Boser B., Vapnik V. (1993) "Automatic capacity tuning of very large VC-dimension classifiers", In S. Hanson et al. (Eds.), *Advances in Neural Information Processing Systems 5*, 147–155, San Mateo CA, Morgan Kaufmann.

[102] Guyon I., Weston J., Barnhill S., Vapnik V. (2002) "Gene selection for cancer classification using support vector machines", *Machine Learning*, **46**, 389–422.

[103] Hamers B., Suykens J.A.K., De Moor B. (2001), "A comparison of iterative methods for least squares support vector machine classifiers", *Internal Report 01-110, ESAT-SISTA, K.U.Leuven (Leuven, Belgium)*.

[104] Hamers B., Suykens J.A.K., De Moor B. (2002) "Compactly supported

RBF kernels for sparsifying the Gram matrix in LS-SVM regression models", *International Conference on Artificial Neural Networks (ICANN 2002)*, Madrid Spain, August 2002, pp. 720–726.

[105] Hampel F.R., Ronchetti E.M., Rousseeuw P.J., Stahel W.A. (1986) *Robust Statistics, the Approach Based on Influence Functions*, John Wiley & Sons, New York.

[106] Hassibi B., Stork D.G. (1993) "Second order derivatives for network pruning: optimal brain surgeon", In Hanson, Cowan, Giles (Eds.) *Advances in Neural Information Processing Systems*, 5, 164-171, San Mateo, CA: Morgan Kaufmann.

[107] Hastie T., Tibshirani R., Friedman J. (2001) *The Elements of Statistical Learning*, Springer-Verlag.

[108] Haussler D. (1999) "Convolution kernels on discrete structures", Technical report UCSC-CRL-99-10.

[109] Haykin S. (1996) *Adaptive Filter Theory*, Third Edition, Prentice-Hall.

[110] Haykin S. (1994) *Neural Networks: a Comprehensive Foundation*, Macmillan College Publishing Company: Englewood Cliffs.

[111] Herbrich R. (2002) *Learning Kernel Classifiers*, MIT Press, Cambridge Massachusetts.

[112] Hoerl A.E., Kennard R.W. (1970) "Ridge regression: biased estimation for nonorthogonal problems", *Technometrics*, $12(1)$, 55–67.

[113] Holte R.C. (1993) "Very simple classification rules perform well on most commonly used datasets", *Machine Learning*, 11, 63–90.

[114] Horn R.A., Johnson C.R. (1985) *Matrix Analysis*, Cambridge University Press.

[115] Hotelling H. (1936) "Simplified calculation of principal components", *Psychometrica*, 1, 27–35.

[116] Hotelling H. (1936) "Relations between two sets of variates", *Biometrica*, 28, 321–377.

[117] Hotelling H. (1957) "The relations of the newer multivariate statistical methods to factor analysis", *British Journal Statistics and Psychology*, 10, 69–79.

[118] Huber P.J. (1964) "Robust estimation of a location parameter", *Ann. Math. Statist.*, 35, 73–101.

[119] Huber P.J. (1981) *Robust Statistics*, John Wiley & Sons, New York.

[120] Isenman A.J. (1991) "Recent developments in nonparametric density estimation", *Journal of the American Statistical Association*, 86, 205–224.

[121] Jaakkola T., Haussler D. (1998) "Exploiting generative models in discriminative classifiers", *Advances in Neural Information Processing Systems*, 11.

[122] Jaeckel L. (1971) "Robust estimation of location: symmetry and asymmetric contamination", *Annals of Mathematical Statistics*, 42, 1020–1034.

[123] Jeffreys H. (1961) *Theory of Probability*, Oxford University Press.

[124] Jobson J.D. (1991) *Applied Multivariate Data Analysis, Vol.I: Regression*

and Experimental Design, Springer-Verlag, New York.

[125] Jobson J.D. (1991) *Applied Multivariate Data Analysis, Vol.II: Categorical and Multivariate Methods*, Springer-Verlag, New York.

[126] John G.H., Langley P. (1995) "Estimating continuous distributions in Bayesian classifiers", *Proceedings of the Eleventh Conference on Uncertainty in Artificial Intelligence*, Montreal, Quebec, Morgan Kaufmann, 338–345.

[127] Johnson N.L., Kotz S. (1970) *Distributions in Statistics: Continuous Univariate Distributions*, Vol.1-2, John Wiley & Sons, New York.

[128] Jolliffe I.T. (1986) *Principal Component Analysis*, Springer Series in Statistics, Springer-Verlag.

[129] Kailath T. (1971) "RKHS approach to detection and estimation problems: Part I: deterministic signals in Gaussian noise", *IEEE Transactions on Information Theory*, **17**(5), 530–549.

[130] Kailath T. (1974) "A view of three decades of linear filtering theory", *IEEE Transactions on Information Theory*, **20**(2), 146–181.

[131] Kalman R.E. (1960) "A new approach to linear filtering and prediction theory", *Transaction American Society Mechanical Engineering, Journal of Basic Engineering*, **82**, 35–45.

[132] Keerthi S.S., Shevade S.K. (2002) "SMO algorithm for Least Squares SVM formulations", *Neural Computation*, to appear.

[133] Kimeldorf G.S., Wahba G. (1971) "A correspondence between Bayesian estimation on stochastic processes and smoothing by splines", *Ann. Math. Statist.*, **2**, 495–502.

[134] Kohonen T. (1990) "The self-organizing map", *Proc. IEEE*, **78**(9), 1464–1480.

[135] Kohonen T. (1997) *Self-Organizing Maps*, Springer Series in Information Sciences, **30**.

[136] Kreyszig E. (1989) *Introductory Functional Analysis with Applications*, John Wiley & Sons, New York.

[137] Krige D.G. (1951) "A statistical approach to some basic mine valuation problems on the Witwatersrand", *J. Chem. Metall. Mining Soc. S. Africa*, **52**(6), 119–139.

[138] Kuh A. (2001) "Adaptive kernel methods for CDMA systems", *Proc. of the International Joint Conference on Neural Networks (IJCNN 2001)*, Washington DC, 1404–1409.

[139] Kwok J.T. (2000) "The evidence framework applied to support vector machines", *IEEE Transactions on Neural Networks*, **10**, 1018–1031.

[140] Le Cun Y., Denker J.S., Solla S.A. (1990) "Optimal brain damage", In Touretzky (Ed.) *Advances in Neural Information Processing Systems*, **2**, 598-605, San Mateo, CA: Morgan Kaufmann.

[141] Lim T.-S., Loh W.-Y., Shih Y.-S. (2000) "A comparison of prediction accuracy, complexity, and training time of thirty-three old and new classification algorithms", *Machine Learning*, **40**(3), 203–228.

[142] Lin C.-J. (2001) "On the convergence of the decomposition method for support vector machines", *IEEE Transactions on Neural Networks.* **12**, 1288–1298.

[143] Lindley D.V. (1980) "Approximate Bayesian methods", In *Bayesian Statistics* (Eds.) J.M. Bernardo, M.H. De Groot, D.V. Lindley, A.F.M. Smith, 223–237, Valencia: Valencia University Press.

[144] Ljung L. (1999) *System Identification: Theory for the User* (2nd ed.), Prentice Hall, New Jersey.

[145] Loeve M. (1948) "Fonctions aléatoires du second ordre", suppl. to P. Lévy, *Processus stochastique et mouvement Brownien,* Paris: Gauthier-Villar.

[146] Lu C., Van Gestel T., Suykens J.A.K., Van Huffel S., Vergote I., Timmerman D. (2003) "Preoperative prediction of malignancy of ovarian tumors using least squares support vector machines", *Artificial Intelligence in Medicine,* **28**(3), 281–306.

[147] Luenberger D.G. (1973) *Introduction to Linear and Nonlinear Programming,* Addison-Wesley, Reading Massachusetts.

[148] Lukas L., Devos A., Suykens J.A.K., Vanhamme L., Van Huffel S., Tate A., Majos C., Arus C. (2002) "The use of LS-SVM in the classification of brain tumors based on magnetic resonance spectroscopy signals", *European Symposium Artificial Neural Networks (ESANN 2002),* Bruges Belgium, 131–136.

[149] MacKay D.J.C. (1992) "Bayesian interpolation", *Neural Computation,* **4**(3), 415–447.

[150] MacKay D.J.C. (1992) "The evidence framework applied to classification networks", *Neural Computation,* **4**, 698–714.

[151] MacKay, D.J.C. (1995) "Probable networks and plausible predictions - a review of practical Bayesian methods for supervised neural networks", *Network: Computation in Neural Systems,* **6**, 469–505.

[152] MacKay, D.J.C. (1999) "Comparison of approximate methods for handling hyperparameters", *Neural Computation,* **11**, 1035–1068.

[153] MacKay, D.J.C. (1998) "Introduction to Gaussian processes". in *Neural networks and machine learning* (Ed. C.M. Bishop), Springer NATO-ASI Series F: Computer and Systems Sciences, Vol.168, 133–165.

[154] Mangasarian O.L. (1994) *Nonlinear Programming,* SIAM, Classics in Applied Mathematics.

[155] Mangasarian O.L., Musicant D.R. (1999) "Successive overrelaxation for support vector machines", *IEEE Transactions on neural Networks,* **10**, 1032–1037.

[156] Marazzi A., Ruffieux C. (1996) *Implementing M-estimators of the Gamma Distribution,* Lecture Notes in Statistics, Vol.109, Springer, Heidelberg.

[157] Marron J.S. (1989) "Automatic smoothing parameter selection: a survey", *Empirical Econom.,* **13**, 187–208.

[158] McCullagh P., Nelder J.A. (1989) *Generalized Linear Models,* Chapman & Hall, London.

[159] Mendelson S., "Rademacher averages and phase transitions in Glivenko-Cantelli classes", *IEEE Transactions on Information Theory*, **48**(1), 251–263.

[160] Mercer J. (1909) "Functions of positive and negative type and their connection with the theory of integral equations", *Philos. Trans. Roy. Soc. London*, **209**, 415–446.

[161] Micchelli C.A. (1986) "Interpolation of scattered data: distance matrices and conditionally positive definite functions", *Constructive Approximation*, **2**, 11–22.

[162] Michel A.N., Herget C.J. (1981) *Applied Algebra and Functional Analysis*, Dover Publications, New York.

[163] Mika S., Rätsch G., Weston J., Schölkopf B., Müller K.-R. (1999) "Fisher discriminant analysis with kernels", In Y.-H. Hu, J. Larsen, E. Wilson, and S. Douglas, editors, *Neural Networks for Signal Processing IX*, 41–48. IEEE.

[164] Mjolness E., DeCoste D. (2001) "Machine learning for science: state of the art and future prospects", *Science*, **293**, 2051–2055.

[165] Moonen M., Vandewalle J. (1993) "A systolic array for recursive least squares computations", *IEEE Transactions on Signal Processing*, **41**(2), 906–912.

[166] Moonen M., Van Dooren P., Vandewalle J. (1993) "A systolic array for SVD updating", *SIAM Journal on Matrix Analysis and Applications*, **14**(2), 353–371.

[167] Moonen M., Golub G.H., De Moor B. (Eds.) (1993) *Linear Algebra for Large-Scale and Real-Time Applications*, Proceedings NATO Advanced Study Institute, Leuven (Belgium), August 3–14 1992.

[168] Moore E.H. (1916) "On properly positive Hermitian matrices", *Bull. Amer. Math. Soc.*, **23**, 59.

[169] Mukherjee S., Tamayo P., Mesirov P., Slonim J.P., Verri A., Poggio T. (1999) "Support vector machine classification of microarray data", *CBCL paper 182/ AI memo 1676*, MIT Cambridge MA.

[170] Narendra K.S., Parthasarathy K. (1991) "Gradient methods for the optimization of dynamical systems containing neural networks", *IEEE Transactions on Neural Networks*, **2**(2), 252–262.

[171] Navia-Vazquez A., Perez-Cruz F., Artes-Rodriguez A., Figueiras-Vidal A.R. (2001) "Weighted least squares training of support vector classifiers leading to compact and adaptive schemes", *IEEE Transactions on Neural Networks*, **12**(5), 1047–1059.

[172] Neal R.M. (1996) *Bayesian Learning for Neural Networks*, Vol. 118 of Lecture Notes in Statistics, Springer, New York.

[173] Nesterov Y., Nemirovskii A. (1993) *Interior-Point Polynomial Algorithms in Convex Programming*, SIAM Vol.13.

[174] Nocedal J., Wright S.J. (1999) *Numerical Optimization*, Springer Verlag.

[175] Osuna E., Freund R., Girosi F. (1997) "Improved training algorithm for

support vector machines", *IEEE Neural Network for Signal Processing (NNSP'97)*, 276–285.

[176] Parisini T., Zoppoli R. (1994) "Neural networks for feedback feedforward nonlinear control systems", *IEEE Transactions on Neural Networks*, **5**(3), 436–449.

[177] Parzen E. (1970), "Statistical inference on time series by RKHS methods", Dep. Statist. Stanford Univ. Tech. Rep.14, Jan. 1970.

[178] Pearson K. (1901) "On lines and planes of closest fit to systems of points in space", *Phil. Mag. (6)*, **2**, 559–572.

[179] Perrone M.P., Cooper L.N. (1993) "When networks disagree: Ensemble method for neural networks", in R.J. Mammone (Ed.) *Neural Networks for Speech and Image processing*, Chapman-Hall.

[180] Pfanzagl J. (1969) "On measurability and consistency of minimum contrast estimates", *Metrika*, **14**, 248–278.

[181] Platt J. (1999) "Probabilistic outputs for support vector machines and comparisons to regularized likelihood methods", In *Advances in Large Margin Classifiers*, Smola, A., Bartlett, P., Schölkopf, B., and Schuurmans, D. (Eds.), MIT Press.

[182] Platt J. (1999) "Fast training of support vector machines using sequential minimal optimization", In Schölkopf B., Burges C.J.C., Smola A.J. (Eds.) *Advances in Kernel methods - Support Vector Learning*, 185–208, MIT Press.

[183] Poggio T., Girosi F. (1990) "Networks for approximation and learning", *Proceedings of the IEEE*, **78**(9), 1481–1497.

[184] Poggio T., Girosi F. (1990) "Regularization algorithms for learning that are equivalent to multilayer networks", *Science*, **247**, 978–982.

[185] Povzner A.Ya. (1950) "A class of Hilbert function spaces", *Dokl. Akad. Nauk USSR*, **68**, 817–820.

[186] Prakasa Rao B.L.S. (1983) *Nonparametric Functional Estimation*, Academic Press.

[187] Principe J., Fisher J., Xu D. (2000) "Information theoretic learning", in S. Haykin (Ed.), *Unsupervised Adaptive Filtering*, John Wiley & Sons, New York.

[188] Quinlan J. (1993) *C4.5 Programs for Machine Learning*, Morgan Kaufmann.

[189] Rencher A.C. (1995) *Methods of Multivariate Analysis*, John Wiley & Sons, New York.

[190] Renegar J. (2001) *A Mathematical View of Interior-Point Methods in Convex Optimization*, MPS/SIAM Series on Optimization 3.

[191] Rice J.A. (1995) *Mathematical Statistics and Data Analysis* (2nd ed.), Duxbury Press.

[192] Ridella S., Rovetta S., Zunino R. (1997) "Circular backpropagation networks for classification", *IEEE Transactions on Neural Networks*, **8**(1), 84–97.

[193] Ripley B.D. (1996) *Pattern Recognition and Neural Networks*, Cambridge: Cambridge University Press.

[194] Ritter H., Martinetz T., Schulten K. (1992) *Neural Computation and Self-Organizing Maps: An Introduction*, Addison-Wesley, Reading, MA.

[195] Rosipal R., Trejo L.J. (2001) "Kernel partial least squares regression in reproducing kernel Hilbert space", *Journal of Machine Learning Research*, **2**, 97–123.

[196] Rousseeuw P.J., Leroy A. (1997) *Robust Regression and Outlier Detection*, John Wiley & Sons, New York.

[197] Sanner R.M., Slotine J.-J. E. (1992) "Gaussian networks for direct adaptive control", *IEEE Transactions on Neural Networks*, **3**(6), 837–863.

[198] Saunders C., Gammerman A., Vovk V. (1998) "Ridge regression learning algorithm in dual variables", *Proc. of the 15th Int. Conf. on Machine Learning (ICML-98)*, Madison-Wisconsin, 515–521.

[199] Sayed A.H., Kailath T. (1994) "A state-space approach to adaptive RLS filtering", *IEEE Signal Processing Magazine*, **11**(3), 18–60.

[200] Schoenberg I.J. (1938) "Metric spaces and completely monotone functions", *Annals of Mathematics*, **39**(3), 811–841.

[201] Schölkopf B., Sung K.-K., Burges C., Girosi F., Niyogi P., Poggio T., Vapnik V. (1997) "Comparing support vector machines with Gaussian kernels to radial basis function classifiers", *IEEE Transactions on Signal Processing*, **45**(11), 2758–2765.

[202] Schölkopf B., Burges C., Smola A. (1998) *Advances in Kernel Methods: Support Vector Learning*, MIT Press, Cambridge, MA.

[203] Schölkopf B., Smola A., Müller K.-R. (1998) "Nonlinear component analysis as a kernel eigenvalue problem", *Neural Computation*, **10**, 1299–1319.

[204] Schölkopf B., Mika S., Burges C., Knirsch P., Müller K.-R., Rätsch G., Smola A. (1999) "Input space vs. feature space in kernel-based methods", *IEEE Transactions on Neural Networks*, **10**(5), 1000–1017.

[205] Schölkopf B., Smola A., Williamson R.C., Bartlett P.L. (2000) "New support vector algorithms", *Neural Computation*, **12**, 1083–1121.

[206] Schölkopf B., Smola A. (2002) *Learning with Kernels*, MIT Press, Cambridge, MA.

[207] Schölkopf B., Shawe-Taylor J., Smola A.J., Williamson R.C. (1999) "Generalization bounds via eigenvalues of the Gram matrix", NeuroCOLT2 Technical Report Series NC2-TR-1999-035.

[208] Schölkopf B., Shawe-Taylor J., Smola A., Williamson R.C. (1999) "Kernel-dependent support vector error bounds", *Proc. of the 9th Int. Conf. on Artificial Neural Networks (ICANN-99)*, 304–309, Edinburgh, UK.

[209] Schwarz G. (1978) "Estimating the dimension of a model", *Ann. Stat.*, **6**, 461–464.

[210] Seber G.A.F. (1977) *Linear Regression Analysis*, John Wiley & Sons, New York.

[211] Serfling R.J. (1984) "Generalized L-, M-, and R-statistics", *Ann. Statist.*,

12, 76–86.

[212] Sezer M.E., Siljak D.D. (1988) "Robust stability of discrete systems", *International Journal of Control*, **48**(5), 2055–2063.

[213] Shao J. (1999) *Mathematical Statistics*, Springer-Verlag, New York.

[214] Shibata R. (1981) "An optimal selection of regression variables", *Biometrica*, **68**, 461–464.

[215] Shiryaev. A.N. (1996) *Probability* (2nd ed.), Reading in Mathematics, Springer-Verlag, New York.

[216] Sjöberg J., Zhang Q., Ljung L., Benveniste A., Delyon B., Glorennec P.-Y., Hjalmarsson H., Juditsky A. (1995) "Nonlinear black-box modeling in system identification: a unified overview", *Automatica*, **31**(12), 1691–1724.

[217] Smale S. (1997) "Complexity theory and numerical analysis", *Acta Numerica*. 523–551.

[218] Smale S. (2000) "Mathematical problems for the next century", *Mathematics: frontiers and perspectives* (V. Arnold, M. Atiyah, P. Lax, B. Mazur, Eds.), AMS, 271–294, CMP 2000:13.

[219] Smola A., Schölkopf B. (1998) "A tutorial on support vector regression", NeuroCOLT2 Technical Report Series NC2-TR-1998-030.

[220] Smola A., Schölkopf B., Müller K.-R. (1998) "The connection between regularization operators and support vector kernels", *Neural Networks*, **11**, 637–649.

[221] Smola A., Schölkopf B. (1998) "On a kernel-based method for pattern recognition, regression, approximation and operator inversion", *Algorithmica*, **22**, 211–231.

[222] Smola A., Schölkopf B., Müller K.-R. (1998) "General cost functions for support vector regression", In T. Downs, M. Frean, and M. Gallagher (Eds.), *Proc. of the Ninth Australian Conf. on Neural Networks,* 79–83, Brisbane, Australia.

[223] Smola A. (1999) *Learning with Kernels*, PhD Thesis, published by: GMD, Birlinghoven.

[224] Smola A.J., Mika S., Schölkopf B., Williamson R.C. (2001) "Regularized principal manifolds", *Journal of Machine Learning Research*, 1, 179–209.

[225] Smola A., Schölkopf B., Rätsch G. (1999) "Linear programs for automatic accuracy control in regression", *International Conference on Artificial Neural Networks*, Conference Publications No. 470, 575–580, London IEE.

[226] Smola A.J., Schölkopf B. (2000) "Sparse greedy matrix approximation for machine learning", in P. Langley (Ed.) *Proc. 17th International Conference on Machine Learning*, 911–918, San Francisco, Morgan Kaufman.

[227] Stigler S.M. (1973) "The asymptotic distribution of the trimmed mean", *Ann. Statist.*, 1, 472–477.

[228] Stone M. (1974) "Cross-validatory choice and assessment of statistical predictions", *J. Royal Statist. Soc. Ser. B*, **36**, 111–147.

[229] Strang G. (1998) *Introduction to Linear Algebra* (2nd ed.), Wellesley-Cambridge Press, Wellesley MA.

[230] Suykens J.A.K., Vandewalle J., De Moor B. (1996) *Artificial Neural Networks for Modelling and Control of Non-Linear systems*, Kluwer Academic Publishers, Boston.

[231] Suykens J.A.K., Vandewalle J. (Eds.) (1998) *Nonlinear Modeling: Advanced Black-box Techniques*, Kluwer Academic Publishers, Boston.

[232] Suykens J.A.K., De Moor B. Vandewalle J. (1994) "Static and dynamic stabilizing neural controllers, applicable to transition between equilibrium points", *Neural Networks*, **7**(5), 819–831.

[233] Suykens J.A.K., De Moor B. Vandewalle J. (1997) "NL_q theory: a neural control framework with global asymptotic stability criteria", *Neural Networks*, **10**(4), 615–637.

[234] Suykens J.A.K., Vandewalle J. (1999) "Training multilayer perceptron classifiers based on a modified support vector method", *IEEE Transactions on Neural Networks*, **10**(4), 907–912.

[235] Suykens J.A.K., Vandewalle J. (1999) "Least squares support vector machine classifiers", *Neural Processing Letters*, **9**(3), 293–300.

[236] Suykens J.A.K., Lukas L., Van Dooren P., De Moor B., Vandewalle J. (1999) "Least squares support vector machine classifiers: a large scale algorithm", *European Conference on Circuit Theory and Design, (ECCTD'99)*, 839–842, Stresa Italy.

[237] Suykens J.A.K., Vandewalle J. (1999) "Multiclass least squares support vector machines", *Proc. of the International Joint Conference on Neural Networks (IJCNN'99)*, Washington DC, USA, CD-Rom.

[238] Suykens J.A.K., Lukas L., Vandewalle J. (2000) "Sparse approximation using least squares support vector machines", *IEEE International Symposium on Circuits and Systems (ISCAS 2000)*, II 757–760, Geneva, Switzerland.

[239] Suykens J.A.K., Lukas L., Vandewalle J. (2000) "Sparse least squares support vector machine classifiers", in *Proc. of the European Symposium on Artificial Neural Networks (ESANN 2000)*, Bruges, Belgium, 2000, 37–42.

[240] Suykens J.A.K., Vandewalle J. (2000) "Recurrent least squares support vector machines", *IEEE Transactions on Circuits and Systems-I*, **47**(7), 1109–1114.

[241] Suykens J.A.K., Vandewalle J., De Moor B. (2001) "Optimal control by least squares support vector machines", *Neural Networks*, **14**(1), 23–35.

[242] Suykens J.A.K., Vandewalle J. (1999) "Chaos control using least squares support vector machines", *International Journal of Circuit Theory and Applications, Special Issue on Communications, Information Processing and Control Using Chaos*, **27**(6), 605–615.

[243] Suykens J.A.K., De Brabanter J., Lukas L., Vandewalle J. (2002) "Weighted least squares support vector machines: robustness and sparse approximation", *Neurocomputing* **48**(1–4), 85–105.

[244] Suykens J.A.K. (2001) "Nonlinear modelling and support vector machines", in *Proc. of the IEEE International Conference on Instrumentation and Measurement Technology (IEEE-IMTC 2001 State-of-the-Art lecture)*, Bu-

dapest, Hungary, May 2001, 287–294.

[245] Suykens J.A.K. (2000), "Least squares support vector machines for classification and nonlinear modelling", *Neural Network World, Special Issue on PASE 2000*, **10**(1-2), 29–48.

[246] Suykens J.A.K. (2001) "Support vector machines: a nonlinear modelling and control perspective", *European Journal of Control, Special Issue on Fundamental Issues in Control*, **7**(2-3), 311–327.

[247] Suykens J.A.K., Van Gestel T., Vandewalle J., De Moor B. (2003) "A support vector machine formulation to PCA analysis and its kernel version", *IEEE Transactions on Neural Networks*, **14**(2), 447–450.

[248] Suykens J.A.K., Vandewalle J., De Moor B. (2001) "Intelligence and cooperative search by coupled local minimizers", *International Journal of Bifurcation and Chaos*, **11**(8), 2133–2144.

[249] Swets J. (1988) "Measuring the accuracy of diagnostic systems", *Science*, **240**, 1285–1293.

[250] Swets J., Dawes R., Monahan J. (2000) "Better decisions through science", *Scientific American*, 82–87.

[251] Tierney L., Kadane J.B. (1986) "Accurate approximations for posterior moments and marginal densities", *Journal of the American Statistical Association*, **81**, 82–86.

[252] Tikhonov A.N., Arsenin V.Y. (1977) *Solution of Ill-Posed Problems*, Winston, Washington DC.

[253] Tipping M.E. (2001) "Sparse Bayesian learning and the relevance vector machine", *Journal of Machine Learning Research*, **1**, 211–144.

[254] Tipping M.E., Bishop C.M. (1999) "Probabilistic principal component analysis", *Journal of the Royal Statistical Society, Series B*, **61**, 611–622.

[255] Tou J.T., Gonzalez R.C. (1974) *Pattern Recognition Principles*, Addison-Wesley Publishing Company, Reading, Massachusetts.

[256] Trefethen L.N., Bau D. (1997) *Numerical Linear Algebra*, SIAM, Philadelphia.

[257] Tresp V. (2000) "A Bayesian committee machine", *Neural Computation*, **12**(11), 2719–2741.

[258] Tresp V. (2001) "Scaling kernel-based systems to large data sets", *Data Mining and Knowledge Discovery*, **5**(3), 197–211.

[259] Tukey J.W. (1960) *A Survey of Sampling of Contaminated Distributions*, In Contributions to Probability and Statistics, Stanford University Press, Stanford CA.

[260] Utschick W. (1998) "A regularization method for non-trivial codes in polychotomous classification", *International Journal of Pattern Recognition and Artificial Intelligence*, **12**, 453–474.

[261] Vandenberghe L., Boyd S. (1996) "Semidefinite programming", *SIAM Review*, **38**, 49–95.

[262] Vanderbei R.J. (1997) *Linear Programming: Foundations and Extensions*, Kluwer Academic Publishers, Hingham MA.

[263] Van Der Vaart A.W. (1998) *Asymptotic Statistics*, Cambridge Series in Statistical and Probabilistic Mathematics, Cambridge University Press.

[264] Van Gestel T., Suykens J.A.K., Baesens B., Viaene S., Vanthienen J., Dedene G., De Moor B., Vandewalle J. (2004) "Benchmarking least squares support vector machine classifiers", *Machine Learning* **54**(1), 5–32.

[265] Van Gestel T., Suykens J.A.K., Baestaens D., Lambrechts A., Lanckriet G., Vandaele B., De Moor B., Vandewalle J. (2001) "Financial time series prediction using least squares support vector machines within the evidence framework", *IEEE Transactions on Neural Networks* (special issue on Neural Networks in Financial Engineering), **12**(4), 809–821.

[266] Van Gestel T., Suykens J.A.K., Lanckriet G., Lambrechts A., De Moor B., Vandewalle J. (2002) "A Bayesian framework for least squares support vector machine classifiers, Gaussian processes and kernel Fisher discriminant analysis", *Neural Computation*, **14**(5), 1115–1147.

[267] Van Gestel T., Suykens J.A.K., Lanckriet G., Lambrechts A., Baestaens D., De Moor B., Vandewalle J. (2001) "Bayesian interpretation of least squares support vector machines for financial time series prediction", in *Proc. of the 5th World Multiconference on Systemics, Cybernetics and Informatics (SCI 2001)*, Orlando, Florida, 254–259.

[268] Van Gestel T., Suykens J.A.K., Lanckriet G., Lambrechts A., De Moor B., Vandewalle J. (2002), "Multiclass LS-SVMs: moderated outputs and coding-decoding schemes", *Neural Processing Letters*, **15**(1), 45–58.

[269] Van Gestel T., Suykens J.A.K., De Moor B., Vandewalle J. (2001), "Automatic relevance determination for least squares support vector machine regression", in *Proc. of the International Joint Conference on Neural Networks (IJCNN 2001)*, Washington DC, USA, 2416–2421.

[270] Van Gestel T., Suykens J.A.K., De Moor B., Vandewalle J. (2001) "Automatic relevance determination for least squares support vector machine regression", *9th European Symposium on Artificial Neural Networks (ESANN 2001)*, 13–18, Bruges Belgium.

[271] Van Gestel T., Suykens J.A.K., De Brabanter J., De Moor B., Vandewalle J. (2001) "Kernel canonical correlation analysis and least squares support vector machines", *Proc. of the International Conference on Artificial Neureal Networks (ICANN 2001)*, Vienna, Austria, 381–386.

[272] Van Gestel T., Suykens J.A.K., De Brabanter J., De Moor B., Vandewalle J. (2001) "Least squares support vector machine regression for discriminant analysis", *Proc. of the International Joint Conference on Neural Networks (IJCNN 2001)*, Washington DC, USA, 2445–2450.

[273] Van Gestel T., Suykens J.A.K., De Moor B., Vandewalle J. (2002) "Bayesian inference for LS-SVMs on large data sets using the Nyström method", *Proc. of the International Joint Conference on Neural Networks (IJCNN 2002)*, Honolulu, Hawaii.

[274] Van Gestel T. (2002) *From Linear to Kernel Based Methods in Classification, Modelling and Prediction*, PhD thesis, K.U. Leuven Department of

Electrical Engineering.

[275] Van Overschee P., De Moor B. (1996) *Subspace Identification for Linear Systems: Theory, Implementation, Applications*, Kluwer Academic Publishers, Boston.

[276] Vapnik V., Lerner A. (1963) "Pattern recognition using generalized portrait method", *Automation and Remote Control*, **24**, 774–780.

[277] Vapnik V., Chervonenkis A. (1964) "A note on one class of perceptrons", *Automation and Remote Control*, **25**.

[278] Vapnik V. (1982) *Estimation of Dependencies based on Empirical Data*, Springer-Verlag, New York.

[279] Vapnik V. (1995) *The Nature of Statistical Learning Theory*, Springer-Verlag, New York.

[280] Vapnik V., Golowich S., Smola A. (1997) "Support vector method for function approximation, regression estimation, and signal processing", In Mozer M., Jordan M., Petsche T. (Eds.) *Advances in Neural Information Processing Systems 9*, 281–287, Cambridge, MA, MIT Press.

[281] Vapnik V. (1998) *Statistical Learning Theory*, John Wiley & Sons, New York.

[282] Vapnik V. (1998) "The support vector method of function estimation", In *Nonlinear Modeling: Advanced Black-box Techniques*, Suykens J.A.K., Vandewalle J. (Eds.), Kluwer Academic Publishers, Boston, 55–85.

[283] Viaene S., Baesens B., Van Gestel T., Suykens J.A.K., Van den Poel D., Dedene D., De Moor B., Vanthienen J. (2001) "Knowledge discovery in a direct marketing case using least squares support vector machines", *International Journal of Intelligent Systems*, **16**(9), 1023–1036.

[284] Vidyasagar M. (1993) *Nonlinear Systems Analysis*, Prentice-Hall.

[285] Vidyasagar M. (1997) *A Theory of Learning and Generalization*, Springer-Verlag.

[286] Wahba G. (1990) *Spline Models for Observational Data*, Series in Applied Mathematics, **59**, SIAM, Philadelphia.

[287] Wahba G. (1998) "Support vector machines, reproducing kernel Hilbert spaces and the randomized GACV", In Schölkopf B., Burges C.J.C., Smola A.J. (Eds.) *Advances in Kernel Methods - Support Vector Learning*, MIT Press, 69–87.

[288] Weigend A.S., Gershenfeld N.A. (Eds.) (1994) *Time Series Prediction: Forecasting the Future and Understanding the Past*, Addison-Wesley.

[289] Werbos P. (1990) "Backpropagation through time: what it does and how to do it", *Proceedings of the IEEE*, **78**(10), 1150–1560.

[290] Wilkinson J.H. (1965) *The Algebraic Eigenvalue Problem*, Oxford University Press, Oxford.

[291] Williams C.K.I., Rasmussen C.E. (1996) "Gaussian processes for regression", In D.S. Touretzky, M.C. Mozer, and M.E. Hasselmo (Eds.), *Advances in Neural Information Processing Systems 8*, 514–520. MIT Press.

[292] Williams C.K.I., Barber D. (1998) "Bayesian classification with Gaussian

processes", *IEEE Transactions on Pattern Analysis and Machine Intelligence*, **20**, 1342–1351.

[293] Williams C.K.I., Seeger M. (2000) "The effect of the input density distribution on kernel-based classifiers", In P. Langley (Ed.) *Proceedings of the Seventeenth International Conference on Machine Learning (ICML 2000)*, Morgan Kaufmann.

[294] Williams C.K.I., Seeger M. (2001) "Using the Nyström method to speed up kernel machines", In T.K. Leen, T.G. Dietterich, and V. Tresp (Eds.), *Advances in neural information processing systems*, **13**, 682–688, MIT Press.

[295] Witten I.H., Frank E. (2000) *Data Mining: Practical Machine Learning Tools and Techniques with Java Implementations*, Morgan Kaufmann, San Francisco.

[296] Wong E. (1971). *Stochastic Processes in Information and Dynamical Systems*, McGraw-Hill.

[297] Xiong M., Fang X., Zhao J. (2001) "Biomarker identification by feature wrappers", *Genome Research*, **11**, 1878–1887.

[298] Xu G., Kailath T. (1994) "Fast estimation of principal eigenspace using Lanczos algorithm", *SIAM Journal on Matrix Analysis and Applications*, **15**(3), 974–994.

[299] Yaglom A.M. (1986) *Correlation Theory of Stationary and Related Random Functions, Part I & II*, Springer Series in Statistics, Springer-Verlag, New York.

[300] Yang Y., Zheng Z. (1992) "Asymptotic properties for cross-validated nearest neighbour median estimates in non-parametric regression: the L_1-view", In *Probability and Statistics*, 242–257. World Scientific, Singapore.

[301] Young D. (1971) *Iterative Solution of Large Linear Systems*, Academic Press, New York.

[302] Zhang T., Golub G.H. (2001) "Rank-one approximation to high order tensors", *SIAM J. Matrix Anal. Appl.*, **30**(2), 534–550.

[303] Zien A., Rätsch G., Mika S., Schölkopf B., Lengauer T., Müller K.-R. (2000) "Engineering support vector machine kernels that recognize translation initiation sites", *BioInformatics*, **16**(9), 799–807.

List of Symbols

x	input vector of network or state of a system
y	output vector of system or network
$\hat{y}$	estimated output vector of system or network
u	input vector of system
θ	parameter vector of model
n	number of inputs or number of state variables
n_y	number of outputs
n_u	number of inputs
n_h	number of hidden units of network
N	number of training data
$[a_1; ...; a_n]$	column vector with components $a_1, ..., a_n$
$[a_1, ..., a_n]$	row vector with components $a_1, ..., a_n$
$\mathrm{diag}([a_1; ...; a_n]))$	diagonal matrix with diagonal elements $a_1, ..., a_n$
1_v	$[1; 1; ...; 1]$
A^T	transpose of matrix A
$A(:)$	scanning of matrix A to a column vector
$A > 0$	positive definite matrix A
$A < 0$	negative definite matrix A
$\det(A), \|A\|$	determinant of matrix A
$K(\cdot, \cdot)$	kernel function
d_{eff}	effective number of parameters
ν, γ, μ, ζ	regularization constants
k, l	indices of training data points
$\{x_k, y_k\}_{k=1}^N$	training set of N data points
J, L	cost function (or loss function)
L_ϵ	Vapnik ϵ-insensitive loss function
J_P, J_D	cost function for primal and dual problem, respectively
w	output layer vector or parameter vector in primal space
b	bias term
α	vector of support values
$\mathcal{L}$	Lagrangian

287

P	probability
$p(\cdot)$	probability density
$F(\cdot)$	distribution function
$F^-(\cdot)$	generalized inverse of F
$\mathcal{D}$	data
$\mathcal{H}_i$	i-th model of a model set
$\mathcal{H}_\sigma$	model with RBF kernel having kernel width σ
$\mathcal{C}$	class
$n_{\mathcal{C}}$	number of classes in multi-class problem
$n_{\mathcal{H}}$	number of models in a model set
$\mathcal{E}[\cdot]$	expected value
σ_e	standard deviation of noise
σ	width of RBF kernel
$\Sigma_{\mathbf{xy}}$	covariance matrix $\mathcal{E}\{(x - \mathcal{E}[x])(y - \mathcal{E}[y])^T\}$
$S_{\mathbf{xy}}$	sample covariance matrix
$\mathrm{Cov}(x, y)$	covariance matrix $\Sigma_{\mathbf{xy}}$
$\mathrm{Var}(x)$	$\mathrm{Cov}(x, x)$
$\mathrm{Corr}(x, y)$	correlation matrix $\mathrm{Cov}(x, y)/(\sqrt{\mathrm{Cov}(x, x)}\sqrt{\mathrm{Cov}(y, y)})$
Ω	kernel matrix (or Gram matrix)
M_c	centering matrix
Ω_c	centered kernel matrix
λ_i	i-th eigenvalue
ϕ_i	i-th eigenfunction
$\varphi(\cdot)$	map from input space to feature space
h	VC dimension
δ_{kl}	Kronecker delta ($\delta_{kl} = 1$ if $k = l$ and 0 otherwise)

Acronyms

1vs1	one versus one
1vsA	one versus all
AA	average accuracy
AR	average rank
ARD	automatic relevance determination
BP	backpropagation
CCA	canonical correlation analysis
CG	conjugate gradient
CV	cross-validation
D	dual
Eff	efficiency
ECOC	error correcting output code
FD	Fisher discriminant
FN	false negative
FP	false positive
GP	Gaussian process
(G)PE	(generalized) prediction error
IB	instance based
IF	influence function
i.i.d.	independently identically distributed
I/O	input/output
KKT	Karush-Kuhn-Tucker
LDA	linear discriminant analysis
LP	linear program
LS-SVM	least squares support vector machine
MLP	multilayer perceptron
MOC	minimum output coding
MSE	mean squared error
NARX	nonlinear autoregressive moving average with exogenous input
NOE	nonlinear output error model
P	primal

PCA	principal component analysis
PLS	partial least squares
Pol	polynomial
QDA	quadratic discriminant analysis
QP	quadratic program
RBF	radial basis function
RKHS	reproducing kernel Hilbert space
RN	regularization network
ROC	receiver operating characteristic
SMO	sequential minimal optimization
SQP	sequential quadratic programming
SRM	structural risk minimization
SV	support vector
SVM	support vector machine
TN	true negative
TP	true positive
VC	Vapnik-Chervonenkis

Index

sparse representation, 181, 186
sparseness, 33, 111
spectral representation, 108
spline network, 23
stability of recurrent networks, 26
state feedback, 233
state space model, 26
stationary random field, 108
statistical learning theory, 44
string kernels, 62
structural risk minimization, 48
successive overrelaxation, 67
support vectors, 33

Takens' embedding theorem, 230
textmining, 62
transductive inference, 183
trimmed mean, 164
Tukey biweight score function, 163

UCI benchmarking results, 89, 136
unbalanced data set, 127
universal approximation, 2

Vapnik ϵ-insensitive loss function, 52
Vapnik's VC bound, 56
variational problem, 104
VC bound, 49
VC dimension, 46

weight decay, 8
weighted LS-SVM, 140, 155
Winsorized mean, 165
within class covariance, 83
Wolfe dual, 66

www.ingramcontent.com/pod-product-compliance
Lightning Source LLC
Chambersburg PA
CBHW050635190326
41458CB00008B/2287